Chemical Engineering Analysis and Optimization Using MATLAB®

Chemical Engineering Analysis and Optimization Using MATLAB®

Weiguo Xie, Sam Toan, and Richard Davis
Department of Chemical Engineering, University of Minnesota
Duluth, MN, USA

WILEY

This edition first published 2025

Registered Office
John Wiley & Sons, Inc., 111 River Street, Hoboken, NJ 07030, USA

For details of our global editorial offices, customer services, and more information about Wiley products visit us at www.wiley.com.

Wiley also publishes its books in a variety of electronic formats and by print-on-demand. Some content that appears in standard print versions of this book may not be available in other formats.

Library of Congress Cataloging-in-Publication Data Applied for:

Hardback ISBN: 9781394205363

Cover Design: Wiley
Cover Image: Courtesy of Weiguo Xie

Set in 9.5/12.5pt STIXTwoText by Straive, Chennai, India

Contents

Preface

The swift evolution of computing technology has revolutionized the landscape of traditional chemical engineering education, ushering in a new era of computer-aided design that is now in high demand. MATLAB, a multi-paradigm numerical computing environment and proprietary programming language developed by MathWorks, is a powerful tool for computer-aided design in chemical engineering. The requirement from employers for newly hired chemical engineering graduates to have computing and data analytics skills is ever-rising.

Our teaching philosophy is to help students acquire useful knowledge, understanding, and skills to build their confidence, which is the key to their future career success. With over 20 years of computer-aided design experience and two books (Davis 2021; Toan et al. 2017), the authors know how students and users get stuck in the coding hurdles and feel frustrated. We also realize that some advanced computational methods can upgrade users' computational skills and bring satisfaction to their work. This book aims to help readers build a fundamental understanding of chemical engineering systems by developing mathematical and statistical models and simulations of these systems using MATLAB. It also aims to develop a systematic understanding with stepwise solution examples and raise a critical awareness of process optimization and analysis.

The book is organized into three major sections: the first is a review section, which includes two chapters introducing modeling in chemical engineering processes and some key MATLAB commands, functions, and toolboxes. The second section has six chapters introducing advanced algorithms and computational methods with case studies in chemical engineering. The third section has two chapters that integrate some algorithms and methods to tackle more complicated case studies in chemical engineering. There are several chemical-engineering-related examples in each chapter at the undergraduate and graduate levels. We hope that this book will prove helpful to a wide variety of users in chemical engineering.

We would also like to acknowledge the valuable discussion, advice, and suggestions of Dr. Mike Croucher and Dr. Aycan Hacioglu from MathWorks Inc.

About the Companion Website

This book is accompanied by a companion website.

www.wiley.com/go/xie/chemicalengineering

This website includes:

Instructor book Companion Website:
Solutions Manuals (Book problems answer key + problem codes + figures and excel sheets)

Student book Companion Website:
Book codes and data files (+ excel sheets)

1

Introduction to Modeling

Modeling in engineering and science transcends the glitz of fashion runways and the meticulous craftsmanship of hobbyists. It embodies a rigorous approach to understanding, analyzing, and optimizing systems, processes, and operations crucial to society's advancement.

Mathematical modeling, a cornerstone of applied mathematics, is more than a mere problem-solving tool – it is a language deeply rooted in mathematical physics. Engineers and scientists leverage mathematical models to conceptualize system behaviors, rigorously testing these hypotheses through meticulous validation and verification.

Engineering modeling entails:

- Crafting mathematical representations of physical reality, expressed through equations and algorithms.
- Employing simulations of these mathematical models as proxies for real-world experimentation.
- Elevating the value of models through rigorous validation processes.

While engineering modeling shares some traits with fashion and hobby modeling, its essence lies in its ability to distill complex phenomena into actionable insights. Like fashion modeling:

- Models serve as idealized depictions of reality, facilitating understanding and analysis.
- They establish relationships between different system states and parameters.

Resonating with hobby modeling:

- Careful construction is paramount to ensure fidelity to the real-world system.
- Models amalgamate empirical laws and constitutive relations to capture system dynamics.
- The distinction between system states and parameters is pivotal in model formulation.

Historically, engineering relied heavily on empirical methods, with designs forged through trial and error experimentation. However, the advent of mathematical modeling, coupled with modern computing tools, has revolutionized this paradigm. Mathematical models serve as powerful guides for equipment and process design, streamlining experimentation, and mitigating risks associated with empirical approaches.

Equation-based models, derived from fundamental principles like mass and energy conservation, thermodynamics, and kinetics, empower chemical engineers to address a myriad of questions, such as:

Chemical Engineering Analysis and Optimization Using MATLAB®, First Edition.
Weiguo Xie, Sam Toan, and Richard Davis.

Companion website: www.wiley.com/go/xie/chemicalengineering

- Will our concepts translate into practical solutions?
- What measurements are essential for validation?
- Which factors significantly influence system behavior?
- Is the proposed solution economically viable and safe?
- Can we effectively control and optimize the system?
- What scale can we feasibly implement?

Engineers and scientists navigate complex technical challenges with precision and confidence by strategically integrating mathematical modeling, propelling innovation, and societal progress. In navigating this terrain, we must balance simplicity and utility. Levenspiel's (2002) concept of "the US$10, US$100, and US$1000 models" serves as a reminder: while complex models may incur significant costs, they do not always offer proportional increases in understanding. Expensive experiments yield to relatively inexpensive computer simulations, offering a cost-effective alternative. Moreover, as depicted in Figure 1.1, the decreasing cost of computing contrasts with the rising costs of traditional experimentation over time.

Models vary in complexity based on their intended purposes. Simple models can often provide sufficient insight to guide decisions on whether to proceed with further development, pause for further investigation, or alter engineering directions. However, the demand for increased model detail grows as the need for accuracy and precision intensifies in design and decision-making processes.

The cost associated with developing a mathematical model hinges on the level of uncertainty deemed acceptable in the model predictions. As depicted in Figure 1.2, generally, the higher the tolerance for uncertainty in the calculated results, the lower the overall cost of the model.

The escalating costs of achieving greater certainty in model predictions often stem from the endeavor to solve the model equations. In preliminary engineering models, there is a propensity toward higher uncertainty – lower precision – particularly when efforts to mitigate uncertainty fail to justify the exponentially increasing modeling expenses.

Contrastingly, scientific models tend to gravitate toward minimizing uncertainty – higher precision – aiming for a comprehensive comprehension of natural phenomena. To navigate the uncertainty, engineers often incorporate design safety factors. For instance, a chemical engineer might design a distillation column by calculating the minimum number of ideal stages, then doubling this figure and adjusting the reflux to attain the desired separation (Walas 1990).

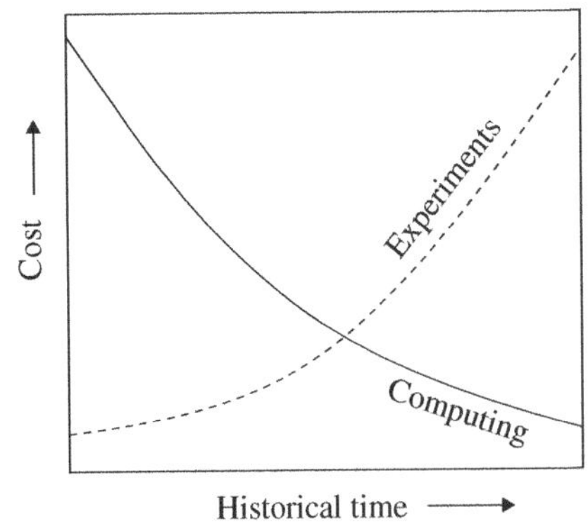

Figure 1.1 Computing costs are decreasing as the cost of experimentation increases over time.

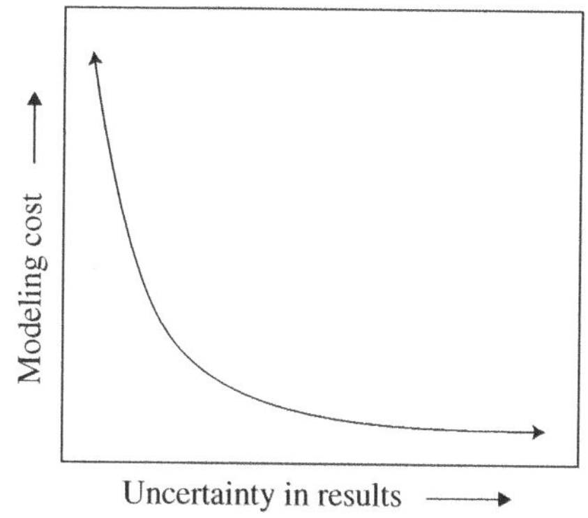

Figure 1.2 Cost of mathematical modeling as a function of the degree of uncertainty in the model predictions.

1.1 Numerical Methods

Numerical methods are computational techniques employed to approximate solutions to mathematical models that may prove inconvenient, challenging, or even impossible to solve through standard analytical methods. Analytical solutions yield symbolic representations, often involving explicit rearrangements of equations to isolate variables. However, when analytical solutions are unattainable, numerical methods offer practical approximations.

Consider calculating the molar volume of a gas using models that correlate molar volume with temperature and pressure. The ideal gas law and the Redlich–Kwong equation of state for nonideal behavior are two potential candidates. We can rearrange the ideal gas law to solve for the molar volume:

$$V = \frac{R_g T}{P} \tag{1.1}$$

where R_g is the ideal gas constant, T is the temperature, and P is the pressure. However, we cannot rearrange the following nonideal Redlich–Kwong model *explicitly* for either molar volume *or* temperature[1]:

$$P = \frac{R_g T}{V - b} - \frac{a}{V(V + b)\sqrt{T}} \tag{1.2}$$

The parameters a and b are functions of the critical temperature and gas pressure. There is no way to rearrange Eq. (1.2) with the molar volume only appearing on one side of the equation. Instead, we determine the Redlich–Kwong equation's molar volume from numerical approximation methods for specific cases of T and P.

Many numerical methods rely on iterative calculations, often facilitated by computers. Compared to analytical solutions, numerical approaches can be easier to implement, saving time and effort that can be directed toward addressing other challenges.

In chemical engineering, typical problems involve large systems of linear equations arising from applying fundamental principles like the conservation of momentum, mass, and energy. Additionally, nonlinear equations abound due to the inherent complexities of physical and chemical properties, transport phenomena, thermodynamics, and chemical reactions.

Some common examples of nonlinear functions encountered in chemical engineering include the temperature-dependent Antoine equation for vapor pressure, the modified

1 Explicit is defined as an expression containing only independent variables.

Arrhenius function for reaction rate constants, and the modified Henri function for enzyme reaction kinetics with substrate inhibition.

$$\text{Antoine} \quad \ln P_v(T) = A - \frac{B}{C+T} \tag{1.3}$$

$$\text{Modified Arrhenius} \quad k(T) = AT^2 \exp\left(-\frac{E_a}{R_g T}\right) \tag{1.4}$$

$$\text{Modified Henri} \quad v(S) = \frac{v_m S}{K_m + aS + bS^2} \tag{1.5}$$

While linear equations possess analytical solutions, determining the solution of a system larger than three or four equations can often prove tedious. Analytical solutions for such large systems entail extensive algebraic manipulations and symbolic bookkeeping. Moreover, many nonlinear equations lack analytical solutions altogether. Hence, methods for obtaining practical approximations of nonlinear functions and large systems of linear equations are essential.

To underscore the necessity of numerical methods, let us consider a simple model from chemical reaction engineering.

The following reaction mechanism is elementary, irreversible, and first order in species A:

$$A \rightarrow B \tag{1.6}$$

Consider the steady-state mole balance for product species B around a perfectly stirred chemical reactor, as illustrated in Figure 1.3, with feed and effluent concentrations of C_{A0} and unreacted C_A and product C_B, respectively.

We start with species conservation to mathematically model this reactor. With no B in the feed, the concentration of B leaving the reactor equals the product of the residence time, τ, with the rate of generation of B per unit volume in the reactor, r_B:

$$C_B = \tau\, r_B \tag{1.7}$$

For elementary, first-order reactions, the rate of production of B is a linear function of concentration:

$$r_B = k(C_{A0} - C_B) \tag{1.8}$$

where $(C_{A0} - C_B)$ is equivalent to the concentration of unreacted A in the reactor in terms of the initial concentration of reactant A, and k is the first-order reaction rate constant. We can rearrange Eqs. (1.7) and (1.8) explicitly for the concentration of product C_B exiting the reactor:

$$C_B = \frac{\tau k C_{A0}}{1 + \tau k} \tag{1.9}$$

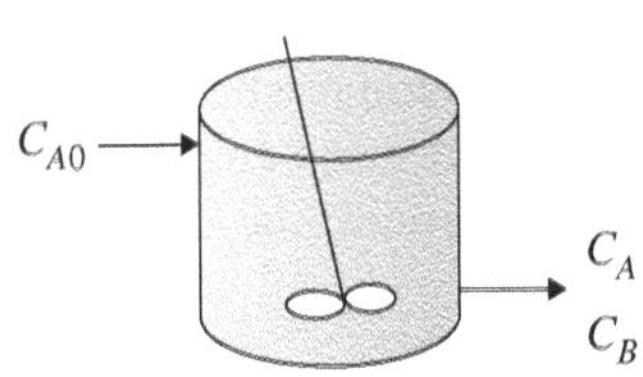

Figure 1.3 Well-mixed, steady-state reactor to produce B from A.

However, we typically describe the generation rate by a nonlinear concentration function for higher-order elementary or nonelementary reactions. Thus, for example, a catalyzed reaction may behave according to some form of Langmuir kinetics:

$$r_B = \frac{k(C_{A0} - C_B)}{\left[1 + K_A(C_{A0} - C_B) + \frac{K_B}{C_B}\right]^2} \tag{1.10}$$

where K_A and K_B are reaction equilibrium constants for reversible steps in the reaction mechanism. Substituting the rate law from Eq. (1.10) into the mole balance of Eq. (1.7) yields a nonlinear equation implicit[2] in C_B:

$$C_B = \frac{\tau k(C_{A0} - C_B)}{\left[1 + K_A(C_{A0} - C_B) + \frac{K_B}{C_B}\right]^2} \tag{1.11}$$

We cannot rearrange Eq. (1.11) explicitly for C_B. Instead, we calculate C_B iteratively by guessing the solution in a "trial and error" approach[3] until our C_B calculations converge on the solution.

The previous illustration demonstrates the need for numerical methods to solve nonlinear equations. Other problems encountered in chemical engineering may require solutions to nonlinear integrals,

$$I(x) = \int f(x)dx \tag{1.12}$$

or coupled systems of nonlinear, nonhomogeneous, "initial-value" first-order differential equations, with f and g representing nonlinear functions of some combination of the independent and dependent variables:

$$\frac{dy}{dx} = f(w, x, y) \text{ and } \frac{dw}{dx} = g(w, x, y) \tag{1.13}$$

Our models may involve nonhomogeneous, second-order "boundary-value"-type differential equations:

$$\frac{\partial^2 C}{\partial x^2} + \frac{\partial^2 C}{\partial y^2} + \lambda r(C) = \frac{\rho}{\alpha}\frac{\partial C}{\partial t} \tag{1.14}$$

C is the concentration, $r(C)$ may be a nonlinear rate law, and α, ρ, and λ are model parameters that may be the functions of x, y, or t.

1.1.1 Linear vs. Nonlinear Equations

We can easily distinguish between linear and nonlinear functions using visual graphical analysis (Figure 1.4):

---

2 Implicit is defined as an expression in which the dependent variable and one or more independent variables are not separated on opposite sides of an equation. http://www.merriam-webster.com/dictionary/implicit.

3 Trial and error methods iterate on two steps: (1) Trial = guess the solution and evaluate the result; (2) Error = use an estimate of the error from the trial to upgrade the guess for a better solution.

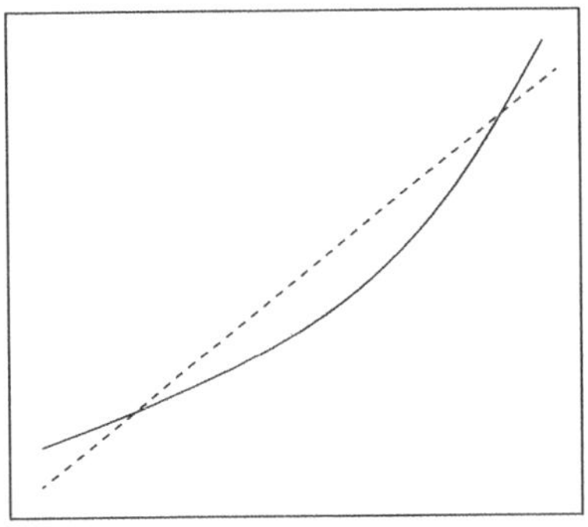

Figure 1.4 Graphical analysis to detect linear vs. nonlinear functions.

Without a graphical analysis, we recognize nonlinear equations from their terms involving exponents, transcendental functions, or combinations of dependent variables. To illustrate nonlinearity, consider the following second-order, ordinary differential equation:

$$\frac{d^2y}{dx^2} + \frac{x}{y}\frac{dy}{dx} + x^2y^2 = 0 \tag{1.15}$$

The first term in Eq. (1.15) is linear in the dependent variable, y. However, the second and third terms are nonlinear in y because they involve the product of two functions of y. Consequently, Eq. (1.15) is a nonlinear ordinary differential equation and may not have an analytical solution.

Of course, we may cite the well-known common exception – the second-order polynomial or quadratic equation:

$$ax^2 + bx + c = 0 \tag{1.16}$$

with two roots calculated by the standard quadratic formula[4]:

$$x_1 = \frac{-b - \sqrt{b^2 - 4ac}}{2a} \text{ and } x_2 = \frac{-b + \sqrt{b^2 - 4ac}}{2a} \tag{1.17}$$

Higher-order polynomials of order $m > 2$ have m roots.

$$\alpha x^m + \beta x^{m-1} + \cdots ax^2 + bx + c = 0 \tag{1.18}$$

In principle, an analytical solution for the m number of roots of a polynomial is possible. However, in practice, analytical solutions of polynomials with m greater than three are not trivial. When analytical solutions are difficult or impossible to achieve, one must resort to numerical approximations as a matter of practicality.

1.1.2 Standard Features of Numerical Methods

Most numerical methods share several standard features:

- They rely on local linear approximations of nonlinear functions to simplify calculations.
- Error approximation is utilized to refine the solution.
- They employ iteration and successive approximation to improve solutions.
- They are susceptible to computational round-off errors, which may lead to instability in the solution method.

4 See Section 1.1.3 to learn about computing problems with the standard quadratic formula.

Many numerical methods necessitate robust linear approximations to nonlinear equations at a local level. As discussed earlier, nonlinear equations can take various forms – algebraic, differential, or integral. Our solution approximation techniques are based on locally linearizing the nonlinear equations through truncated Taylor series expansions for each problem type. Novice practitioners of numerical methods soon discover that a surprisingly small set of numerical techniques, based on the Taylor series, can solve a wide range of nonlinear problems.

Linear approximations for nonlinear functions enable approximate analytical solutions to nonlinear problems. We gain access to all the analytical tools required to solve linear equations by employing good linear approximations of nonlinear functions. We utilize linear approximation error estimates iteratively to refine solutions until the error falls below our specified threshold. Each refinement step is considered one iteration. While a rudimentary approach to finding an approximate solution involves guessing and checking, we aim to systematically improve the approximation through numerical methods, removing guesswork from the process.

Once we grasp these fundamental numerical computation features, we can delve into the specifics of individual methods.

Problems involving solutions across a range of time or space necessitate a sequence of calculations. We successively approximate the solution at different points in time or space based on the solution at previous points. For instance, we might approximate solutions across spatial and temporal dimensions by discretizing the problem's time and space domains. Whether aiming for a precise solution at a single point or traversing through a function to obtain a range of solutions, iteration, and successive approximation principles remain relevant.

A common point of confusion arises when implementing iteration in a computer program, where we replace a variable's value with a mathematical operation involving that variable. For instance, from a mathematical standpoint, we might stipulate that the following equation is only valid when $\phi = 0$:

$$x = x + \phi \tag{1.19}$$

When we state that ϕ must equal zero, we imply that the direction of equality goes both ways (e.g. $x + \phi = x$). However, when setting up a sequence of iterative calculations using a computer, we mean that the current value of x is *replaced* by $x + \phi$, such that the flow of information is in one direction: $x \leftarrow x + \phi$. For example, suppose we need to calculate the sum of a series $\phi_1, \phi_2 \ldots \phi_n$. We start by initializing $x = 0$ and $i = 0$, then, iterate on Eq. (1.20) until $i = n$:

$$i \leftarrow i + 1 \quad x_i \leftarrow x_{i-1} + \phi_i \tag{1.20}$$

where the subscript i is an index variable for ϕ and the sum of the values in the series is the final value of x when $i = n$. We make extensive use of this type of operation in numerical methods. As an illustration, if $x_0 = 0$ and $\phi = (1, 2, 3, 4, 5 \ldots)$, then $x_1 = 0 + 1 = 1, x_2 = 1 + 2 = 3$, $x_3 = 3 + 3 = 6, x_4 = 6 + 4 = 10, x_5 = 10 + 5 = 15 \ldots.$

1.1.3 Significant Figures and Computer Round-Off Error

In scientific and engineering practice, numerical results are typically reported with only the significant figures accounted for. Significance is attributed to digits deemed reliable within a number, with convention dictating that only the rightmost digit retains any significant

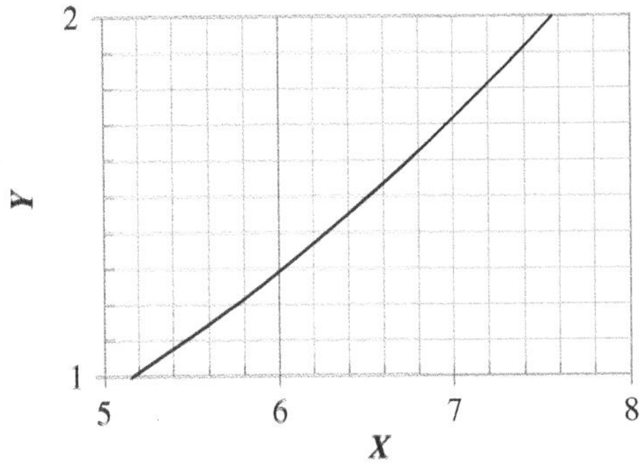

Figure 1.5 Reading values from a graph.

uncertainty. Computers store numbers in either single-precision format, allowing for seven significant digits, or double-precision format, accommodating 15 significant digits. When a single-precision number extends to eight significant digits, the computer rounds it to seven significant digits.

Consider reading a value from the Y-axis of the following graph, where the smallest division is 1/10 or 0.1. If a value at $X = 6.4$ correlates to $Y = 1.46$, only three significant figures are appropriate – the last digit, "6," being uncertain (Figure 1.5).

Consider the addition of two numbers in *the MATLAB* programming environment using single vs. double-precision values:

- *Single-precision*: 123456789 + 987654321 = 1111111168 (round to seven significant figures 1111111000, **incorrect**)
- *Double-precision*: 123456789 + 987654321 = 1111111110 (**correct**)

The single-precision result is accurate to just seven significant figures, whereas the double-precision result gives the correct answer. This example teaches us not to trust single-precision results involving more than seven significant figures. How far can we go with double precision?

$$123456789123456789 + 987654321987654321 = 1111111111111110000$$

The result has 15 significant figures! What do we conclude? Use double precision.

By default, MATLAB uses double precision. However, simply using double precision does not eliminate all round-off errors. Avoid adding or subtracting numbers with extreme differences in their orders of magnitude. For example, in double precision, $123456789 \pm 0.00000001 = 123456789$, with no change! Generally, we should sum up the small positive numbers first and sum the small negative numbers before combining them with large numbers.

Example 1.1 ***Computer Round-Off Errors in Derivative Approximations***

We use the following definition of the calculus derivative of a function to demonstrate computer round-off errors and the limits of numerical approximation:

$$\frac{df}{dx} = \lim_{\Delta x \to 0} \frac{f(x + \Delta x) - f(x)}{\Delta x} \tag{1.21}$$

Our test function with its analytical result for the derivative evaluated at $\pi/4$:

$$f(x) = \sin(x) \qquad \frac{df}{dx} = \cos(x) \qquad \cos\left(\frac{\pi}{4}\right) = \frac{1}{\sqrt{2}} = 0.70711 \tag{1.22}$$

Substitute the test function into Eq. (1.21) for a first-order approximation:

$$\frac{df}{dx} \cong \frac{f\left(\frac{\pi}{4}+\Delta x\right)-f\left(\frac{\pi}{4}\right)}{\Delta x} = \frac{\sin\left(\frac{\pi}{4}+\Delta x\right)-\sin\left(\frac{\pi}{4}\right)}{\Delta x} \tag{1.23}$$

Solution:

```
%MATLAB code for Example 1.1
% Define the function to be plotted
%f = @(dx) (sin(pi/4 + dx) - sin(pi/4)) ./ dx;
syms dx
f = (sin(pi/4+dx)-sin(pi/4))/dx;
g = abs((cos(pi/4)-(sin(pi/4+dx)-sin(pi/4))/dx)/cos(pi/4));
% Plot f(dx)
figure
fplot(f, [1e-16, 1], 'LineWidth', 1)
set(gca, 'XScale', 'log', 'YScale', 'log') % Set both axes to
% logarithmic scale
xlabel('dx')
ylabel('df/dx')
grid on
% Plot g(dx)
figure
fplot(g, [1e-16, 1], 'LineWidth', 1)
set(gca, 'XScale', 'log', 'YScale', 'log') % Set both axes to
% logarithmic scale
xlabel('dx')
ylabel('|% Error|')
grid on
```

The first *semi*-log plot in Figure 1.6 shows the numerical derivative approximation of the test function using Eq. (1.23) approaching the true solution near 0.71 as Δx decreases before the error becomes noticeable at small Δx. Note that the plots show decreasing values for Δx moving from left to right for convenience.

The second log–log plot shows the absolute percent error in the numerical approximation of the test function's derivative function as $\Delta x \rightarrow 0$. The error in the derivative approximation decreases exponentially as expected until $\Delta x \approx 10^{-8}$, then starts to climb again to noticeable values because of round-off error. These results have implications for controlling computational round-off error when we use derivative approximations for several numerical methods to find roots of algebraic equations, optimization, and uncertainty analysis.

Note that the abscissa (Δx-axis) scale is reversed for values in decreasing magnitude.

Example 1.2 ***Avoid Round-Off Errors in the Quadratic Formula***
One of the two standard roots to a quadratic equation, defined by Eqs. (1.16) and (1.17), potentially loses precision due to round-off error through subtraction. To avoid a loss of precision, we keep the root involving addition (x^+) and find the second root (x^-) from the relationships:

$$x^+ \cdot x^- = \frac{c}{a} \text{ and } x^- = \frac{c}{a \cdot x^+} \tag{1.24}$$

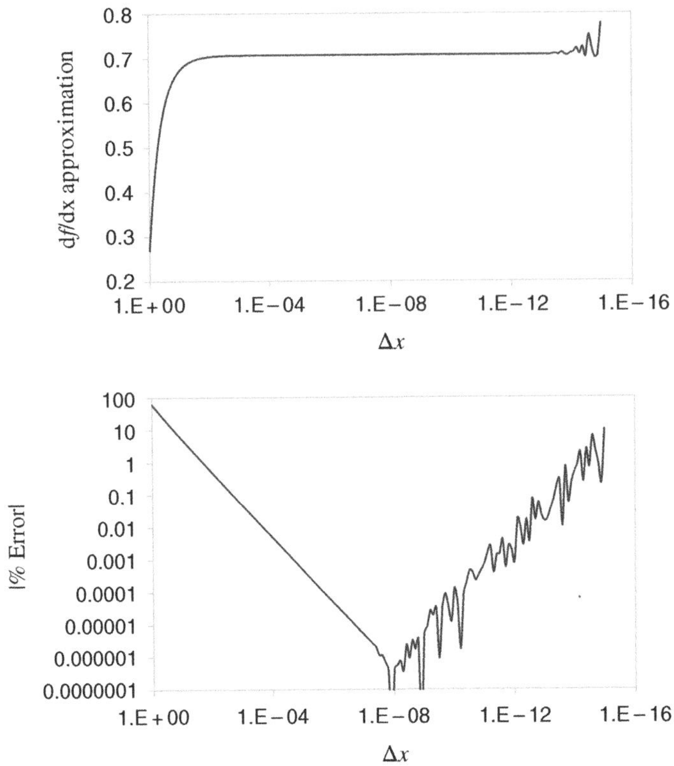

Figure 1.6 Semi-log plot of the numerical derivative approximation in Eq. (1.23) vs. Δx. The numerical approximation approaches the true solution before the round-off error becomes noticeable at $\Delta x = 10^{-13}$. Log–log plot displaying the percent error in the numerical derivative approximation of sin(x) evaluated at $\pi/4$. Round-off errors in the computer cause the approximation error to increase for small numbers.

Thompson's (1987) example of a quadratic root involves a buffer solution of 0.0010 M chloroacetic acid and 0.0100 M sodium chloroacetate. The chemical equilibrium relationship in terms of the hydrogen ion concentration x is

$$1.4 \times 10^{-3} = \frac{x(0.0100 + x)}{0.0010 - x}$$

which reduces to a quadratic function in x:

$$x^2 + 0.0114x - 1.4 \times 10^{-6} = 0$$

The standard roots using 10^{-4} precision are

$$x = \frac{-0.0114 \pm 0.0116}{2}$$

The first root involves a subtraction to give an incorrect answer $x^- = (0.0116 - 0.0114)/2 = 0.0001$. The second root involves an addition (of two terms with the same sign) to give a correct value of $x^+ = (-0.0114 - 0.0116)/2 = -0.0115$. Use x^- in Eq. (1.24) to get the other correct root:

$$x^- = \frac{-1.4 \times 10^{-6}}{-0.0115} = 1.22 \times 10^{-4}$$

Press et al. (1992) recommend the following alternative form of the quadratic rule that eliminates round-off error by avoiding subtraction between two potentially small terms:

$$q = -0.5\left[b + \text{sign}(b)\sqrt{b^2 - 4ac}\right] \tag{1.25}$$

where

$$x_1 = \frac{q}{a} \text{ and } x_2 = \frac{c}{q} \tag{1.26}$$

and

$$\text{sign}(b) = \begin{cases} -1 \text{ for } b < 0 \\ +1 \text{ for } b \geq 0 \end{cases} \tag{1.27}$$

To illustrate further, compare the calculation with the roots in the following equation:

$$f(x) = 0.321x^2 + 214x + 0.00123 = 0 \tag{1.28}$$

Solution:

```
%MATLAB code for Example 1.2
% Coefficients of the quadratic equation
a = 0.321; % Quadratic coefficient
b = 213;   % Linear coefficient
c = 1.23e-3; % Constant term
% Standard quadratic formula
x1_standard = (-b + sqrt(b^2 - 4*a*c)) / (2*a);
x2_standard = (-b - sqrt(b^2 - 4*a*c)) / (2*a);
% Alternative method for numerical stability
q = -0.5 * (b + sign(b) * sqrt(b^2 - 4*a*c));
x2_correct = q / a;
x1_correct = c / q;
% Function evaluations to check correctness
f1_standard = a*(x1_standard^2) + b*x1_standard + c; % Should be
% close to 0
f1_correct = a*(x1_correct^2) + b*x1_correct + c; % Should be
% close to 0
% Display results
fprintf('Standard method roots: x1 = %.16f, x2 = %.16f\n',...
 x1_standard, x2_standard);
fprintf('Alternative method roots: x1 = %.16f, x2 = %.16f\n',...
 x1_correct, x2_correct);
fprintf('Function values at standard roots: f(x1) = %.16e,...
 f(x2) = %.16e\n', f1_standard, a*(x2_standard^2) +...
 b*x2_standard + c);
fprintf('Function values at alternative roots: f(x1) = %.16e,...
 f(x2) = %.16e\n', f1_correct, a*(x2_correct^2) + b*x2_correct...
 + c);
```

Comparing the two solution methods for the root in Table 1.1 reveals the problem with the standard version of the quadratic formula. The bottom row shows the standard solution's

Table 1.1 Comparing the standard quadratic formula with an alternative to eliminate round-off errors using $a = 0.321$, $b = 213$, and $c = 1.23 \times 10^{-3}$.

Standard quadratic formula	Alternative quadratic formula
—	$q = -212.999998146338000000$
$x_1 = -0.00000577464792771470$	$x_1 = c/q = -0.00000577464793757862$
$x_2 = -663.551396094511000000$	$x_2 = q/a = -663.551396094511000000$
$f(x_1) = 0.00000000000210101511$	$f(x_1 = c/q) = 0.00000000000000000000$

inaccuracy when substituting the first root into Eq. (1.28). From this example, we learn to be careful with arithmetic involving the addition or subtraction of an extremely large number and an extremely small number.

1.2 Mathematical Model Building

Henri's equation is a classic example of empirical model building, which he discovered for describing the rate of enzyme-catalyzed reactions as a function of the reacting substrate concentration. A typical plot in Figure 1.7 of reaction rate, r, vs. substrate concentration, $[S]$ reveals the asymptotic nature observed in many catalysis reactive systems.

The curve plotted in Figure 1.7 represents a *graphical* model. Therefore, we may interpret the function's behavior directly from the plot. Henri observed that the reaction rate is proportional to the substrate concentration, $[S]$, for small substrate concentrations. For large $[S]$, the rate approaches an asymptotic maximum value, $r \rightarrow r_{max}$. He proposed the following mathematical equation, or model, to match the observations of the data:

$$r = \frac{a[S]}{b + [S]} \tag{1.29}$$

where a and b are the constants specific to the reaction mechanism. For a small substrate concentration, $[S] \ll b$, the rate expression approaches a linear function in $[S]$ with proportionality constant a/b. In the limit of large $[S] \gg b$, the reaction rate approaches a constant, maximum value, $a = r_{max}$:

$$r \cong \begin{cases} \frac{a}{b} & [S] \ll b \\ a & [S] \gg b \end{cases} \tag{1.30}$$

We may also determine the value of parameter b by recognizing that parameter a disappears from Eq. (1.29) when $r = a/2$:

$$\frac{a}{2} = \frac{a[S]}{b + [S]} \rightarrow \frac{1}{2} = \frac{[S]}{b + [S]} \rightarrow b = [S] \quad \text{at } r = \frac{a}{2} \tag{1.31}$$

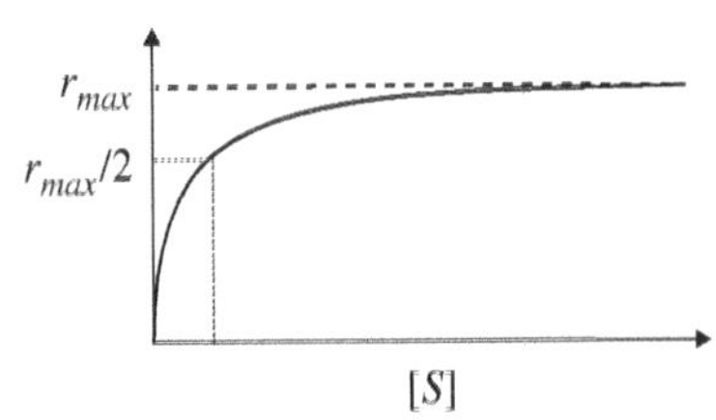

Figure 1.7 Henri plot of enzyme reaction rate, r, as a function of substrate concentration s.

Henri's model successfully fits enzyme reaction data for interpolation but does not explain *how* or *why* the reaction kinetics behave this way. Menten and Michaelis (1913) proposed a two-step reaction mechanism to explain how Henri's model works. Their model involves an initial, reversible step that forms a complex intermediate molecule of enzyme–substrate. Then, the enzyme–substrate complex reacts irreversibly to form the reaction product in the second step:

$$S + E \underset{k_{-1}}{\overset{k_1}{\rightleftharpoons}} ES \tag{1.32}$$

$$ES \xrightarrow{k_2} E + P \tag{1.33}$$

E is the free enzyme, ES is the enzyme–substrate complex, and k_1, k_{-1}, and k_2 are the forward and reverse reaction rate constants for the two steps. If the second step is rate limiting, the reaction rate is proportional to the concentration of the complex $[ES]$:

$$r = k_2[ES] \tag{1.34}$$

while the first step reaches a pseudo-equilibrium state:

$$\frac{[ES]}{[S][E]} = \frac{k_1}{k_{-1}} \tag{1.35}$$

The conservation of enzyme catalyst in the reaction yields the following balance:

$$[E] = [E]_0 - [ES] \tag{1.36}$$

where $[E]_0$ is the initial enzyme concentration in the reactor. Combining Eqs. (1.34) through (1.36) to eliminate $[E]$ and $[ES]$ gives the well-known Michaelis–Menten expression for the reaction rate:

$$r = \frac{k_2[E]_0[S]}{\frac{k_{-1}}{k_1} + [S]} = \frac{r_{\max}[S]}{K_M + [S]} \tag{1.37}$$

where K_M is the Michaelis equilibrium constant. We should confirm that Eq. (1.37) has the same basic functionality of reaction rate with substrate concentration as the Henri equation (1.29) (i.e. what are a and b?) and comes with an explanatory mechanism in Eqs. (1.32) and (1.33). We encounter the alternative Briggs–Haldane mechanism in the next section.

Like Henri's equation, empirical models lack a foundation in the fundamental principles of momentum, energy, or mass conservation. Nevertheless, they can incorporate mathematical functionality that captures trends in the data. Polynomials and rational functions, such as Eq. (1.29), serve as convenient models for interpolating experimental or process data. However, they may struggle to elucidate the underlying mechanisms of the system.

We introduce various approaches to mathematical modeling, each based on underlying assumptions about the system. We commence with a review of the inviolable law of conservation, as our models are rendered ineffective if they fail to account for all mass and energy inputs. Subsequently, we explore pairs of opposing principles (Rasmuson et al. 2014), with underlying assumptions denoted in brackets: mechanistic (based on physics) vs. empirical, steady (time-independent) vs. unsteady-state, lumped (uniform or well-mixed) vs. distributed systems, and stochastic (incorporating uncertainty) vs. deterministic modeling.

1.2.1 Laws of Conservation

Mathematical models bridge the descriptive language of conservation equations and the symbolic language of mathematics. While not mandatory, expressing governing equations in symbolic mathematical form proves significantly more convenient and efficient. To illustrate, attempt to describe the derivative of the Henri equation's reaction rate (1.29) concerning substrate concentration without employing mathematical terms. It is undoubtedly challenging!

Constructing a mathematical model involves establishing conservation laws for momentum, mass, and energy. Initially, boundaries are defined around the system's control volume. Subsequently, rate balances are applied to the system and across the defined boundaries for each conservation law:

$$(\text{Rate of } \xi \text{ in}) - (\text{Rate of } \xi \text{ out}) + (\text{Net rate of } \xi \text{ generation}) = (\text{Rate of } \xi \text{ accumulation}) \tag{1.38}$$

where ξ may represent mass, energy, or momentum. Conservation laws often yield nonlinear algebraic, integral, or differential equations. Operations on these equations are straightforward, using standard analytical and numerical techniques. Mathematical models replace each term in Eq. (1.38) with its equivalent symbolic mathematical representation of measurable, quantifiable parameters. As an illustration, the volumetric flow rate product with density, $Q \times \rho$, may replace the mass (rate in). Check that the dimensions of each term in Eq. (1.38) are mutually consistent. The dimensions are usually in velocity units, mass (or moles), or energy per time. Thus, if we describe the (rate in) term by the flux (rate per unit area), multiply by the normal cross-sectional area to convert the flux into a rate. Table 1.2 contains typical constitutive rates for transport phenomena.

By definition, a batch reactor has neither inlet nor outlet streams, which leaves only the generation and accumulation terms in the mass balance:

$$0 - 0 + (\text{Product generation rate}) = (\text{Product accumulation rate}) \tag{1.39}$$

Steady-state systems have no accumulation terms. A steady-state batch process has reached an equilibrium where the net generation rate is zero. This leads to one form of the chemical equilibrium constant as a ratio of reaction rate constants for reversible reactions. For example, consider the first-order reversible reaction $A \rightleftharpoons B$. At steady-state for this reaction, Eq. (1.39)

Table 1.2 Constitutive expressions of transport rates.

	Rate	Diffusive	Advection	Convection
Newton's law of viscosity	Momentum transfer	$\tau = -\mu \dfrac{\partial u}{\partial x}$	$V\rho u$	—
Fourier's law of heat conduction	Heat transfer	$q = -A\kappa \dfrac{\partial T}{\partial y}$	$q = \dot{m}\left(H^0 + \int_{T^0}^{T} c_p dT\right)$	$q = hA\Delta T$
Fick's law of molecular diffusion	Mass transfer	$N = -AD\dfrac{\partial C}{\partial y}$	$\dot{m} = A\rho u$	$\dot{m} = kA\Delta C$

A = area, c_p = specific heat, h = heat transfer coefficient, H = enthalpy, k = mass transfer coefficient, m = mass flow rate, T = temperature, u = velocity, V = volume, x = length dimension, κ = thermal conductivity, ρ = density, and μ = dynamic viscosity.

reduces to the following algebraic expression:

$$-r = k_f C_A - k_r C_B = 0 \tag{1.40}$$

where k_f and k_r represent the forward and reverse reaction rate constants, respectively. Rearrange Eq. (1.40) for the reaction equilibrium constant:

$$K = \frac{C_B}{C_A} = \frac{k_f}{k_r} \tag{1.41}$$

Equation (1.35) is another example of an equilibrium constant defined in terms of rate constants.

The generation term in Eq. (1.39) requires a reaction rate law describing a molecular species' net generation (generation–consumption). Chemical reactions conserve mass, but not necessarily molecules. The Michaelis–Menten reaction rate law derived from Eq. (1.37) is an example of chemical species generation. Consider an alternative model of enzyme reactions proposed by Briggs and Haldane to develop further the steady-state, equilibrium, accumulation, and generation concepts. Instead of assuming an equilibrium step, as in the case of Michaelis and Menten, Briggs and Haldane proposed conservation equations for each species, including the enzyme–substrate complex:

Rate of generation = Rate of accumulation

$$k_1[S]\cdot[E] - k_{-1}[ES] - k_2[ES] = \frac{d[ES]}{dt} \text{ and } k_2[ES] = \frac{d[P]}{dt} \tag{1.42}$$

Briggs and Haldane assumed that the enzyme–substrate complex concentration conservation quickly reaches a pseudo-steady-state condition, $d[ES]/dt \cong 0$. Thus, their model combines Eqs. (1.36) and (1.42) to eliminate $[E]$ and $[ES]$ from the enzyme-catalyzed reaction rate expression:

$$r = \frac{k_2[E]_0[S]}{\dfrac{k_{-1} + k_2}{k_1} + [S]} \tag{1.43}$$

Equation (1.43) maintains the same functionality as the Henri and Michaelis–Menten equations but contains an alternative form of the Michaelis constant.

Another approach solves the conservation equations without assumptions of pseudo-equilibrium or steady state for the reaction steps. However, this approach does not yield a simple algebraic equation. Instead, we must solve a simultaneous system of differential and algebraic equations:

$$\frac{d[ES]}{dt} = k_1[S]\cdot[E] - k_{-1}[ES] - k_2[ES] \text{ and } \frac{d[P]}{dt} = k_2[ES] \tag{1.44}$$

$$[E]_0 = [E] + [ES] \text{ and } [S]_0 = [S] + [P] \tag{1.45}$$

The initial concentrations of the enzyme–substrate complex $[ES]$ and product $[P]$ are assumed to be zero.

Indeed, the heat-generation term in a thermal energy conservation equation becomes essential when there are endo- or exothermic chemical reactions or electrical heating within the system boundaries. It is crucial to note that the heat-generation term can be either negative or positive, depending on the nature of the processes involved.

For instance, a chemical reaction that consumes molecular species typically results in a negative heat-generation term. Conversely, an exothermic reaction, conventionally characterized by a negative heat of reaction, leads to a positive heat-generation term. Conversely, an endothermic reaction, where the reaction absorbs heat from its surroundings, results in a negative heat-generation term. This occurs when the rate of consumption of the reactant (r_A) is negative.

Understanding the sign convention for heat generation is vital for accurately modeling and analyzing thermal systems involving chemical reactions or other sources of heat generation. It allows engineers to effectively predict and control heat transfer processes, ensuring optimal performance and safety in various industrial applications.

$$\text{Chemical reaction energy: } q = V_s r_A \Delta H_{rxn,A} \tag{1.46}$$

where V_s is the system volume, r_A is the reaction rate of species A, and ΔH is the heat of the reaction.

Thermal energy generation may involve heat from an electrical resistor:

$$\text{Electrical energy: } q = \frac{I^2 R}{V} \tag{1.47}$$

where I, R, and V are the electrical current, resistance, and voltage, respectively.

1.2.2 Lumped vs. Distributed Systems

Indeed, engineers rely on mathematical models for design, simulation, and gaining insights into processes. These models often necessitate simplifying assumptions to tackle complexities arising from the system's geometry, flow patterns, or other properties. While these assumptions and approximations can introduce errors and limit the model's applicability, they are often indispensable for obtaining tractable solutions to mathematical equations.

Let us explore six examples demonstrating how to relax simple modeling assumptions to enhance descriptive capabilities:

1. *Incorporating nonlinearities*: Initially, assuming linear relationships between variables may simplify the model, but relaxing this assumption to account for nonlinearities can provide a more accurate representation of real-world behavior.
2. *Considering three-dimensional geometry*: Many models assume two-dimensional geometry for simplicity. However, relaxing this assumption to include three-dimensional effects can offer a more comprehensive understanding of the system's behavior.
3. *Accounting for turbulence*: Simplified models often assume laminar flow, neglecting turbulence effects. Relaxing this assumption to incorporate turbulent flow phenomena can improve the accuracy of predictions, particularly in fluid dynamics applications.
4. *Introducing time dependency*: Some models assume steady-state conditions, disregarding temporal variations. Engineers can more effectively capture dynamic behavior and transient effects by relaxing this assumption and incorporating time dependency.
5. *Incorporating spatial variability*: Simplified models often assume spatial uniformity, overlooking variations across different regions of the system. Relaxing this assumption to account for spatial variability allows a more nuanced understanding of localized phenomena.
6. *Considering non-Newtonian behavior*: Many models assume Newtonian fluid behavior due to its simplicity. However, relaxing this assumption to accommodate non-Newtonian

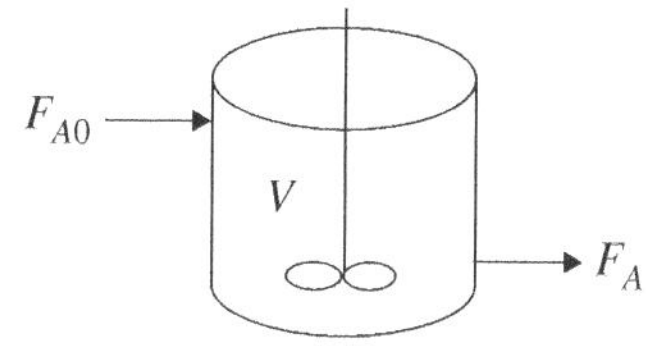

Figure 1.8 Well-mixed flow reactor.

behavior can be essential for accurately representing the rheological properties of complex fluids.

By relaxing these simplifying assumptions, engineers can develop models that better capture the intricacies of real-world systems, leading to more reliable predictions and insights for design, simulation, and troubleshooting processes.

We begin with a well-mixed reactor's steady-state isothermal operation, pictured in Figure 1.8. F_{A0} and F_A are species A's inlet and effluent molar flow rates, respectively, and V is the reactor volume.

The conservation equation for a molecular species in a steady-state reactor is simply the difference between the molar flow rates in and out of the reactor vessel and generation due to reaction:

$$F_{A0} - F_A + Vr_A = 0 \tag{1.48}$$

where r_A is the reaction rate of formation of A per unit volume of the reactor. Define conversion in a flow reactor in terms of molar flow rates in and out of the reacting species:

$$X_A = \frac{F_{A0} - F_A}{F_{A0}} \tag{1.49}$$

We solve Eq. (1.49) for F_A, then substitute it into the conservation Eq. (1.48) and rearrange it for alternative forms of species conservation in terms of conversion of reactant A:

$$\frac{V}{F_{A0}} = \frac{X_A}{-r_A} \text{ or } \frac{X_A}{V} = \frac{-r_A}{F_{A0}} \tag{1.50}$$

Numerical methods for finding roots to nonlinear algebraic equations may be required to solve Eq. (1.50) for X_A when the reaction rate law for r_A is nonlinear.

The straightforward design Eq. (1.53) for a well-mixed reactor overlooks the influence of mixing on the conversion of a chemical reaction. This lumped model assumes that the reactor's composition is uniform throughout, mirroring the composition of the exit stream. We require a model that considers spatial or temporal variations to represent mixing effects accurately.

For example, we can devise a model that incorporates mixing effects by simulating a reactor behaving as if it consists of two mixing zones, akin to two stirred reactors-in-series, as illustrated in the flow diagram depicted in Figure 1.9. This approach allows us to account for the gradual mixing of reactants and products within the reactor, providing a more realistic representation of the chemical reaction's conversion. We can better understand and optimize

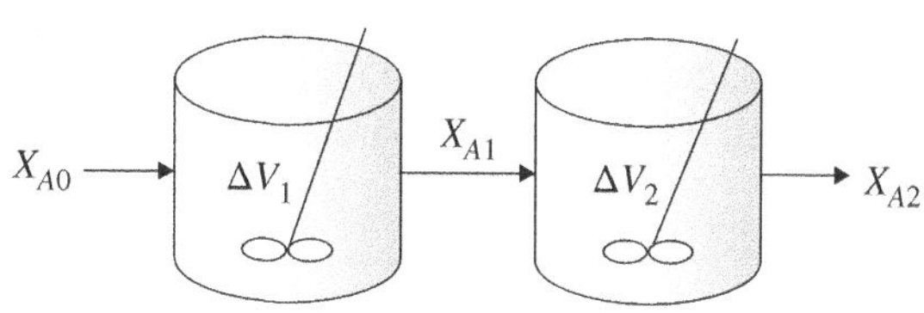

Figure 1.9 Two well-mixed tanks in series represent the reactor.

reactor performance in real-world applications by incorporating such spatial or temporal variations into our model.

The steady-state material balances for species A around each tank are:

$$\frac{X_{A1}}{\Delta V_1} = \frac{-r_{A1}}{F_{A0}} \text{ and } \frac{X_{A2} - X_{A1}}{\Delta V_2} = \frac{-r_{A2}}{F_{A0}} \tag{1.51}$$

where the sum of volumes for the mixing zones equals the total volume of the reactor:

$$V = \Delta V_1 + \Delta V_2 \tag{1.52}$$

Optimization tools may help determine the best split between the two volumes that match model predictions with experimental results for conversion.

Eventually, as the number of tanks-in-series increases to infinity, the volume of each tank becomes infinitely small, and we replace the ratio of conversion to volume with a derivative:

$$\lim_{\Delta V \to 0} \frac{X_{A,V+\Delta V} - X_{A,V}}{\Delta V} = \frac{dX_A}{dV} \tag{1.53}$$

In this case, the model approaches the distributed design equation for a continuous plug flow reactor (PFR), pictured in Figure 1.10:

$$\frac{dX_A}{dV} = \frac{-r_A}{F_{A0}} \tag{1.54}$$

The solution to Eq. (1.54) may require numerical methods to approximate solutions to integrals or ordinary differential equations when the reaction rate law, r_A, is nonlinear.

By adding increasing degrees of complexity, we transformed the simple algebraic design equation for a well-mixed reactor into the differential *PFR* design equation. The *PFR* equation accounts for changes in the axial direction of flow but ignores mixing effects in the radial direction (hence, "plug flow"). One accounting method for deviations from plug flow without a three-dimensional model introduces a dispersion term into the *PFR* balance. Dispersion accounts for uneven mixing in the radial direction by assuming that the gradient in the concentration along the flow axis drives the mixture away from a flat concentration profile in the radial direction, akin to Fick's law. The dispersion flux equation is

$$J_A = -S^2 E \frac{dC_A}{dV} \tag{1.55}$$

E is the dispersion coefficient, like a diffusion coefficient, and S is the cross-sectional area of the PFR. The molar flow in and out terms now takes the form.

$$\text{Molar flow rate in or out} = F_{A0} X_A + S J_A \tag{1.56}$$

Substitute the results from Eqs. (1.55) and (1.56) into the differential *PFR* design Eq. (1.54), divide each term by the differential volume, and take the limit as the differential volume goes to zero:

$$\lim_{\Delta V \to 0} \frac{\Delta \left[F_{A0} X_A + C_{A0} S^3 E \left(\frac{dX_A}{dV} \right) \right]}{\Delta V} = F_{A0} \frac{dX_A}{dV} + C_{A0} S^3 E \frac{d^2 X_A}{dV^2} = -r_A \tag{1.57}$$

The second-order differential Eq. (1.57) may require appropriate boundary conditions and a numerical method for approximating the solution to this boundary-value problem.

Figure 1.10 Plug flow reactor (PFR) with uniform conditions in the radial dimension.

Indeed, adding higher levels of rigor can transform the original algebraic model into a second-order differential equation, introducing increased complexity. Consequently, a simple reactor model utilizing the well-mixed design equation becomes more rigorous to accommodate imperfect mixing effects in a continuous flow reactor. However, if perfect mixing is a reasonable assumption, the simple algebraic well-mixed design equation yields the same result as more complex plug flow or dispersion models. This convergence occurs, for instance, as the dispersion coefficient approaches zero ($E \to 0$).

Generally, when constructing a mathematical model, making as many reasonable simplifying assumptions as possible is advisable. Only incorporate higher levels of rigor into the model as dictated by the problem's requirements. From a practical standpoint, there is no need to develop an expensive, highly rigorous mathematical model to describe a process if the solution to the model equations is intractable or if a more straightforward model provides satisfactory results (Figure 1.2). Often, simple models can offer sufficient information and insights to fulfill preliminary design and economic analysis needs.

The famous Bernoulli equation from fluid mechanics is another classic example of a lumped system that relates the fluid velocity to pressure in a pipe:

$$\frac{\Delta P}{\rho} + g\Delta z + \frac{\Delta u^2}{2} = 0 \tag{1.58}$$

where ΔP is the pressure difference between the fluid inlet and outlet of a system, g is the acceleration due to gravity, Δz is the vertical change in height between the fluid inlet and outlet points, u is the average fluid velocity, and ρ is the fluid density. When ignoring friction, the only information needed to determine the pipe flow is the conditions at the ends. The model requires neither the length nor the pipe geometry between the inlet and outlet, as displayed in Figure 1.11.

When we include a term in Eq. (1.58) to account for friction, we need to understand the complete geometry to calculate the friction term:

$$\frac{\Delta P}{\rho} + g\Delta z + \frac{\Delta u^2}{2} = F = \frac{2fLu^2}{d} \tag{1.59}$$

where f is the Fanning friction factor, and L and d are the tube's length and diameter, respectively. The nonlinear Colebrook equation gives the friction factor for flow in rough circular pipes as a function of the Reynolds number, $Re = \rho uD/\mu$, where ρ, u, D, and μ are the fluid density, velocity, pipe diameter, and viscosity, respectively. With this information, we can calculate the velocity or pressure at any position along the pipe's length using a numerical root-finding method.

Consider the mass and energy balances around a mixing tank shown in Figure 1.12. To simplify our analysis, we assume all the fluid mixing and heat transfer behaviors are lumped together. In this case, we cool a full tank of hot water with volume V and initial temperature T_0, with a stream of cold liquid with a constant volumetric flow rate Q and temperature T_{in}.

We derive an expression for calculating the water temperature as a function of time by assuming constant properties for water, a constant feed rate, and a well-mixed tank. Our model system is the water in the tank. Because the tank is full, the water exiting the tank must

Figure 1.11 Lumped equations ignore geometric considerations.

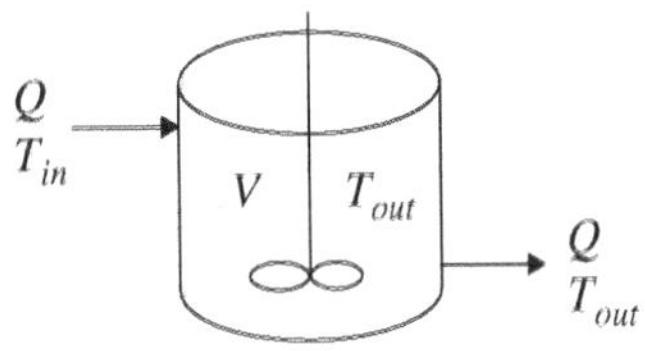

Figure 1.12 Mixed tank with volume V and constant inlet and outlet flow rate Q.

equal the entering rate. Thus, there is no mass accumulation or generation in the tank, i.e.:

$$\text{Rate of mass in} = \text{Rate of mass out} \tag{1.60}$$

For the case of constant density, the mass of water in the tank remains constant, such that:

$$\rho \frac{dV}{dt} = 0 \tag{1.61}$$

where ρ is the fluid density.

Next, we apply the energy conservation principle to the water in the tank. Because the system generates no energy within the water in the tank, the energy balance reduces to

$$(\text{Rate of energy in}) - (\text{Rate of energy out}) = (\text{Rate of energy accumulation}) \tag{1.62}$$

We replace the terms with their mathematical equivalents from thermodynamics:

$$Q\rho c_p(T_{in} - T) = V\rho c_p \frac{dT}{dt} \tag{1.63}$$

where c_p is the average specific heat of the water. We separate variables and integrate Eq. (1.63) assuming constant Q and V:

$$\frac{Q}{V}\int_0^t dt = \int_{T_0}^{T} \frac{dT}{(T_{in} - T)} \tag{1.64}$$

then solve Eq. (1.64) for the temperature in the tank at time t:

$$T = T_{in} - (T_{in} - T_0)\exp\left(-\frac{t}{\tau}\right) \tag{1.65}$$

where the residence time is $\tau = Q/V$. The function in Eq. (1.65) behaves asymptotically: $T \to T_{in}$ as $t \to \infty$.

Study heat convection from a hot aluminum sphere with diameter d suddenly plunged into an infinite volume of colder fluid (also known as quenching). Our model system is a solid sphere. If the diameter of the sphere is small, and the thermal conductivity of the material is large (true for aluminum) such that the temperature distribution in the solid is practically uniform, the energy balance gives:

$$0 - (\text{Rate of energy out}) = (\text{Rate of energy accumulation})$$

$$-h(T - T_\infty)\frac{\pi d^2}{4} = \frac{\pi d^3}{6}\rho c_p \frac{dT}{dt} \tag{1.66}$$

with the initial condition $T = T_0$ at $t = 0$.

We assume the temperature in Eq. (1.66) is independent of position in the sphere (lumped).

On the other hand, it may be necessary to consider variations in the system's physical space without uniform properties. We treat this type of problem at the local level of the system to distribute the properties of viscosity, temperature, or concentration. The modeling approach for a distributed system applies the conservation equations in a shell balance to a local finite volume element. We describe the local *in* and *out* terms of a finite volume element with

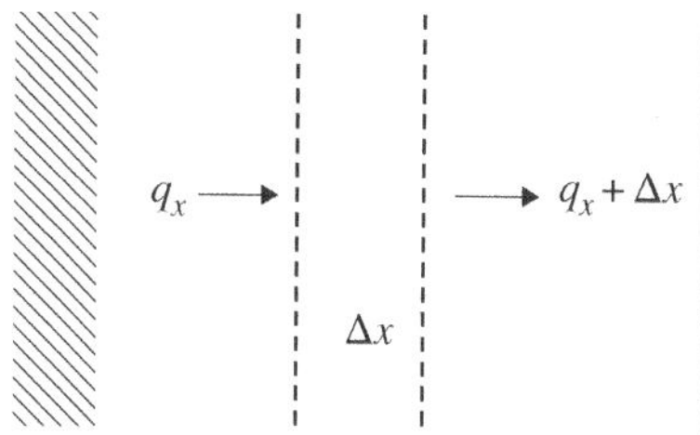

Figure 1.13 Energy balance around a finite wall element.

constitutive equations, such as Newton's law of viscosity, Fick's law of diffusion, or Fourier's law of heat conduction from Table 1.2. For example, consider the illustration of heat conduction through a wall in Figure 1.13. The left side of the wall is heated. The right side of the wall loses heat by convection to a neighboring fluid. Start with a differential element having a thickness Δx at any point in the wall along the x-direction, as illustrated in Figure 1.13.

An energy balance around the differential length Δx gives:

(Rate of energy in) – (Rate of energy out) = (Rate of energy accumulation)

$$-A\kappa \left.\frac{\partial T}{\partial x}\right|_x - A\left(-\kappa \left.\frac{\partial T}{\partial x}\right|_{x+\Delta x}\right) = A\Delta x \rho c_p \frac{\partial T}{\partial t} \tag{1.67}$$

where the terms on the left side represent heat conduction in and out by Fourier's law. The term on the right side accounts for the accumulation of thermal energy. Divide Eq. (1.67) by the differential volume of the wall represented by the product of the cross-sectional area and differential thickness, $A\Delta x$, and then take the limit as the differential length goes to zero:

$$\lim_{\Delta x \to 0} \frac{-A\kappa \left.\frac{\partial T}{\partial x}\right|_x - A\left(-\kappa \left.\frac{\partial T}{\partial x}\right|_{x+\Delta x}\right) = A\Delta x \rho c_p \frac{\partial T}{\partial t}}{A\Delta x} \to \frac{\partial}{\partial x}\left(\kappa \frac{\partial T}{\partial x}\right) = \rho c_p \frac{\partial T}{\partial t} \tag{1.68}$$

The result is a second-order partial differential equation subject to the boundary conditions at the left and right sides of the wall where $x = 0$ and $x = L$.

Apply energy conservation at the walls for the boundary conditions. On the left side, the thermal heating rate equals the conduction rate from the surface into the wall. The heat conduction into the surface at the right-hand side of the wall equals the rate of thermal convection to the fluid:

Rate of energy in = Rate of energy out

$$q = -\kappa \left.\frac{\partial T}{\partial x}\right|_{x=0} \quad \text{and} \quad -\kappa \left.\frac{\partial T}{\partial x}\right|_{x=L} = h(T_L - T_\infty) \tag{1.69}$$

Thermodynamic conditions at system boundaries may require additional constitutive models, such as vapor–liquid equilibrium. We derive Fick's *second* law in one dimension from the local species balance around a finite volume element of length Δx, width Δz, and height Δy, as shown in Figure 1.14.

N represents the molar flux for the diffusing species. The rate balance in Figure 1.14 accounts for the flux in and out at positions x and $x + \Delta x$. Similar expressions apply to the y and z dimensions:

$$(\text{Rate in} - \text{out})_x + (\text{Rate in} - \text{out})_y + (\text{Rate in} - \text{out})_z + (\text{Rate in} - \text{out})$$
$$= (\text{Rate of accumulation})$$

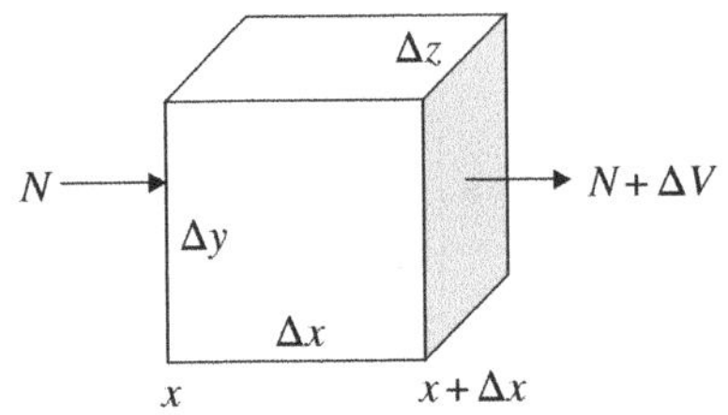

Figure 1.14 Finite volume element in a Cartesian coordinate system with dimensions Δx, Δy, and Δz.

$$\Delta y\Delta z(N_x - N_{x+\Delta x}) + \Delta x\Delta z(N_y - N_{y+\Delta y}) + \Delta x\Delta y(N_z - N_{z+\Delta z}) = \Delta x\Delta y\Delta z\frac{dC}{dt} \tag{1.70}$$

where the product of the lengths of two sides $\Delta y\Delta z$ is the area of flux normal to the x-direction, and so forth. The product of all three sides $\Delta x\Delta y\Delta z$ is the finite element volume. The element volume multiplied by the time derivative of concentration represents the accumulation term. The equation takes the form of a second-order *partial differential* equation (space and time dimensions) in the zero limits of the volume of the finite element $\Delta x\Delta y\Delta z \to 0$:

$$\lim_{\Delta x\to 0}\left\{\frac{\Delta y\Delta z(N_x - N_{x+\Delta x}) + \Delta x\Delta z(N_y - N_{y+\Delta y}) + \Delta x\Delta y(N_z - N_{z+\Delta z})}{\Delta x\Delta y\Delta z}\right\}$$
$$= -\left(\frac{\partial}{\partial x}N_x + \frac{\partial}{\partial y}N_y + \frac{\partial}{\partial z}N_z\right) = \frac{\partial C}{\partial t} \tag{1.71}$$

If we replace the molar flux terms in Eq. (1.71) by Fick's first law from Table 1.2, Eq. (1.71) becomes:

$$-\left[\frac{\partial}{\partial x}\left(-D\frac{\partial C}{\partial x}\right) + \frac{\partial}{\partial y}\left(-D\frac{\partial C}{\partial y}\right) + \frac{\partial}{\partial z}\left(-D\frac{\partial C}{\partial z}\right)\right] = \frac{\partial C}{\partial t} \tag{1.72}$$

subject to an initial condition: $C = C_0$ at $t = 0$ and two boundary conditions for each dimension. For example, the x-dimension boundary conditions are:

Rate in = Rate out

$$\beta_2(C - C_\infty) = \beta_1\frac{\partial C}{\partial x} \text{ at } x = x_0 \text{ and } \beta_3\frac{\partial C}{\partial x} = \beta_4(C - C_\infty) \text{ at } x = x_1 \tag{1.73}$$

where β_1 through β_4 are diffusion and convection parameters. The y and z dimensions require similar boundary conditions. The initial condition is the equation's solution at $t = 0$ and is the calculation's starting point. Boundary conditions impose the unique, physical nature of the system into the model. We derive boundary conditions from applying equilibrium thermodynamics or conservation equations to the boundary surfaces.

For constant diffusivity and in the limit $V \to 0$, Eq. (1.72) reduces to

$$D\left(\frac{\partial^2 C}{\partial x^2} + \frac{\partial^2 C}{\partial y^2} + \frac{\partial^2 C}{\partial z^2}\right) = \frac{\partial C}{\partial t} \tag{1.74}$$

To complicate the matter further, reaction rates other than first-order give nonlinear rate expressions, which give nonlinear differential equations when substituted for the generation terms. For example, consider the one-dimensional problem with a nonlinear reaction term for the generation part of the mole balance:

$$D\frac{\partial^2 C}{\partial x^2} - \frac{kC}{K + C} = \frac{\partial C}{\partial t} \tag{1.75}$$

We can approximate solutions to this partial differential equation using the numerical methods introduced in Chapter 14.

1.2.3 Degrees of Freedom

Understanding the relationship between the number of equations (E) and the number of variables (V) in a system is crucial for finding solutions. Here is a breakdown of the different scenarios and their implications:

$E = V$: In an independent system of equations, where the number of equations equals the number of variables, one unique solution exists. This is characteristic of linear, independent equations. However, nonlinear equations may have multiple roots.

$E \neq V$: An ill-conditioned system of equations has an unequal number of equations and unknowns, leading to challenges in finding solutions.

$E < V$: In this scenario, the solution has $F = V - E$ degrees of freedom. Specifying the values of F unknowns allows for solving equations for a unique solution. An underspecified problem may present an opportunity for optimization.

$E > V$: If there are more equations than unknowns, the system is over-specified and has no unique solution. In such cases, the set of solutions to the system of equations is infinite. However, a subset of V-independent equations may have a unique solution. Utilizing uncertainty in the additional equations can help find a solution that best satisfies the system of equations.

When dealing with an extensive system of equations, it is crucial to specify it correctly before attempting a solution. Simply matching the number of variables with the number of equations may not suffice, as some linear systems may be singular and lack a unique solution. For nonlinear problems, multiple solutions may exist, but identifying the correct one involves ensuring that the results align with the system's physics. For instance, solutions that yield negative mole fractions or concentrations greater than one should be eliminated, as absolute temperatures and concentrations must be positive by definition.

1.2.4 Dimensionless Equations

We may recast model equations in dimensionless terms to identify *significant* vs. *negligible* terms and minimize the effects of round-off error in numerical solutions. Follow Aris' (1993) rules when making equations dimensionless:

1. When we are interested in the effect of a certain quantity, that quantity should appear in only one dimensionless parameter, not in a dimensionless variable.
2. Dimensionless dependent variables should range from zero to one.
3. Dimensionless independent variables should range over finite intervals, e.g. (0, 1) or $(-p, +p)$.
4. Dimensionless parameters show the relative magnitudes of terms in the model equations. Consider ignoring relatively small terms.

Making the model equations dimensionless usually reduces the number of parameters by at least one. Nondimensionalization also reveals the minimum number of parameters needed to describe the physical nature of the system.

To illustrate, make Eq. (1.57) dimensionless following the Aris rules:

$$\frac{dX_A}{d\overline{V}} + \varepsilon \frac{d^2X_A}{d\overline{V}^2} = \phi \tag{1.76}$$

where the conversion X_A is already dimensionless, with a range of zero to one. Define the dimensionless independent variable and parameters as follows:

$$\bar{V} = \frac{V}{V_0} \quad \varepsilon = \frac{C_{A0}A^2E}{F_{A0}V_0} \quad \phi = \frac{-r_A V_0}{F_{A0}}$$

Note the reduction in the total number of parameters. The three parameter groups F_{A0}, $C_{A0}S^3E$, and $-r_A$ are combined into two parameters ε and ϕ. Aris' fourth rule of dimensionless numbers applies to the parameter ε. For small ε (for cases of large $F_{A0}V_0$ or small $C_{A0}S^3E$), Eq. (1.76) reduces to the PFR result in Eq. (1.54).

For some problems, there are advantages of scaling the independent variable to range between the integration limits of zero to one:

$$x^* = \frac{x - x_0}{x_n - x_0} \rightarrow x = x_0 + x^*(x_n - x_0) \tag{1.77}$$

where x_0 and x_n are the integration limits or initial and final values for x. The advantages include:

- Increased computational stability with smaller integration steps
- Improved accuracy with smaller integration steps
- Higher computational efficiency with fewer integration steps
- Ability to integrate both directions of x (positive from x_0 to x_n or negative from x_n to x_0)

Recall that x^* is a function of x according to Eq. (1.77). To apply variable scaling in our numerical methods, we multiply each function for the differential equations by the difference in the limits of integration $(x_n - x_0)$ and define x in terms of x^* in the expressions for f. As a result, the first-order differential equation becomes dimensionless in the independent variable:

$$\begin{aligned} \frac{dy}{dx} &= \frac{1}{(x_n - x_0)}\frac{dy}{dx^*} = f(x^*, y) \quad \rightarrow \quad \frac{dy}{dx^*} \\ &= (x_n - x_0)f(x^*, y)x^* = \frac{x - x_0}{x_n - x_0} \quad \rightarrow \quad x = x_0 + x^*(x_n - x_0) \end{aligned} \tag{1.78}$$

1.2.5 Stochastic vs. Deterministic Modeling

Unlike deterministic models, which generate outcomes based on fixed numerical inputs, stochastic models take into account the inherent uncertainty of the system by producing a probability distribution of results. This distribution is derived from random inputs drawn from unique probability distributions of uncertainty. Stochastic models predict the expected outcome and provide confidence intervals, which serve as reliability measures. Smaller confidence intervals indicate higher reliability of the probable outcome. Given the unpredictable nature of forecasting, models in this domain often rely on stochastic inputs.

This book also delves into empirical and stochastic modeling, exploring techniques such as evolutionary optimization and uncertainty propagation through data analysis and least-squares regression. These approaches broaden the scope of modeling by incorporating real-world data and addressing the inherent uncertainty present in many systems.

1.2.6 Model Verification and Validation

The utility of mathematical models is indisputable, yet a prudent approach is essential. Models are invaluable tools for comprehending systems, identifying correlations among variables, refining processes, and facilitating design. However, it is crucial to recognize that while mathematical modeling enhances our understanding, it cannot wholly replace empirical evidence.

Advancements in computing power, accuracy, and accessibility have transformed the landscape. In chemical property prediction, reliance on computer models instead of experimentation is becoming commonplace. Notably, esteemed journals such as the *ACS Journal of Chemical and Engineering Data* now welcome submissions based on computational experiments. In harsh conditions where conventional measurements are impractical, computer simulations serve as "soft sensors," providing unattainable insights through traditional means.

Despite these strides, the pinnacle of chemical process simulation still necessitates a blend of bench-scale experiments and pilot-scale studies before unequivocal trust is bestowed upon simulations. The modeling journey is iterative, involving continuous refinement:

1. Define the physical system and assess existing knowledge.
2. Formulate simplified models to render the system mathematically tractable.
3. Iteratively solve model equations across varying parameters to elucidate relationships.
4. Validate the model against empirical data, refining assumptions and understanding as needed.

Computers dazzle us with their computational and graphical prowess. However, it is imperative not to succumb to complacency by unquestioningly trusting computer-generated results. Rigorous verification and validation processes are paramount. Robust models should yield accurate solutions and demonstrate predictive efficacy and uncertainty resilience.

The adage "Garbage in, garbage out" underscores the importance of input quality in determining model output reliability. Computational tools like MATLAB offer immense potential when coupled with validated models, enabling rapid exploration of "what-if" scenarios without disrupting operations. The true value of modeling lies not merely in its creation but in its ongoing utilization to probe process dynamics and uncover latent opportunities. Failure to leverage models for such exploration squanders the effort invested in their development, solving, and validation.

1.3 Expert Problem-Solvers

Engineers derive satisfaction from solving problems and moving on to the next challenge. However, it is crucial to resist the temptation to skip the critical step in problem-solving: documenting and reflecting on our results. Our proven problem-solving methodology comprises four principal components: starting with basic equations, clearly stating all assumptions, performing the analysis, and checking the results for engineering consistency and common sense.

The legacy of the US NASA space shuttle program serves as a sobering reminder of the profound impact and consequences of both practical and flawed problem-solving processes, evident in the triumphs and tragic accidents of the Challenger in 1986 and Columbia in 2003 (Gehman 2003). Indeed, history underscores the adage that "there are no small mathematical errors in engineering work." (Kowalczyk 1963) The loss of NASA's Mars Orbiter due to overlooked unit conversions and the collapse of the Quebec Cantilevered Bridge due to failure to redesign for added length are poignant examples of costly engineering mistakes.

The accuracy of our results and the clarity of our presentation are matters of ethics and professionalism, as lives may depend on the reliability of our work. It is emphasized that a solution checked by an engineer significantly enhances credibility and trust, and this psychological gain must be matched with technical sufficiency. Checking calculations is an essential layer of safety protection, and verification and validation are not trivial parts of the solution process. The problem-solving format recommended here promotes engineering documentation, proper numerical precision, accuracy, sound engineering judgment, and critical evaluation of results.

Mathematical models empower engineers to explore parameter space, understand processes, uncover relationships, and identify improvement opportunities. These models may stem from first principles (laws of conservation) (Farhadi et al. 2009) or convenient functions such as polynomials that capture observed system behavior. It is advisable to start with simple expressions and add rigor only when necessary. In many cases, solving model equations requires numerical approximation techniques. The subsequent chapters introduce MATLAB as a platform for building models and provide a comprehensive suite of tools for solving common modeling problems.

2

MATLAB Skill Preparation

MATLAB skill preparation reviews MATLAB commands, functions, and essential concepts to refresh the readers' MATLAB experience. It can also accelerate the MATLAB skills of new users to assist them in building up a foundation for further study in later chapters.

The chemical engineering discipline applies chemistry, physics, biology, and computation to efficiently and economically transform energy and materials to benefit society. There are many state variables (such as mass, enthalpy, internal energy, Gibbs free energy, entropy, pressure, temperature, volume, chemical composition, specific volume or its reciprocal density, and particle number) and process variables (such as mechanical work and heat) in chemical engineering processes. Furthermore, there are many kinds of unit operations with interconnections and different kinds of fluids with different phases. All of those form a large amount of data. MATLAB (https://www.mathworks.com), an abbreviation of "matrix laboratory," was initially designed to calculate, process, and store data using matrix form, which is perfectly suitable for handling data and performing computer-aided design for chemical engineering processes.

This chapter introduces useful MATLAB commands and functions and discusses vector and matrix operations. It briefly reviews solving equations using existing MATLAB solvers and the optimization toolbox to support the advanced chapters. Next, some simple examples demonstrate the coding skills of using loops & conditional statements as building blocks.

2.1 Introduction

MATLAB applications typically involve the "Command Window" as an interactive mathematical shell, or executing text files containing MATLAB code.

2.1.1 MATLAB User Interface

As the user interface, MATLAB Desktop consists of many components and windows that can be reorganized and turned on/off by mouse actions or via Desktop ➤ Layout menu options in the HOME tab. Figure 2.1 shows some commonly used windows, such as Current Folder Window, Editor Window, Figure Window, Workspace Window, Command Window, and Command History Window. Note the following features:

Chemical Engineering Analysis and Optimization Using MATLAB®, First Edition.
Weiguo Xie, Sam Toan, and Richard Davis.

Companion website: www.wiley.com/go/xie/chemicalengineering

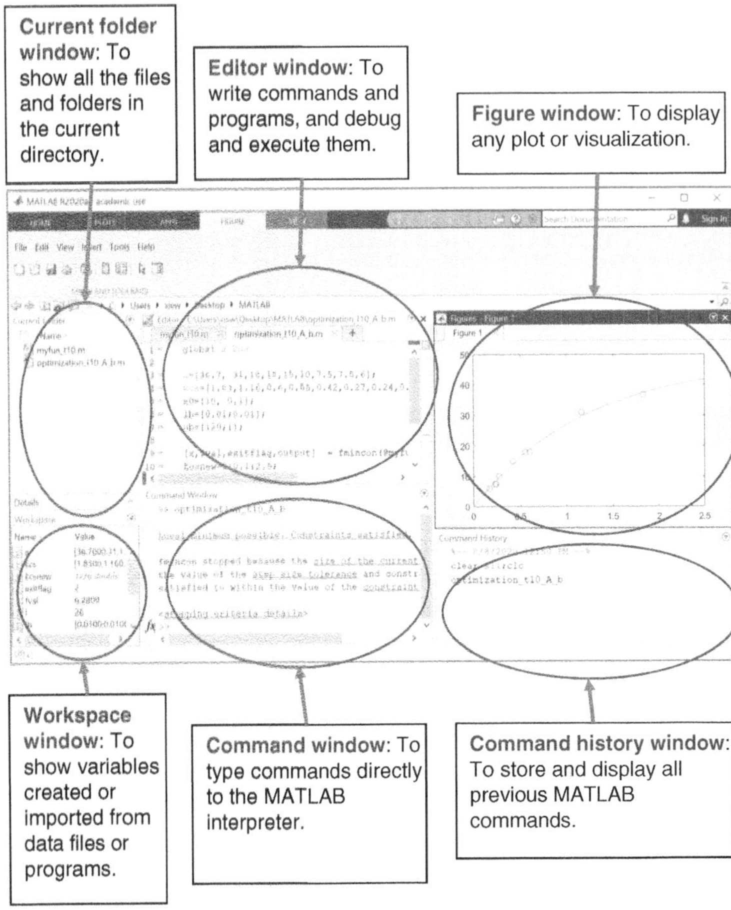

Figure 2.1 MATLAB user interface.

1. The .m files in the Editor Window may not be listed in your Current Folder Window due to their different locations. However, to run the .m file, it must be in the Current Folder Window. Otherwise, you need to click on "Change Folder" button.
2. Workspace Window shows the format or values of variables either before a Breakpoint in a running current .m file or just finished .m file or any direct input from Command Window. Any variables from subroutines and not marked with global are not listed in Workspace Window.
3. The "clear all" function removes all variables, globals, functions, and MEX links. At the Command Window, "clear all" also clears the base import list.
4. The "close all" deletes all figures without hidden handles. The "close all hidden" deletes all figures, including those with hidden handles. The "close all force" deletes all figures, including GUIs (graphical user interfaces) for which CloseRequestFcn was altered not to close the window. However, figures with the "HandleVisibility" property set to "off" are not closed with "close all."
5. The "clc" clears everything in the Command Window.

> **MATLAB – Learning Tips**
>
> - Do not be afraid of trying
> - "More practice makes perfect"
> - Read online user help documents
> - Take online training
> - Search free download codes

2.1.2 MATLAB File Extensions

(1) Script m-files

To use MATLAB code for later usage, the code can be typed in the Editor Window and saved with a .m file extension, a script file (m-file). Write an m-file in the Editor Window of the MATLAB user interface. Figure 2.2 shows an m-file example.

Use an m-file to:

- Edit the code
- Save the work
- Debug the code
- Call other m-files and/or function files

(2) Functions

In MATLAB, functions are used to call an expression, a set of expressions, or tasks multiple times without repeating the coding.

There are three ways to define a function:

(a) Use a function handle "@"

- In the function definition, the handle is followed by brackets containing all input variables.

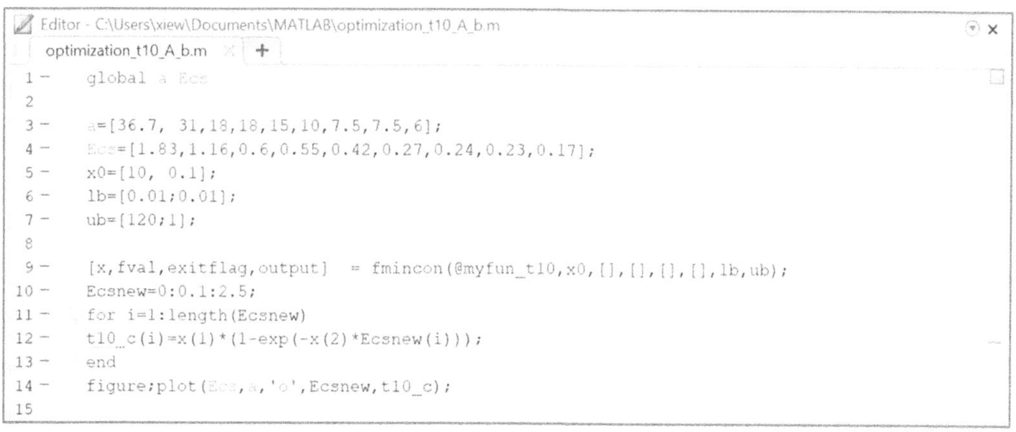


```
global a Ecs

a=[36.7, 31,18,18,15,10,7.5,7.5,6];
Ecs=[1.83,1.16,0.6,0.55,0.42,0.27,0.24,0.23,0.17];
x0=[10, 0.1];
lb=[0.01;0.01];
ub=[120;1];

[x,fval,exitflag,output]  = fmincon(@myfun_t10,x0,[],[],[],[],lb,ub);
Ecsnew=0:0.1:2.5;
for i=1:length(Ecsnew)
t10_c(i)=x(1)*(1-exp(-x(2)*Ecsnew(i)));
end
figure;plot(Ecs,a,'o',Ecsnew,t10_c);

```

Figure 2.2 Example of a script m-file.

For example, calculate the function $x_1^2 - 2x_1x_2 + 2x_2^2 - 6x_1$, when $x_1 = 2$ and $x_2 = 3$

```
>> fun = @(x)x(1)^2-2*x(1)*x(2)+2*x(2)^2-6*x(1);
>> x = [2 3]; b=fun(x)
b =
      -2
```

(b) A symbolic definition

- Define the input variables as symbolic variables using the *syms* command. **Note**: List the input variables after the *syms* command without separating them with a comma.

For example, calculate the Reynolds number for a 1.0 l/min water flow rate in a pipe ($ID = 2$ inches) at 20°C (68°F).

```
>> syms d u rho mu Re;
>> Re(d,u,rho,mu)= d*u*rho/mu;
>> Re1= double(Re(0.05,0.0085,1000,0.001))
Re1 =
    425
```

(c) Use a function file. We must write a function file in the Editor Window if a function requires several lines or many output variables/arrays. The syntax of the first executable line to declare function name, inputs, and outputs follows:

```
function [y1,...,yN] = myfun(x1,...,xM)
```

The function named myfun has input arguments x1, … ,xM and returns outputs assigned to variables y1, … ,yN. Valid function names must begin with an alphabetic character and can contain letters, numbers, or underscores. Figure 2.3 shows an example of function m-file.

(3) Live script and live functions

MATLAB introduced live scripts starting in R2016a and live functions in R2018a. The live scripts and live functions are stored using the Live Code file format in a file with a .mlx extension. These interactive documents combine MATLAB code with formatted text, equations, and images in the Live Editor. Live scripts store and display output next to the code. Figure 2.4 shows a live script mlx-File.

The Live Editor does not support some MATLAB features:

- *Classes*: The Live Editor does not support classes. Instead, create classes as plain code files (.m). Classes become available for use in live scripts or functions.

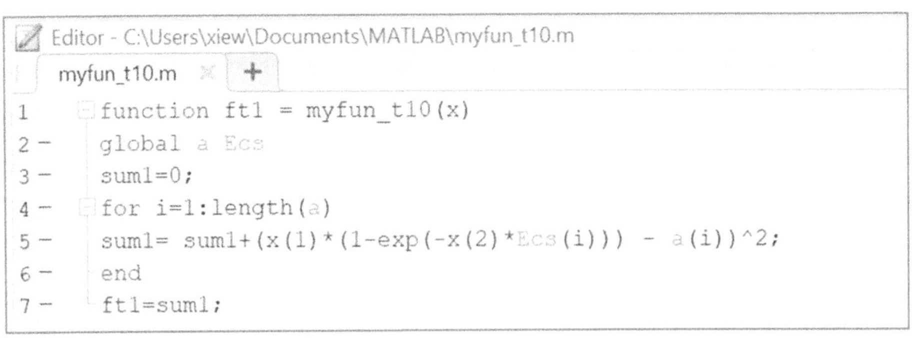

Figure 2.3 Example of a function m-file.

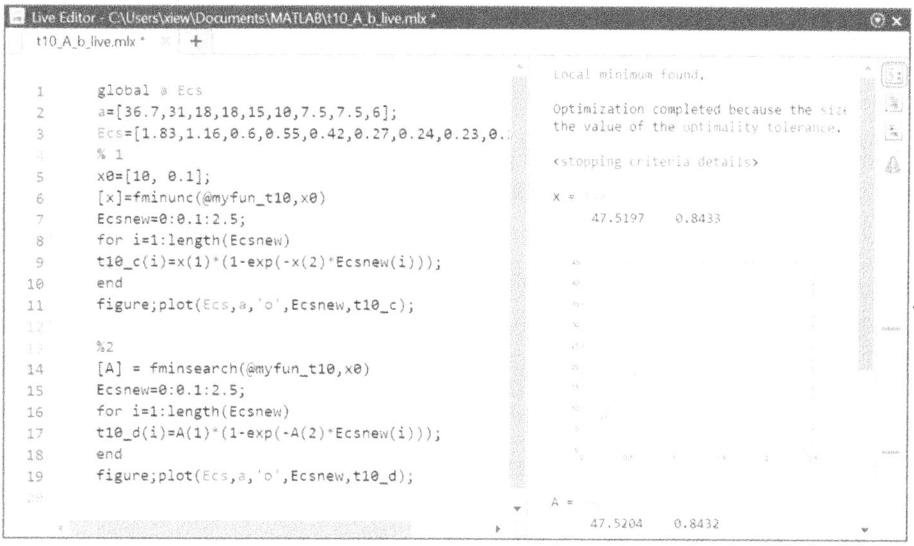

Figure 2.4 Example of a live script mlx-file with output displayed alongside the code.

- *MATLAB preferences*: The Live Editor ignores some MATLAB preferences, including custom keyboard shortcuts and Emacs-style keyboard shortcuts.

(4) The fig-files

A fig-file in MATLAB contains a figure created by a code command, which stores one or more plots or graphs to visualize data. The figure function creates a figure (window), which allows users to insert plots and graphs.

The fig-file can be saved through the figure window and give it a file name or through typing *savefig(filename)* in Command Window to save the current figure to a fig-file named filename.fig.

The figure can also be saved into other formats as shown in Figure 2.5. However, only a fig-file contains data structure and figure edit information (Figure Properties, Axes

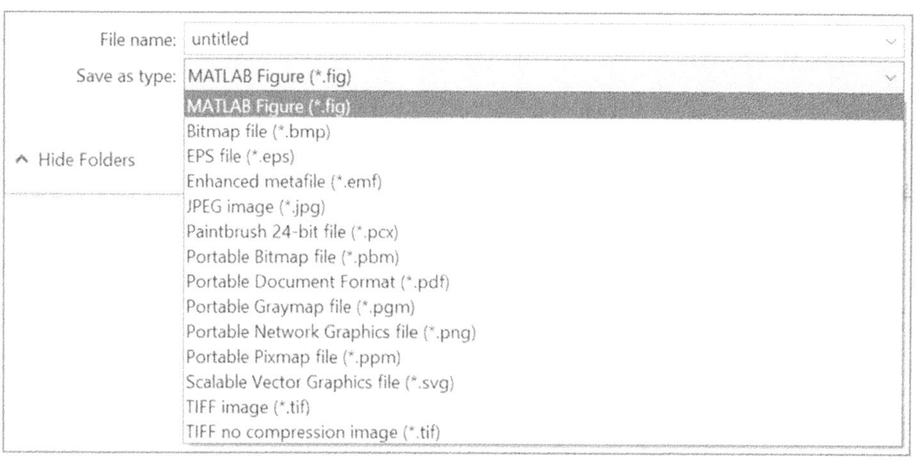

Figure 2.5 All possible formats for saving a figure.

Properties, Current Object Properties, and Colormap), which can be extracted and used for future applications. The following command line can extract the fig-file data:

```
a = load('figurename.fig','-mat')
```

Figure 2.6 shows four commonly used plots in MATLAB, such as

(a) 2-D line plot: plot(X1,Y1,LineSpec1, … ,Xn,Yn,LineSpecn);
(b) 3-D surface plot: surf(X,Y,Z,C);
(c) Contour plot: contour(X,Y,Z);
(d) Vector plot: quiver(X,Y,U,V).

(5) The MAT-files

MAT-files are binary MATLAB files that store current workspace variables. All saved operations create Version 7 MAT-files by default. The only exception is to new MAT-files

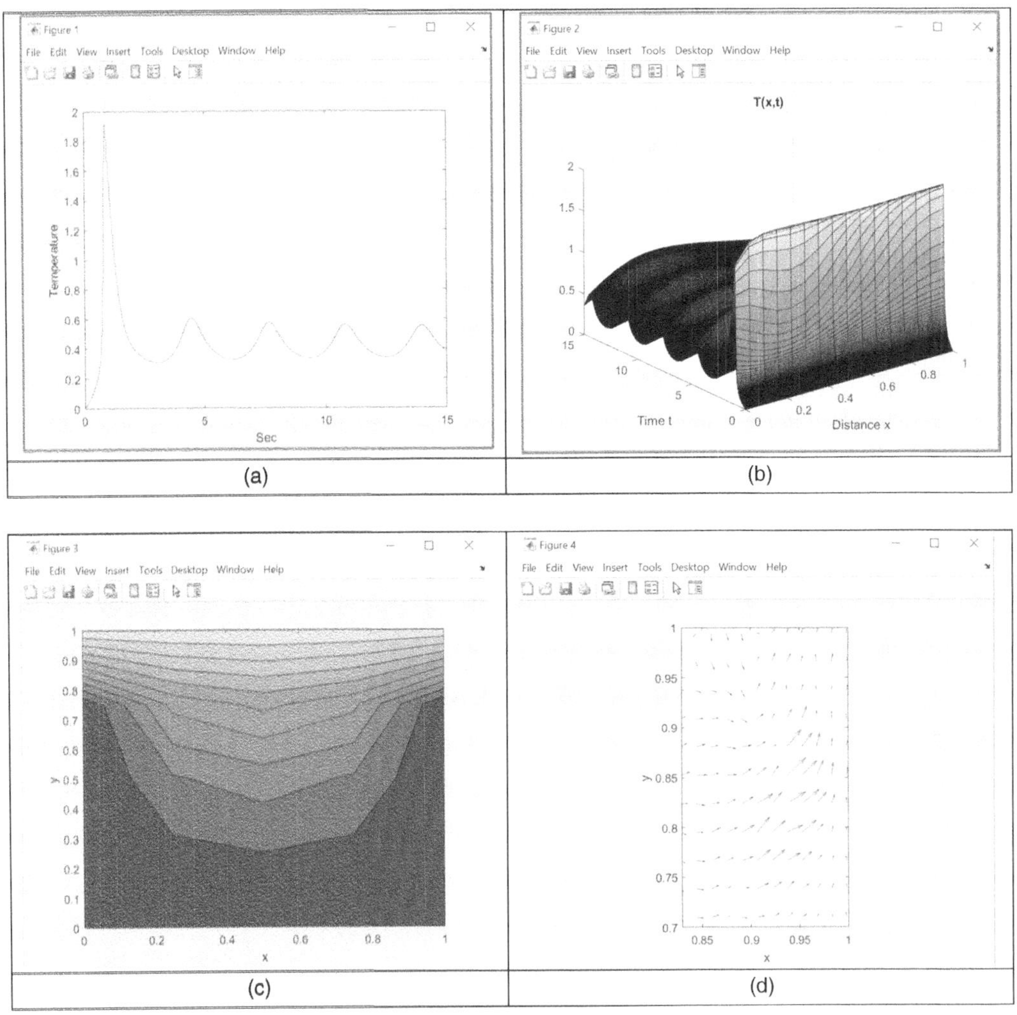

Figure 2.6 Four commonly used plots in MATLAB: (a) 2-D line plot; (b) 3-D surface plot; (c) contour plot; (d) vector plot.

using the *matfile* function, by which the default MAT-file version is 7.3. MATLAB compresses data written to MAT-files to use storage space efficiently. Unfortunately, data compression and decompression slow down save and load processes. However, reducing the file size is worth the additional processing time.
The *save* function can save workspace variables to a MAT-file:

```
save(filename,variables,fmt)
```

Apply the following command to add new variables to an existing MAT-file:

```
save(filename,variables,'-append')
```

The above command overwrites an existing variable in a MAT-file with the value in the current workspace.
Use the *matfile* function to access and change variables in MAT-file without loading the file into memory:
matObj = matfile(filename) creates a matlab.io.MatFile object connected to the MAT-file specified by filename.
matObj = matfile(filename, 'Writable', isWritable) enables or disables write access to the file. Specify isWritable as true or false.

2.2 Some Useful MATLAB Commands and Functions

This section lists useful MATLAB commands and functions, as shown in Tables 2.1–2.8, to facilitate the readers' coding skills.

2.2.1 MATLAB Commands

(1) General purpose commands (https://www.mathworks.com).

(2) Input/output and formatting commands (https://www.mathworks.com).

(3) Vector, matrix, and array commands (https://www.mathworks.com)

2.2.2 MATLAB Functions

Example 2.1
To type the following functions in Command Window to see the answers:

Solution:

```
≫ x=linspace(0, 1, 9)*pi
x =
0   0.3927   0.7854   1.1781   1.5708   1.9635   2.3562   2.7489   3.1416
≫ y1=sin(x)
y1 =
0   0.3827   0.7071   0.9239   1.0000   0.9239   0.7071   0.3827   0.0000
≫ y2=exp(x)
y2 =
1.0000   1.4810   2.1933   3.2482   4.8105   7.1242   10.5507   15.6253   23.1407
```

Table 2.1 Operators and special characters in MATLAB.

	Operators and special characters
+	Addition operator
–	Subtraction operator
*	Scalar and matrix multiplication operator
.*	Element-wise multiplication operator
^	Scalar and matrix power operator
.^	Element-wise power operator
\	Left-division operator
/	Right-division operator
.\	Element-wise left-division operator
./	Element-wise right-division operator
:	Generates regularly spaced elements and represents an entire row or column
( )	Encloses function arguments and array indices; overrides precedence
[ ]	Brackets; enclosures array elements
.	Decimal point
...	Ellipsis; line-continuation operator
,	Separates statements and elements in a row
;	Separates columns and suppresses display
%	Percent sign; designates a comment and specifies formatting
.'	Transpose operator
=	Assignment (replacement) operator

Table 2.2 Logical operators in MATLAB.

	Logical operators
>	Greater than
>=	Greater than or equal to
<	Less than
<=	Less than or equal to
~=	Not equal
==	Equal
&&	Logical AND (for scalars) with short-circuiting
&	Logical AND
||	Logical OR (for scalars) with short-circuiting
|	Logical OR
~	Logical NOT
xor	Logical exclusive or
all	True if all elements of a vector are nonzero
any	True if any element of a vector is nonzero

Table 2.3 Special variables and constants in MATLAB.

	Special variables and constants
ans	Most recent answer
eps	Floating-point relative accuracy
i,j,1i, 1j	Imaginary unit
Inf	Infinity
NaN	Undefined numerical result (not a number)
pi	The number π

Table 2.4 Input/output commands in MATLAB.

	Input/output commands
disp	Displays contents of an array or string
fscanf	Reads formatted data from a file
format	Controls screen-display format
fprintf	Performs formatted writes to screen or file
input	Displays prompts and waits for input
;	Suppresses screen printing

Table 2.5 Format codes for fprintf and fscanf in MATLAB.

	Format codes for fprintf and fscanf
%s	Format as a string
%d	Format as an integer
%f	Format as a floating point value
%e	Format as a floating point value in scientific notation
%g	Format in the most compact form: %f or %e
\n	Insert a new line in the output string
\t	Insert a tab in the output string

```
≫ y3=log(x)
y3 =
-Inf  -0.9347  -0.2416  0.1639  0.4516  0.6747  0.8570  1.0112  1.1447
Note: log(0) is non-existent and gives a -Inf result.
≫ y4=log10(x)
y4 =
-Inf  -0.4059  -0.1049  0.0712  0.1961  0.2930  0.3722  0.4392  0.4971
Note: log10(0) is non-existent and gives a -Inf result.
≫ y5=sqrt(x)
y5 =
```

Table 2.6 Numeric display formats in MATLAB.

	Numeric display formats
format short	Four digits after the decimal point (default)
format long	15 digits after the decimal point for double-precision values and 7 digits after the decimal point for single-precision values.
format shortE	Short scientific notation with four digits after the decimal
format longE	Long scientific notation with 15 digits after the decimal point for double values and 7 digits after the decimal point for single values
format Long6	Returns the more compact display among long, fixed-decimal format or scientific notation, with 15 digits for double-precision values and 7 digits for single-precision values
format shortEng	Short engineering notation (exponent is a multiple of 3) with four digits after the decimal point
format longE	Long engineering notation (exponent is a multiple of 3) with 15 significant digits
format bank	Two decimal digits
format +	+, −, or blank characters displayed for positive, negative, or zero elements
format rat	Ratio of small integers
format compact	Suppresses the display of blank lines
format loose	Resets to less compact display mode
format hex	Hexadecimal representation of a binary double-precision number

Table 2.7 Array commands in MATLAB.

	Array commands
cat	Concatenates arrays
find	Finds indices of nonzero elements
length	Computers number of elements
linspace	Creates a regularly spaced vector
logspace	Creates a logarithmically spaced vector
max	Returns the largest element
min	Returns smallest element
prod	Product of each column
reshape	Change size
size	Computes array size
sort	Sorts each column
sum	Sums each column

Table 2.8 Functions in MATLAB (https://www.mathworks.com/The MathWorks, Inc.).

sin(x), asin	Sine and inverse (argument in radians)
sind(x), asind	Sine and inverse (argument in degrees)
sinh(x), asinh	Hyperbolic sine and inverse (argument in radians)
Analogous for the other trigonometric functions: cos, tan, csc, sec, and cot	
abs(x)	Absolute value of x, complex magnitude
exp(x)	Exponential of x
sqrt(x)	Square root
nthroot(x,n)	Real nth root of real numbers
log(x)	Natural logarithm of x
log2(x), log10	Logarithm with base 2 and 10, respectively
factorial(n)	Factorial of n
sign(x)	Sign of x
mod(x,d)	Remainder after division (modulo)
ceil(x), fix, floor	Round toward +inf, 0, −inf
round(x)	Round to the nearest decimal or integer

```
0  0.6267  0.8862  1.0854  1.2533  1.4012  1.5350  1.6580  1.7725
≫ y6=sum(x)
y6 =
14.1372
≫ y7=round(x)
y7 =
0  0  1  1  2  2  2  3  3
≫ y8=ceil(x)
y8 =
0  1  1  2  2  2  3  3  4
≫ y9=floor(x)
y9 =
0  0  0  1  1  1  2  2  3
≫ y10=fix(x)
y10 =
0  0  0  1  1  1  2  2  3
≫ y11=abs(x)
y11 =
0  0.3927  0.7854  1.1781  1.5708  1.9635  2.3562  2.7489  3.1416
≫ m1=length(x)
m1 =
9
≫ m2=size(x)
m2 =
1     9
≫ format long;y6
y6 =
14.137166941154069
≫ format bank;y6
```

```
y6 =
14.14
≫ format short;y6
y6 =
14.1372
≫ format rat;y6
y6 =
3195/226
≫ format short e;y6
y6 =
1.4137e+01
≫ format long e;y6
y6 =
1.413716694115407e+01
≫ format;y6
y6 =
14.1372
≫ plot(x,y1,'k.-',x,y2,'*')
```

A plot is shown in Figure 2.7.

MATLAB – Learning Tips

- Most error messages come from the mismatch of dimensions: "Matrix dimensions must agree." Or "Subscripted assignment dimension mismatch"
  - To check dimensions in Debug process
- The error messages of dimension mismatch may be displayed when the data is updated or overwritten.
  - To clear the old data (may need to save data with other name before clearing it)

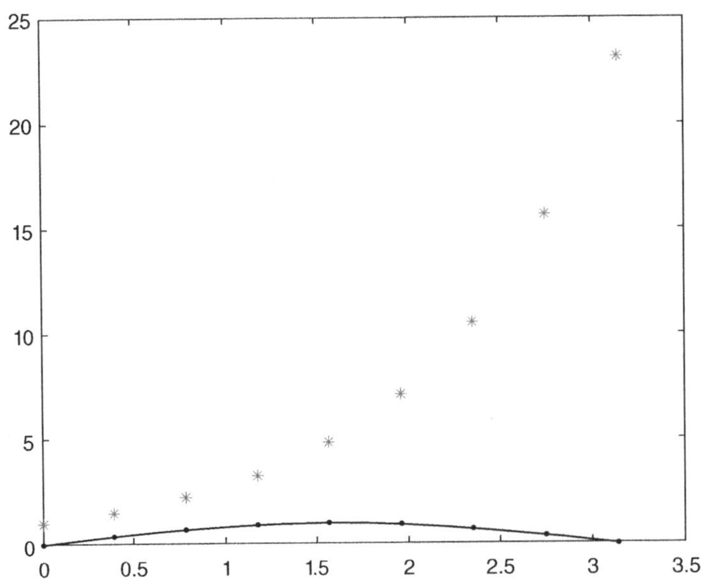

Figure 2.7 Example of a plot.

2.3 Vector and Matrix Operations

MATLAB is based on vectors and matrices and their operators, as shown in Tables 2.9 and 2.10.

2.3.1 Vector Operations

Dot Arithmetic

- It requires that the vectors are of the same size.
- It performs element -wise operation.
- Dot arithmetic only applies to the following operators: "./ ", ".\" " .* ", " .^ ".

Example 2.2
To type the following functions in Command Window to see the answers:

Solution:

```
≫ A = [1 0.5 3]; B = [2 7 1.2]; C1 = A.*B
C1 =
2.0000    3.5000    3.6000
≫ C2 = A./B
C2 =
0.5000    0.0714    2.5000
≫ C3 = A.\B
C3 =
2.0000   14.0000    0.4000
≫ C4 = A.^B
```

Table 2.9 Vector operations in MATLAB (https://www.mathworks.com/The MathWorks, Inc.).

.'	Transpose operator
+	Addition operator
−	Subtraction operator
dot(A,C)	The dot product of vectors A and C
cross(A,C)	The cross-product of vectors A and C
sum(D)	The sum of elements of a vector D
min(D)	The minimum value found in a vector D
max(D)	The maximum value found in a vector D
mean(D)	The mean value of all entries in a vector D
median(D)	The median value of the elements in a vector D
std(D)	The standard deviation of a vector D around the mean value
var(D)	The variance of a vector D around the mean value

```
C4 =
1.0000      0.0078      3.7372
>> C5=dot(A,B)
C5 =
9.1000
>> C6=cross(A,B)
C6 =
-20.4000      4.8000      6.0000
```

2.3.2 Matrix Operations

Example 2.3

Use the function M = magic(n) to create a magic square. This function constructs an n-by-n matrix with equal row and column sums from integers 1 through n^2. Note that n must be greater than or equal to 3. To type the following functions in Command Window to see the answers:

Solution:

```
>> M=magic(4)
M =
```

Table 2.10 Matrix operations in MATLAB (https://www.mathworks.com/The MathWorks, Inc.).

size(A)	The sizes of each dimension of matrix A
rank(A)	Rank of matrix A
trace(A)	Sum of diagonal elements of matrix A
det(A)	Determinant of matrix A
poly(A)	Characteristic polynomial of matrix A
eig(A), eigs	Eigenvalues and vectors of matrix A (subset)
inv(A), pinv	Inverse and pseudo-inverse of matrix A
norm(A)	Norm of vector or matrix A
expm(A), logm	Matrix exponential and logarithm
cross(A,B)	Cross product of matrices A and B
dot(A,B)	Dot product of matrices A and B
kron(A,B)	Kronecker tensor product of matrices A and B
null(A)	Null space of matrix A
orth(A)	Orthonormal basis for the range of matrix A
tril(A), triu	Lower and upper triangular parts of matrix A
linsolve(A,B)	Solves a linear system of the form AX=B
lsqminnorm(A,B)	Least-squares solution to a linear equation
qr(A), lu, chol	Matrix decompositions of matrix A
svd(A)	Singular value decomposition of matrix A
gsvd(A,B)	Generalized SVD
rref(A)	Reduced row echelon form of matrix A

```
16     2     3    13
 5    11    10     8
 9     7     6    12
 4    14    15     1
≫ rank(M)
ans =
3
≫ trace(M)
ans =
34
≫ det(M)
ans =
5.1337e-13
≫ poly(M)
ans =
1.0e+03 *
0.0010   -0.0340   -0.0800    2.7200    0.0000
≫ eig(M)
ans =
34.0000
8.9443
-8.9443
-0.0000
≫ inv(M)
Warning: Matrix is close to singular or badly scaled. Results
 may be inaccurate.
RCOND =  4.625929e-18.
ans =
1.0e+15 *
-0.2649   -0.7948    0.7948    0.2649
-0.7948   -2.3843    2.3843    0.7948
0.7948    2.3843   -2.3843   -0.7948
0.2649    0.7948   -0.7948   -0.2649
≫ pinv(M)
ans =
0.1011   -0.0739   -0.0614    0.0636
-0.0364    0.0386    0.0261    0.0011
0.0136   -0.0114   -0.0239    0.0511
-0.0489    0.0761    0.0886   -0.0864
≫ norm(M)
ans =
34
≫ null(M)
ans =
-0.2236
-0.6708
0.6708
```

```
0.2236
≫ orth(M)
ans =
-0.5000    0.6708    0.5000
-0.5000   -0.2236   -0.5000
-0.5000    0.2236   -0.5000
-0.5000   -0.6708    0.5000
≫ qr(M)
ans =
-19.4422  -10.5955  -10.9041  -18.5164
0.1411  -16.0541  -15.7259   -0.9848
0.2539    0.1504    1.9486   -5.8458
0.1129    0.4976   -0.5257    0.0000
≫ lu(M)
ans =
16.0000    2.0000    3.0000   13.0000
0.2500   13.5000   14.2500   -2.2500
0.5625    0.4352   -1.8889    5.6667
0.3125    0.7685    1.0000    0.0000
≫ svd(M)
ans =
34.0000
17.8885
4.4721
0.0000
```

2.4 Loops & Conditional Statements

Within the MATLAB program, loops & conditional statements are used to control the program flow, as shown in Table 2.11. Loops use a "for" or "while" keyword. Conditional statements use "if" or "switch" keywords.

MATLAB Language Syntax:

Example 2.4

(1) Assign values to every matrix A element using "for" loops. To type the following functions in Command Window to see the answers:

Solution:

```
≫ n = 5;
A = zeros(n);
for c = 1:n
  for r = 1:n
    A(r,c) = 1/r+1/c;
  end
end
≫ A
```

Table 2.11 Loops & conditional statements in MATLAB (https://www.mathworks.com/The MathWorks, Inc.).

if, elseif, else	Execute statements if a condition is true
switch, case, otherwise	Execute one of several groups of statements
for	For loop to repeat a specified number of times
while	While loop to repeat when a condition is true
try, catch	Execute statements and catch resulting errors
break	Terminate execution of for or while loop
return	Return control to invoking script or function
continue	Pass control to the next iteration of for or while loop
pause	Stop MATLAB execution temporarily
parfor	Parallel for-loop
end	Terminate a block of code or indicate the last array index

```
A =
2.0000    1.5000    1.3333    1.2500    1.2000
1.5000    1.0000    0.8333    0.7500    0.7000
1.3333    0.8333    0.6667    0.5833    0.5333
1.2500    0.7500    0.5833    0.5000    0.4500
1.2000    0.7000    0.5333    0.4500    0.4000
```

(2) Using "if, elseif, else" statements to use the "floor" function for the upper triangle of matrix A and the "ceil" function for the lower triangle of matrix A. To type the following functions in Command Window:

```
≫ for c = 1:n
     for r = 1:n
          A(r,c) = 1/r+1/c;
          if c>r A(r,c)=floor(A(r,c));
          elseif c<r A(r,c)=ceil(A(r,c));
          end
     end
end
≫ A
A =
     2.0000    1.0000    1.0000    1.0000    1.0000
     2.0000    1.0000         0         0         0
     2.0000    1.0000    0.6667         0         0
     2.0000    1.0000    1.0000    0.5000         0
     2.0000    1.0000    1.0000    1.0000    0.4000
```

2.5 Solving Equations

There are some functions to solve equations in MATLAB, including algebraic equations, differential equations, and differential algebraic equations (DAEs), as shown in Table 2.12.

Table 2.12 Linear and nonlinear equations and systems in MATLAB (https://www.mathworks.com/The MathWorks, Inc.).

Roots	Solve polynomial equations of the form $p_1x^n + \ldots + p_nx + p_{n+1} = 0$
Fsolve	Solve a system of nonlinear equations
Fzero	Find a point x where fun(x) = 0
linsolve	Solve linear equations in matrix form
Poles	Poles of expression or function
Solve	Equations and systems solver
vpasolve	Solve symbolic equations numerically

2.5.1 Linear and Nonlinear Equations and Systems

Example 2.5

(1) MATLAB represents a polynomial by a vector of coefficients. If vector P describes a polynomial ax^3+bx^2+cx+d, then P(1)=a, P(2)=b, P(3)=c and P(4)=d. P=[1 0 -2 1] means polynomial x^3-2x+1, and then r=roots(P) is to solve the equation:

$$x^3 - 2x + 1 = 0 \tag{2.1}$$

Solution:

```
>> P=[1 0 -2 1] ; r=roots(P);
r =
-1.6180
 1.0000
 0.6180
```

(2) The fsolve is to be used to solve the equation.
   (a) To use initial value x0=0.

```
>>  x = fsolve(@(x)x^3-2*x+1,0)
Equation solved.
fsolve completed because the vector of function values is near
 zero as measured by the value of the function tolerance, and
 the problem appears regular as measured by the gradient.
<stopping criteria details>
x =
    0.6180
```

(b) To use initial value x0=1.

```
>> x = fsolve(@(x)x^3-2*x+1,1)
Equation solved at the initial point.
fsolve completed because the vector of function values at the
  initial point is near zero as measured by the value of the
  function tolerance, and the problem appears regular as
```

```
 measured by the gradient.
<stopping criteria details>
x =
     1
```

(c) To use initial value x0=-2.

```
≫ x = fsolve(@(x)x^3-2*x+1,-2)
Equation solved.
fsolve completed because the vector of function values is
 near zero as measured by the value of the function
 tolerance, and the problem appears regular as measured
 by the gradient.
<stopping criteria details>
x =
   -1.6180
```

(3) The fzero is to be used to solve the equation.
   (a) To use initial value x0=0.

```
≫ x = fzero(@(x)x^3-2*x+1,0)
x =
 0.6180
```

(b) To use initial value x0=1.

```
≫ x = fzero(@(x)x^3-2*x+1,1)
x =
     1
```

(c) To use initial value x0=-2.

```
≫ x = fzero(@(x)x^3-2*x+1,-2)
x =
   -1.6180
```

(4) Use the "solve" with the equation.

```
≫ syms x; eqn = x^3-2*x+1==0; S = solve(eqn,x)
S =
                    1
- 5^(1/2)/2 - 1/2
  5^(1/2)/2 - 1/2
```

(5) The "vpasolve" is to be used to solve the equation.

```
≫ syms x; eqn = x^3-2*x+1==0; S = vpasolve(eqn,x)
S =
-1.6180339887498948482045868343656
0.61803398874989484820458683436564
                              1.0
```

MATLAB – Learning Tips

- The "roots", "solve" and "vpasolve" solvers give the full solutions.
- The "fsolve" and "fzero" solvers only give a solution because of their local minimization of the sum of squares of the components.

2.6 Optimization Toolbox

Use MATLAB's optimization toolbox to tackle various optimization problems. The toolbox offers several solvers, including linear (LP), mixed-integer linear (MILP), quadratic (QP), second-order cone (SOCP), nonlinear (NLP) programming, as well as constrained linear and nonlinear least squares, and nonlinear equations.

MathWorks (https://www.mathworks.com/) provides a downloadable comprehensive Optimization Toolbox User's Guide for readers to learn algorithms in more detail.

2.6.1 Minimization vs. Maximization Problems

Minimizers formulate optimization problems in the form:

$$\min_{x} f(x) \tag{2.2}$$

subject to constraints. $f(x)$ is called an objective function. In general, $f(x)$ is a scalar function of type double, and x is a vector or scalar of type double. However, multiobjective optimization, equation solving, and some sum-of-squares minimizers can have vector or matrix objective functions $F(x)$ of type double. **All solvers in MATLAB are based on the minimization of an objective function**.

To use Optimization Toolbox solvers for a maximization problem, convert to a minimization problem via multiplication of (-1), e.g. convert the form

$$\max_{x} f(x) \tag{2.3}$$

by defining

$$g(x) = -f(x) \tag{2.4}$$

and minimize

$$\min_{x} g(x) \tag{2.5}$$

2.6.2 Optimization Solvers

There are some commonly used optimization solvers as shown in Table 2.13:

Example 2.6

(1)

$$f(x) = \cos(x) - 2\ \ln(x) \tag{2.6}$$

with x range (0.01, 20). To plot the curve of f(x) vs. x

Table 2.13 Optimization solvers in MATLAB (https://www.mathworks.com/The MathWorks, Inc.).

fminsearch	Find a minimum of an unconstrained multivariable function using a derivative-free method
fminunc	Find a minimum of an unconstrained multivariable function
fminbnd	Find a minimum of a single-variable function on a fixed interval
fmincon	Find a minimum of a constrained nonlinear multivariable function
fseminf	Find a minimum of a semi-infinitely constrained multivariable nonlinear function
linprog	Solve linear programming problems
quadprog	Quadratic programming
lsqlin	Solve constrained linear least-squares problems
lsqnonlin	Solve nonlinear least-squares (nonlinear data-fitting) problems
lsqnonneg	Solve nonnegative linear least-squares problem
lsqcurvefit	Solve nonlinear curve-fitting (data-fitting) problems by least-squares regression.
coneprog	Second-order cone programming solver

Solution:

```
>> x=linspace(0.01,20,200);
y=cos(x) - 2*log(x);plot(x,y);
```

The curve is plotted as shown in Figure 2.8.

(2) To use fminsearch to find the minimum point with the initial x0=1.

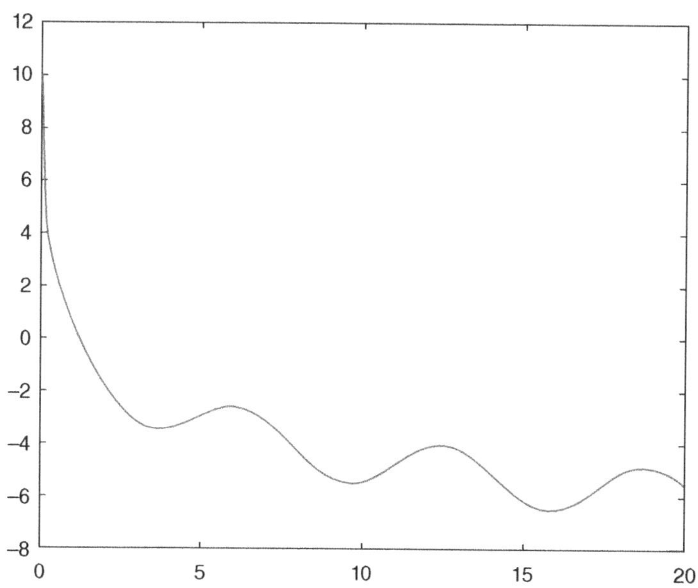

Figure 2.8 The plot of Example 2.6.

```
≫ [x, y, exitflag]=fminsearch(@(x)cos(x)-2*log(x), 1)
x =
    3.7108
y =
   -3.4648
exitflag =
     1
```

***Note**: exitflag – Reason fminsearch stopped:

1	The function converged to a solution x.
0	Number of iterations exceeded options.MaxIter or number of function evaluations exceeded options.MaxFunEvals.
−1	The algorithm was terminated by the output function.

The exitflag is also used for other solvers and provides very useful information on whether the optimization successfully terminated. However, in our book, we will look at more details on result messages from the MATLAB, rather than just rely on the exitflag values.

(3) To use fminunc to find the minimum point with the initial x0=1.

```
≫ [x, y]=fminunc(@(x)cos(x)-2*log(x), 1)
Local minimum found.
Optimization completed because the size of the gradient is
 less than the value of the optimality tolerance.
<stopping criteria details>
x =
    3.7108
y =
   -3.4648
```

(4) To use fminbnd to find the minimum point with the interval [0.01 20].

```
≫ [x, y]=fminbnd(@(x)cos(x)-2*log(x), 0.01, 20)
x =
   15.8346
y =
   -6.5164
```

(5) To use fmincon to find the minimum point with the initial x0=1, lower bound lb=0.01, and upper bound ub=20.

```
≫ [x, y]=fmincon(@(x)cos(x)-2*log(x), 1, [], [], [], [],...
 0.01, 20)
Local minimum found that satisfies the constraints.
Optimization completed because the objective function is
 non-decreasing in feasible directions, to within the value of
 the optimality tolerance, and constraints are satisfied within
 the value of the constraint tolerance.
```

```
<stopping criteria details>
x =
    3.7108
y =
   -3.4648
```

MATLAB – Learning Tips

- The "fminsearch", "fminunc" and "fmincon" solvers give the local optimal points.
- The "fminbnd" solver gives a global optimal point for this example.

Problems

2.1 (1) Write matrix $A = \begin{bmatrix} 1.2 & 5 & 4.5 \\ 0 & 3.1 & 2.1 \\ 0 & 2 & 1 \end{bmatrix}$ in the Command Window.

(2) Write matrix $B = \begin{bmatrix} 1.2 & 5 & 4.5 & 8.5 \\ 0 & 3.1 & 2.1 & 4 \\ 0 & 2 & 1 & 3.5 \end{bmatrix}$ using matrix A in the Command Window.

(3) Write matrix $C = \begin{bmatrix} 1.2 & 5 & 4.5 & 8.5 \\ 0 & 3.1 & 2.1 & 4 \\ 0 & 2 & 1 & 3.5 \\ 3.6 & 1.7 & 0 & 9.9 \end{bmatrix}$ using matrix B in the Command Window.

2.2 (1) Find an operator to change data from 5.15 to integer 5.
(2) Find an operator to change data from 5.15 to integer 6.
(3) Try a random number generator: rand (1, 6).
(4) Generate 5 random numbers between 2.5 and 3.6.
(5) Generate 8 random integers between 10 and 100.

2.3 (1) Equally distribute "x1" data with 0.5 intervals from 0 to 20, and function y1=sin(x1).
(2) Equally distribute "x2" data with 0.1 intervals from 0 to 20, and function y2=sin(x2).
(3) Plot y1 vs. x1, and y2 vs. x2 in same figure.
(4) Save it in a fig file and a jpg file.

2.4 To calculate:

$$a = 2.5\pi \times \frac{5e^{-2}}{2^3 + \sqrt{3} + \frac{3}{4}} \times \cos\left(\frac{5}{6}\right) \tag{2.7}$$

2.5 To evaluate the following logical (Boolean) expressions (true or false):
(1) $10 > 7$
(2) $9 = 3 * 3$ (**Note**: It should be written as 9==3*3 in MATLAB for a logical operation)
(3) $\sin(1) > \cos(1)$
(4) $\sin(100) > \cos(100)$

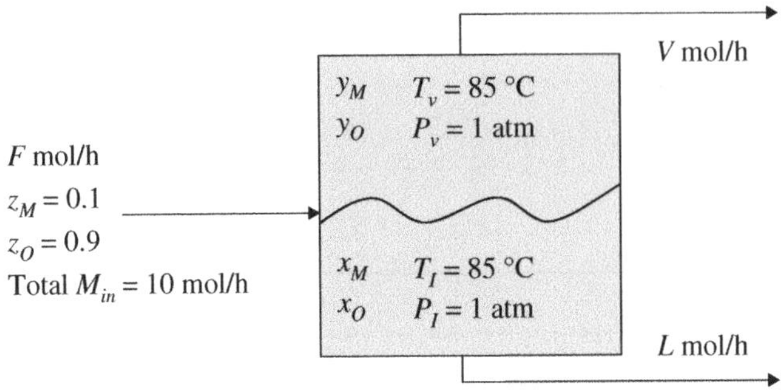

Figure 2.9 A nonadiabatic flash separation drum for methane and octane.

(5) If A is true and B is true, then (A and not (B)) and (A or B)?
(6) If A is true and B is false, then (A and not (B)) and (A or B)?

2.6 Use optimization solvers to find an x with the minimum value of the following function:

$$f(x) = (\cos(e^x) + x^2 - 1)^2 \tag{2.8}$$

(1) Use fminsearch to find the minimum point with the initial x0=1.
(2) Use fminunc to find the minimum point with the initial x0=1.
(3) Use fminbnd to find the minimum point with the interval [0.01 20].
(4) Use fmincon to find the minimum point with the initial x0=1, lower bound lb=0, and upper bound ub=20.
(5) Use fsolve to solve the equation $f(x) = \cos(e^x) + x^2 - 1 = 0$ with initial x0=1.
(6) Use fzero to solve the equation $f(x) = \cos(e^x) + x^2 - 1 = 0$ with initial x0=1.

2.7 A nonadiabatic flash separation drum for methane and octane is operated so that at equilibrium, two phases exist (liquid and vapor), as shown in Figure 2.9. The drum pressure and temperature can be maintained at some specified values. Under these conditions, methane has a boiling point of −162°C and octane 126°C. Virtually all the methane leaves through the vapor stream, and there is a distribution of octane in both product streams. With equilibrium constants:

$$y_M/x_M = K_M(T,P) = 220 \tag{2.9}$$

$$y_O/x_O = K_O(T,P) = 0.5 \tag{2.10}$$

3

Model Fitting and Spline Functions

Model fitting and spline functions in this chapter use algebraic equations to develop mathematical models for a system of interest. Chapter 9 on Complex Equation Systems introduces differential models.

MATLAB has a built-in toolbox specifically for curve fitting. However, this toolbox has limitations on certain equations and does not work with more than three variables. Although the "Custom Equation" option in MATLAB can widen the capability of its curve-fitting toolbox, it is insufficient for Chemical Engineers or students working with complex phenomena in chemical processing systems. We introduce new methodologies for model fitting via optimization algorithms to overcome the general MATLAB toolbox shortcomings.

MATLAB has spline functions, including cubic spline interpolants for curves or surfaces, smoothing splines, and shape-preserving cubic spline interpolants only for curves and thin-plate splines only for surfaces. Splines are piecewise polynomials that provide a smooth representation of functions over large intervals. They are helpful when using a single approximating polynomial would be impractical.

This chapter introduces the curve fitting toolbox in MATLAB and then discusses model fitting via optimization algorithms. Then spline functions for complex models are introduced with chemical engineering examples.

3.1 Introduction

Algebraic equations are used widely for model fitting and spline functions in chemical engineering due to their relatively easy to solve and excellent approximations to the real system data. In this chapter, only steady-state systems are of interest, in which the partial derivatives with respect to (w.r.t.) time are zero, as in Eq. 3.1.

$$\frac{\partial p}{\partial t} = 0 \tag{3.1}$$

Chapter 9 discusses complex equation systems with differential terms.

The models in this chapter are based on scalar or vector systems:

(1) A scalar equation can be, e.g.

$$f(x_1, x_2, \ldots, x_N) = 0 \tag{3.2}$$

The function f can be a linear or a nonlinear function of inputs $x_1, x_2, \ldots, x_N$. The set of inputs can be split into two subsets: state variables and parameters (degrees of freedom).

Chemical Engineering Analysis and Optimization Using MATLAB®, First Edition.
Weiguo Xie, Sam Toan, and Richard Davis.

Companion website: www.wiley.com/go/xie/chemicalengineering

(2) Vector systems are based on sets of equations, which can be linear or nonlinear, e.g.

$$\overline{F}(\overline{y},\overline{z}) = 0 \tag{3.3}$$

where

the functions $\overline{F} = \begin{bmatrix} F_1 \\ F_2 \\ \vdots \\ F_N \end{bmatrix}$, the vector of the state variables $\overline{z} = \begin{bmatrix} z_1 \\ z_2 \\ \vdots \\ z_N \end{bmatrix}$, and the set of parameters $\overline{y} = \begin{bmatrix} y_1 \\ y_2 \\ \vdots \\ y_N \end{bmatrix}$.

3.1.1 Linear Equations

3.1.1.1 Linear Models

Linear models are simplest with the relationships of the state to a linear function of one parameter. This model type can be generated empirically from sufficiently linear experimental data or through data statistics. For example, the electrical resistance of platinum sensor can be modeled as a linear function of temperature between the measured temperature range (20–100 °C) (Kim et al. 2020):

$$R(T) = 3.33 \times 10^{-3} R_0(T - 20) + R_0 \tag{3.4}$$

where $R(T)$ is the resistance at temperature T (°C) and R_0 is the resistance at 20 °C.

The model can also be inverted by measuring resistance to determine temperature.

3.1.1.2 Model Linearization

Linear models can be constructed by linearization of the nonlinear models.

To consider a nonlinear function:

$$f(x) = 0 \tag{3.5}$$

Linearization can be conducted via a Taylor series at point $x = x_o$:

$$f(x) \approx f(x_o) + \left.\frac{df}{dx}\right|_{x_0} (x - x_o) + h.o.t = 0 \tag{3.6}$$

where $h.\ o.\ t.$ means "higher order terms" around x_o.

Based on Eq. 3.6, neglecting $h.o.t.$, it yields the linear equation to approximate the nonlinear function $f(x)$, i.e.

$$f(x) \approx ay + b \tag{3.7}$$

where $y = x - x_o$ and $b = f(x_o)$.

For a special case when $a = 0$, we obtain an infinite number of solutions for the linearized equation. However, the nonlinear equation might still have a unique solution because the linearization approximation is not exact as the original function.

Learning Tips

- Linear models should be first considered due to their simplest feature.
- Model linearization can be very useful for many complex systems.

3.1.2 Algebraic Nonlinear Models

3.1.2.1 Nonlinear Models

In the real world, nonlinear systems are often encountered. Consider the following nonlinear problem:

$$\overline{F}(\overline{x}) = 0, F : R^n \rightarrow R^n \tag{3.8}$$

Often there are no direct methods to find some solutions for a nonlinear problem, e.g.

$$\cos(e^x) + x^2 - 1 = 0 \tag{3.9}$$

3.1.2.2 Solution Methods for Nonlinear Problems

Fitting models with nonlinear parameters involves solving nonlinear problems. However, it usually happens that it is impossible to have analytical solutions or direct methods to find solutions for nonlinear problems. In common practice, we can use the following iterative methods to find some solutions for a nonlinear problem.

(1) Bisection method

Bisection is a root-finding method that applies to any continuous function with two values with opposite signs.

To consider the one-variable problem as an example:

$$f(x) = 0 \tag{3.10}$$

Given a, b such that $f(a) \cdot f(b) < 0$, by the intermediate value theorem, the continuous function f must have at least one root in the interval (a, b). An iterative algorithm performs as follows.

Until termination, do the following:

- Set $c \equiv \frac{a+b}{2}$;
- If $f(c) \cdot f(b) < 0$, $a \leftarrow c$; else, $b \leftarrow c$.

After the kth step, the difference between c_{k+1} and a solution x^* is bounded by $|c_{k+1} - x^*| \leq \frac{|b-a|}{2^{k+1}}$.

The bisection method only needs one f-evaluation per iteration. It guarantees convergence (with strong assumption) but slow convergence (q-linear) compared to other methods.

(2) Newton's method

The Newton–Raphson method, created by Isaac Newton and Joseph Raphson, is a highly effective algorithm for approximating a real-valued function's roots (or zeroes). Its linear approach enables it to accurately and reliably find these roots, making it a valuable tool in mathematics and beyond. It is based on Eq. 3.6, the Taylor series at point $x = x_0$, and omit the *h. o. t.*, then

$$x = x_o - \left(\frac{df}{dx}\Big|_{x_o} \right)^{-1} \cdot f(x_o) = x_o - (f'(x_o))^{-1} \cdot f(x_o) \tag{3.11}$$

Given an initial x_o, an iterative algorithm is performed as follows:
Do the following until termination:

- $x_{new} \leftarrow x_o - (f'(x_o))^{-1} \cdot f(x_o)$;
- $x_o \leftarrow x_{new}$.

If f is continuously differentiable near x^* such that $f(x^*) = 0$ and $f'(x^*) \neq 0$, then for an initial guess x_o sufficiently near x^*. After the kth step, the difference between x_{k+1} and a solution x^* is bounded by $|x_{k+1} - x^*| \leq M \cdot |x_{k+1} - x^*|^2$, where $M > \frac{|f''(x^*)|}{2|f'(x^*)|}$.
The Newton method needs one f-evaluation and one f'-evaluation per iteration. It can obtain only local convergence in general and may diverge without a good initial guess. It can have very fast local convergence (q – quadratic), q means quotient.

(3) Secant method
The secant method is a highly effective algorithm crafted specifically for root-finding purposes. By utilizing a sequence of secant lines, it can accurately approximate the root of any real-valued function with ease. It can be considered a finite-difference approximation of Newton's method; however, it predates Newton's method by over three thousand years.
It needs two initial guesses, x_0 and x_{-1}. An iterative algorithm follows:
Do the following until satisfying the termination criterion:

- $x_{new} \leftarrow x_0 - \left(\frac{f(x_0)-f(x_{-1})}{x_0-x_{-1}}\right)^{-1} \cdot f(x_0)$;
- $x_{-1} \leftarrow x_0$; $x_0 \leftarrow x_{new}$,

where the subscript -1 denotes the previous iteration step.
If f is continuously differentiable near the root x^* such that $f(x^*) = 0$ and $f'(x^*) \neq 0$ then for x_0 and x_{-1} sufficiently near x^*, after the kth step, the difference between x_{k+1} and a solution x^* is bounded by $|x_{k+1} - x^*| \leq C \cdot |x_k - x^*| \cdot |x_{k-1} - x^*|$.
The secant method needs one f-evaluation per iteration and only local convergence in general. It can have fast local convergence (q – superlinear).

(4) Hybrid method – Brent's algorithm
The motivation to develop hybrid methods is to retain desirable properties and eliminate undesirable ones of a particular method by combining existing algorithms. Brent's algorithm is an example of a hybrid method, which combines aspects of the bisection method, the secant method, and inverse quadratic interpolation. It guarantees convergence against the worst case of the secant method. It is an "Enclosure" method with one f-evaluation per iteration and no need to have f' evaluations.
Brent's algorithm usually converges as fast as the secant method. Given a tolerance $\delta > 0$ terminates with an approximate solution within 2δ of the actual solution.

Learning Tips

- The choice of the solution methods really depends on the features of the problem.
- There are also other solution methods such as Quasi-Newton Methods, Secant update methods, Continuation methods, Newton's method with iterative linear solvers, and so on.

3.1.3 Model Fitting

3.1.3.1 Least Squares Procedure

Solving numerical equations typically requires the following steps:

(1) Define a residual $r(x)$
For a set of linear equations $Ax = b$, we can define the residual as:

$$r = Ax - b \tag{3.12}$$

where the residual has the exact dimensions as b.

(2) Minimize this residual $r(x)$ by choosing appropriate values of the variables x
- What we want to minimize is the Euclidean norm of r: $||r|| = \sqrt{r^T r}$
- Therefore:

$$\min_x ||r||^2 = \min_x ||(Ax - b)||^2 \tag{3.13}$$

where

$$||(Ax - b)||^2 = (Ax)^T(Ax) - (Ax)^T b - b^T(Ax) + b^T b \tag{3.14}$$

The derivatives w.r.t. x vanishes at the extrema of the norm of the residual:

$$\frac{\partial(||(Ax - b)||^2)}{\partial x} = \frac{\partial((Ax)^T(Ax) - (Ax)^T b - b^T(Ax) + b^T b)}{\partial x} = 0 \tag{3.15}$$

The residual minimization is, therefore, equivalent to finding the solution to:

$$A^T A x = A^T b \tag{3.16}$$

Note: $A^T A$ gives a square symmetric array of the same length as x.

3.1.3.2 Regression Analysis

(1) Optimization problem
Obtaining better results via model fitting performance is an optimization problem to determine the fitting coefficients, i.e.
- To find $\{a_i\}$ such that $\min_i f(y_{obs} - y_{pred})$

where a_i is the set of adjustable parameters; y_{obs} is the observed value; y_{pred} is the value predicted by the model.
The choice of the objective function may depend on the problem. Two commonly used objective functions include:
(a) Least squares difference between the observed value and model predicted value:

$$\min_i \left[\sum (y_{obs} - y_{pred})^2\right] \tag{3.17}$$

(b) Normalized least squares difference between the observed and model predicted values, as shown in Eq. 3.18. It is particularly useful for data that span a large range to reduce the coefficients' dependence on larger absolute value data.

$$\min_i \left[\sum \left(\frac{y_{obs} - y_{pred}}{y_{obs}}\right)^2\right] \tag{3.18}$$

(2) Error variance and standard error
The error variances quantify the uncertainty about the predictor's values and the response.

$$\sigma^2 = S^2 = \frac{\sum_{i=1}^{N}(y_i - \hat{y}_i)^2}{N - P} \tag{3.19}$$

where y_i is the observed value of the sample items, $\hat{y}_i$ is the model predicted value of these observations, P is the number of parameters in the regression model, and N is the total number of data points.
The standard error, also called root mean squared error (RMSE), can be calculated from the error variances:

$$S = \sqrt{\frac{\sum_{i=1}^{N}(y_i - \hat{y}_i)^2}{N - P}} \tag{3.20}$$

(3) Goodness of fit
Sum of squares due to error (SSE) is a reliable statistical measure that accurately reflects the extent of deviation between the response values and the fit to those values. It is also referred to as the summed square of residuals. A value closer to zero indicates that the model has a smaller random error and that the fit is more useful for prediction.

$$SSE = \sum_{i=1}^{N} w_i(y_i - \hat{y}_i)^2 \tag{3.21}$$

where w_i is the weight coefficient of the ith value.
The coefficient of determination, R^2, or correlation coefficient (R) evaluates the goodness of fit:

$$R^2 = 1 - \frac{SSE}{SST} = 1 - \frac{\sum_{i=1}^{N} w_i(y_i - \hat{y}_i)^2}{\sum_{i=1}^{N} w_i(y_i - \bar{y})^2} \tag{3.22}$$

$$R = \sqrt{1 - \frac{SSE}{SST}} = \sqrt{1 - \frac{\sum_{i=1}^{N} w_i(y_i - \hat{y}_i)^2}{\sum_{i=1}^{N} w_i(y_i - \bar{y})^2}} \tag{3.23}$$

where SST is the sum of the squared difference from the mean of the dependent variable $\bar{y}$ (total sum of squares).

The adjusted R-square is based on the residual degrees of freedom. The residual degrees of freedom are defined as the number of response values n minus the number of fitted coefficients m estimated from the response values.

$$v = n - m \tag{3.24}$$

v indicates the number of independent pieces of information involving the n data points required to calculate the sum of squares. Note that if parameters are bounded, and one or more estimates are at their bounds, those estimates are regarded as fixed.

$$\text{Adjusted } R^2 = 1 - \frac{SSE(n-1)}{SST(v)} \tag{3.25}$$

The R-square and the adjusted R-square statistic can take on any value between zero and one, with a value closer to one indicating a better fit. Including noncontributing terms in the model leads to negative values in predicting the response.

Learning Tips

- $R^2 = 1$, the best fit of the presented model and that the curve passes through every data point.
- $R^2 = 0$, the regression model does not fit the data.
- Note: Just because R is close to 1 does not guarantee that the fit or the model is correct.

(4) Commonly used equations
Some types of equations we often use:
(a) Linear equations, e.g.

$$y = f(x) = a + bx \tag{3.26}$$

(b) Polynomial equations, e.g.

$$C_P = a + bT + cT^2 + dT^3 + eT^4 + \cdots \tag{3.27}$$

(c) Exponential equations, e.g.

$$k = k_o e^{-E_A/RT} \tag{3.28}$$

$$c = c_o(1 - e^{-kt}) \tag{3.29}$$

(d) Logarithmic equations, e.g.

$$y = a \log x \tag{3.30}$$

3.1.4 Spline Functions

3.1.4.1 Linear Interpolation

Linear interpolation uses a linear function connecting two data points. If an intermediate data point (x, y) is between these two given data points (x_a, y_a) and (x_b, y_b), then the following linear equation can be used to calculate the intermediate data point:

$$\frac{y - y_a}{y_b - y_a} = \frac{x - x_a}{x_b - x_a} \tag{3.31}$$

3.1.4.2 Polynomial Interpolation

Polynomial interpolation involves finding the polynomial with the minimum degree that can pass through the given dataset. This method can be used to estimate complex curves.

There is an interpolation theorem, which states there exists a unique polynomial of degree at most n that interpolates the $n + 1$ data points, $\{(x_0, y_0), (x_1, y_1), \cdots, (x_n, y_n)\} \in R^2$, where no two x_j are the same.

Suppose that the interpolation polynomial is in the form:

$$p(x) = a_n x^n + a_{n-1}x^{n-1} + \cdots + a_2 x^2 + a_1 x + a_0 \tag{3.32}$$

where p interpolates the data points: $p(x_i) = y_i$ for all $i \in \{0, 1, \ldots, n\}$

Lagrange polynomials have the form as

$$p(x) = \frac{(x - x_1)(x - x_2)\cdots(x - x_n)}{(x_0 - x_1)(x_0 - x_2)\cdots(x_0 - x_n)} y_0 + \frac{(x - x_0)(x - x_2)\cdots(x - x_n)}{(x_1 - x_0)(x_1 - x_2)\cdots(x_1 - x_n)} y_1 + \cdots + \frac{(x - x_0)(x - x_1)\cdots(x - x_{n-1})}{(x_n - x_0)(x_n - x_1)\cdots(x_n - x_{n-1})} y_n = \sum_{i=0}^{n} \left(\prod_{0 \le j \le n, j \ne i} \frac{x - x_j}{x_i - x_j} \right) y_i \tag{3.33}$$

3.1.4.3 Cubic Spline Interpolation

A cubic spline comprises third-order polynomials pieced together to ensure they pass through a set of control points. A cubic spline is often the best approach to estimate an intermediate data point in a nonlinear data trend.

The general expression for the ith C^2 interpolating cubic spline at a point x with the natural condition can be found using the following formula:

$$S_i(x) = \frac{z_i(x-t_{i-1})^3}{6h_i} + \frac{z_{i-1}(t_i-x)^3}{6h_i} + \left[\frac{f(t_i)}{h_i} - \frac{z_i h_i}{6}\right](x-t_{i-1}) + \left[\frac{f(t_{i-1})}{h_i} - \frac{z_{i-1}h_i}{6}\right](t_i - x) \tag{3.34}$$

where

$z_i = f''(t_i)$ is the value of the second derivative at the ith knot.

$h_i = t_i - t_{i-1}$

$f(t_i)$ is the value of the function at the ith knot.

3.2 Curve Fitting Toolbox in MATLAB

If you are working with data, MATLAB has a curve-fitting toolbox that can help you fit curves and surfaces to your data. You can use the toolbox to explore your data, preprocess and post-process it, compare different models, and even remove outliers. You will have access to a library of linear and nonlinear models, or you can create custom equations. The toolbox also has optimized solver parameters and starting conditions to help you get the best possible results. And if you need to use nonparametric modeling techniques like splines, interpolation, or smoothing, the toolbox has you covered. Once you have created a fit, you can use the toolbox's post-processing tools to plot your data, estimate confidence intervals, calculate integrals and derivatives, and more.

3.2.1 Two Ways to Curve Fitting

There are two ways to perform curve fitting in MATLAB: Interactive Curve Fitting and Programmatic Curve Fitting.

- For Interactive Curve Fitting, after load data, you must open the Curve Fitting app in Command Window:

```
>> cftool
```

After selecting X and Y data, you can choose a different model type using the fit category drop-down list.

- For Programmatic Curve Fitting, after load data, you must create a fit using the fit function, specifying the variables and a model type:

```
>> fitobject = fit(x,y,fitType)
```

It creates the fit to the data in x and y with the model specified by fitType.

More details in MATLAB online materials: https://www.mathworks.com/products/curvefitting.html.

3.2.2 An Example of Using Curve Fitting App

Example 3.1

This example of particle size distribution uses the curve fitting app in the following section. Table 3.1 shows the particle size distribution. The data are saved in the file "Particle size data.xlsx."

Use the following command to load data in Command Window:

Solution:

```
≫ A=xlsread('Particle size data.xlsx','B3:C27');
≫ x=A(:,1);y=A(:,2);
```

3.2.2.1 Open the Curve Fitting App in Command Window

After loading data, you can open the Curve Fitting app (as shown in Figure 3.1) in Command Window by typing

```
≫ cftool
```

Table 3.1 The particle size distribution of passing (%) in terms of size (mm).

Size (mm)	Passing (%)
300	100
212	99.48
150	95.99
106	89.42
75	81.04
53	72.45
38	64.98
27	58.11
19	51.48
13.5	44.95
9.6	38.13
6.8	31.13
5	25.28
3.4	19.14
2.4	15
1.75	12.14
1.2	9.43
0.87	7.597
0.62	6.03
0.44	4.583
0.31	3.137
0.22	1.847
0.16	0.977
0.11	0.373
0.08	0.123

Figure 3.1 The window of the Curve Fitting app.

3.2.2.2 Selecting Data and the Default Linear Model

Select X and Y data as shown in Figure 3.2 with a red highlight circle.

After selecting X and Y data, a default result will be shown in Figure 3.2. The default model in the Curve Fitting app is a linear model (i.e. Polynomial with Degree 1), as shown in the middle of Figure 3.2. As seen at the bottom graph window of Figure 3.2, a linear model is not a good fit to the particle size distribution data in this case. The bad curve fitting result can also be seen from the left-bottom Results window "Goodness of fit" in Figure 3.2. SSE (the summed square of residuals) is 9986, which is very large compared to the measured passing data and far away from 0. R-square of 0.6716 and adjusted R-square of 0.6573 are far from 1, indicating the bad fit using the linear model. The RMSE is 20.84, which also shows a bad fit because it is far from 0.

3.2.2.3 Selecting Better Models

Choose a different model type using the fit category drop-down list, which includes Custom Equation, Exponential, Fourier, Gaussian, Interpolant, Linear Fitting, Polynomial, Power, Rational, Smoothing Spline, and Weibull. Here, we show some of them and leave the rest of the models to all interested readers.

(a) Figure 3.3 shows the model fitting using Exponential with two terms. It can be seen from the graph that it is much better than the linear model fit in Figure 3.2. The model fitting results are also shown in the left-bottom Results window as follows:

```
General model Exp2:
     f(x) = a*exp(b*x) + c*exp(d*x)
```

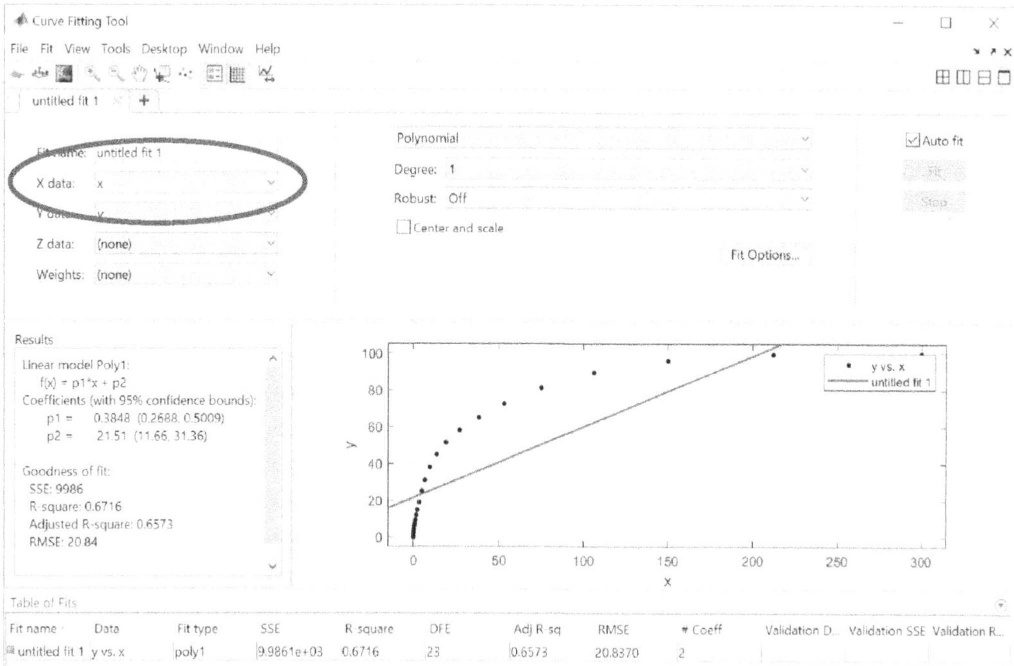

Figure 3.2 Select X and Y data with a red highlight circle.

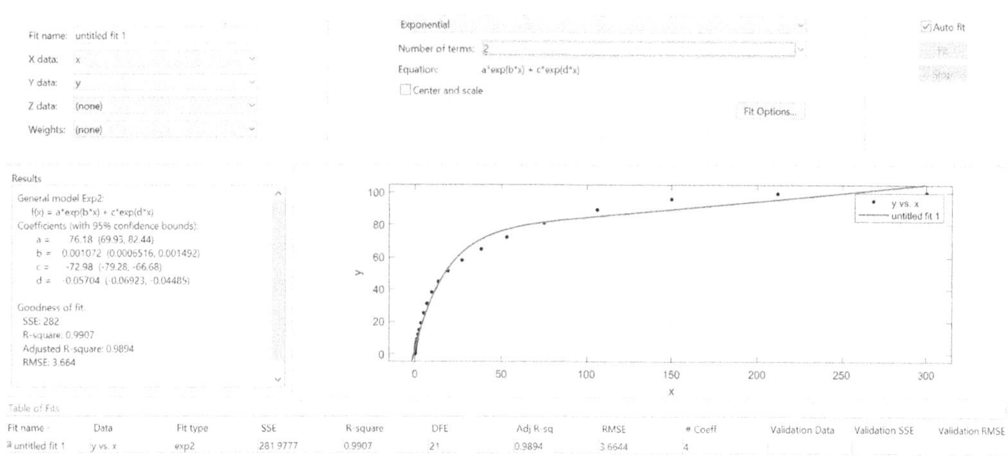

Figure 3.3 Model fitting using exponential with number of terms 2.

```
Coefficients (with 95% confidence bounds):
       a =       76.18  (69.93, 82.44)
       b =    0.001072  (0.0006516, 0.001492)
       c =      -72.98  (-79.28, -66.68)
       d =    -0.05704  (-0.06923, -0.04485)
Goodness of fit:
  SSE: 282
```

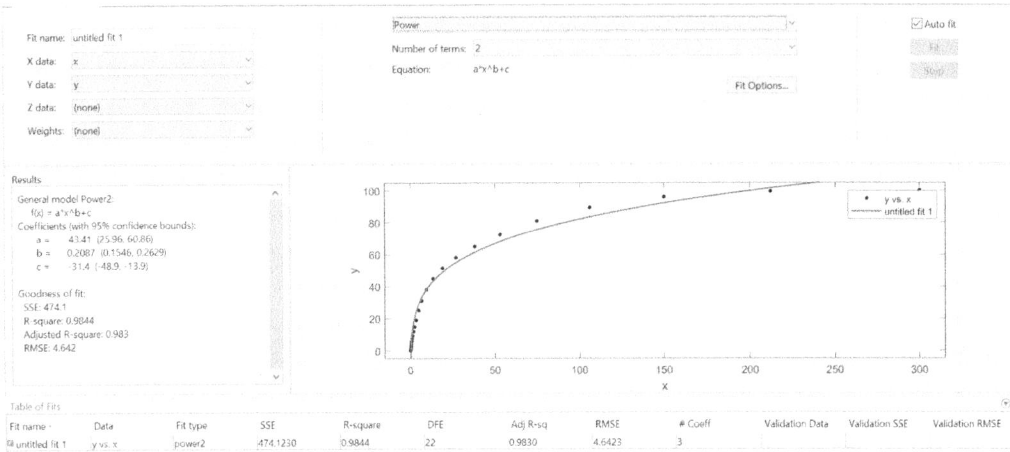

Figure 3.4 Model fitting using Power with two terms.

```
R-square: 0.9907
Adjusted R-square: 0.9894
RMSE: 3.664
```

R-square and Adjusted R-square are close to one, which means the model fits the data well.

(b) Figure 3.4 shows the model fitting using Power with number of terms as 2. It can be seen from the graph that it is much better than the linear model fit in Figure 3.2. The model fitting results are also shown in the left-bottom Results window as follows:

```
General model Power2:
     f(x) = a*x^b+c
Coefficients (with 95% confidence bounds):
       a =        43.41  (25.96, 60.86)
       b =       0.2087  (0.1546, 0.2629)
       c =        -31.4  (-48.9, -13.9)
Goodness of fit:
  SSE: 474.1
  R-square: 0.9844
  Adjusted R-square: 0.983
  RMSE: 4.642
```

It can be seen that both R-square and adjusted R-square are close to 1, which means the model is a good fit for the data. It is much better than the linear model. However, it is worse than the Exponential model comparing the R-square and adjusted R-square values. The good thing is that the Power model uses three parameters (a, b, and c), which is one less than the Exponential model. This feature is helpful for a large dataset in practical applications.

(c) Figure 3.5 shows the model fitting using Rational with a numerator degree of two and a denominator degree of two. It can be seen from the graph that it is much better than the linear model fit in Figure 3.2. The model fitting results are also shown in the left-bottom Results window as follows:

```
General model Rat22:
     f(x) = (p1*x^2 + p2*x + p3) / (x^2 + q1*x + q2)
Coefficients (with 95% confidence bounds):
       p1 =       105.8  (102, 109.5)
       p2 =      -25.31  (-103, 52.42)
       p3 =      -37.85  (-69.24, -6.457)
       q1 =          20  (16.8, 23.21)
       q2 =      -15.78  (-23.08, -8.472)
Goodness of fit:
  SSE: 128
  R-square: 0.9958
  Adjusted R-square: 0.9949
  RMSE: 2.53
```

It can be seen that both R-square and adjusted R-square are close to 1, which means the model is a good fit for the data. It is much better than a linear model. It is better than the above Exponential and Power models comparing the R-square and adjusted R-square values. However, the Rational model uses five parameters (p1, p2, p3, q1, and q2), which is a disadvantage compared to the above Exponential and Power models.

(d) All the above models are just based on random selection or with some experience from the data trend. However, the chosen models may only work for the data used to fit the model and are unsuitable for other experimental data. Rosin–Rammler distribution model with two parameters (a and b) is a good model to fit particle size distribution in the industry (Yu et al. 2018):

$$y = 100 * \left(1 - 0.2^{\left(x/a\right)^b}\right) \tag{3.35}$$

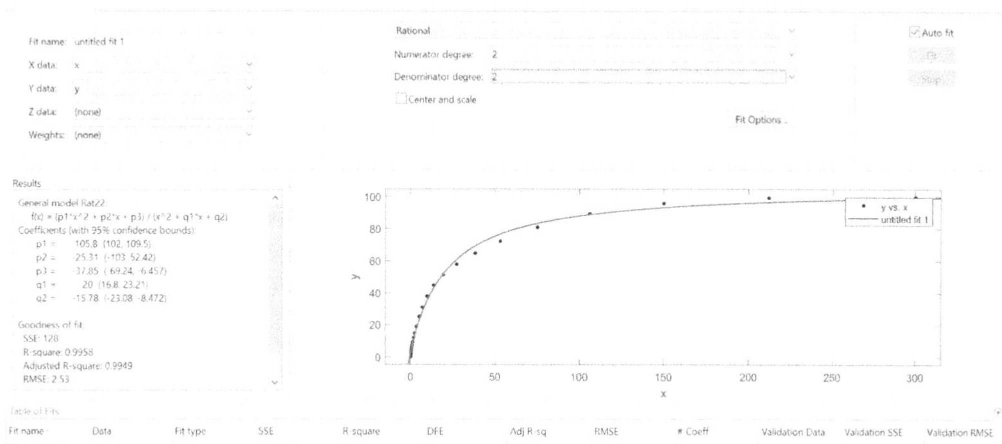

Figure 3.5 Model fitting using Rational with numerator degree of two and denominator degree of two.

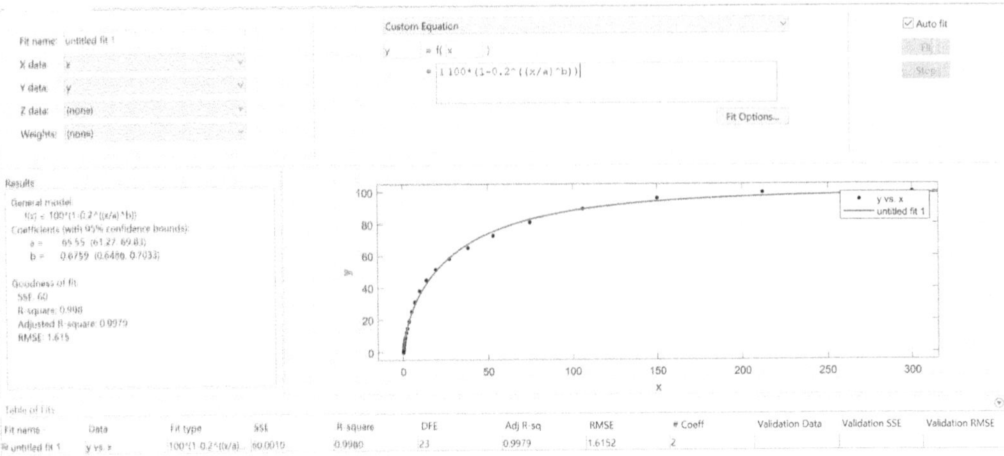

Figure 3.6 Model fitting using Custom Equation (Rosin–Rammler distribution model).

Figure 3.6 shows the model fitting using Custom Equation (Rosin–Rammler distribution model). It can be seen from the graph that it is much better than the linear model fit in Figure 3.2. The model fitting results are also shown in the left-bottom Results window as follows:

```
General model:
     f(x) = 100*(1-0.2^((x/a)^b))
Coefficients (with 95% confidence bounds):
       a =        65.55  (61.27, 69.83)
       b =       0.6759  (0.6486, 0.7033)
Goodness of fit:
  SSE: 60
  R-square: 0.998
  Adjusted R-square: 0.9979
  RMSE: 1.615
```

It can be seen that both R-square and adjusted R-square are very close to 1, which means the model is a very good fit for the data. It is much better than a linear model. It is better than all the above Exponential, Power, and Rational models comparing the R-square and adjusted R-square values. Furthermore, the Rosin–Rammler distribution model has one significant advantage of only two parameters (the least number of parameters compared to all other models). This feature is handy, especially for very large datasets in industrial applications.

Learning Tips

- The knowledge of the trend of some models will be very useful in the model fitting.
- Some empirical models or models with theoretical background may give much better fit to the experimental data.

3.3 Model Fitting via Optimization Algorithms

In the previous section, a curve fitting toolbox was introduced for model fitting, which is quite efficient for two variables in the following form.

$$y = f(x) \tag{3.36}$$

There is also the surface fitting for three variables in the following form.

$$z = f(x, y) \tag{3.37}$$

However, in real-world practice, such as chemical engineering, many cases with multivariable (≥3 variables) are out of the MATLAB curve/surface fitting capability.

For model fitting for cases with multivariable (≥3 variables), optimization algorithms can be applied to determine the fitting coefficients:

To find $\{a_i\}$ such that the least squares difference between the observed value and model predicted value:

$$\min_{i} \left[\sum (y_{obs} - y_{pred})^2 \right] \tag{3.38}$$

where y_{obs} is the observed value; y_{pred} is the value predicted by the model with multivariable (≥3 variables) a_i is the set of adjustable parameters in function g:

$$y_{pred} = g(x_1, x_1, \cdots x_N) \tag{3.39}$$

The function g can be a linear or a nonlinear function of the multivariable (i.e. the inputs x_1, $x_2, \cdots\cdots, x_N$). The set of inputs can be split into two subsets: state variables and parameters (degrees of freedom).

Optimization algorithms are used to determine the fitting coefficients $\{a_i\}$. Some commonly used optimization solvers in MATLAB include fminbnd, fmincon, fminsearch, fminunc, and fsolve. The fminbnd solver is used for cases with single $\{a_i\}$, while other solvers can be used for any number of $\{a_i\}$. The optimization toolbox in MATLAB is introduced in Chapter 4, and examples of how to use optimization algorithms for model fitting are demonstrated in Chapter 7.

3.4 Spline Functions for Complex Models

A lack of specific data points is common in real-world engineering practice, especially with complex models. For example, in the characterization of product particle size, P80 is defined as the size at which 80% of the product passes due to the measurement of particle size using sieves, which is discretized data and may not be able to get the P80 just using a limited number of sieves. The P80 is often obtained using spline functions (or from empirical models).

Example 3.2
Calculate P80 of the data from Example 3.1 using the spline function in MATLAB.

To load data of Example 3.1 in Command Window and plot the data:

Solution:

```
>> A=xlsread('Particle size data.xlsx','B3:C27');
>> y=A(:,1);x=(:,2);
>> plot(y,x)
>> xlabel('Particle size (mm)');ylabel('Passing (%)')
```

3.4.1 Spline Interpolation in MATLAB – *interp1*

The *interp1* is used for 1-D data interpolation (table lookup) in MATLAB. Its syntax is listed in the Help Center from MathWorks as follows:

```
vq = interp1(x,v,xq)
vq = interp1(x,v,xq,method)
vq = interp1(x,v,xq,method,extrapolation)
vq = interp1(v,xq)
vq = interp1(v,xq,method)
vq = interp1(v,xq,method,extrapolation)
pp = interp1(x,v,method,'pp')
```

Table 3.2 shows the commonly used interpolation methods.
In Example 3.2, to use 'linear' to calculate P80:

```
>> vq = interp1(y,x,80,'linear')
vq =
   82.3916
```

To use 'nearest' to calculate P80:

```
>> vq = interp1(y,x,80,'nearest')
vq =
   81.0400
```

Table 3.2 Common spline interpolation methods as one of the options using the *interp1* (https://www.mathworks.com/The MathWorks, Inc.).

Method	Description	Comments
'linear'	Linear interpolation (default).	Requires at least two points
'nearest'	Nearest neighbor interpolation.	Requires at least two points; Fastest computation time
'pchip'	Shape-preserving piecewise cubic interpolation.	Requires at least four points; Requires more memory and computation time than 'linear'
'cubic' or 'v5cubic'	Cubic convolution used in MATLAB.	Requires at least three points; Points must be uniformly spaced; This method falls back to 'spline' interpolation for irregularly spaced data.
'makima'	Modified Akima cubic Hermite interpolation. The interpolated value at a query point is based on a piecewise function of polynomials with degree at most three. The Akima formula is modified to avoid overshoots.	Requires at least two points; Produces fewer undulations than 'spline', but does not flatten as aggressively as 'pchip'.
'spline'	Spline interpolation using not-a-knot end conditions.	Requires at least four points.

To use 'pchip' to calculate P80:

```
>> vq = interp1(y,x,80,'pchip')
vq =
   82.6267
```

To use 'cubic' to calculate P80:

```
>> vq = interp1(y,x,80,'cubic')
Warning: The 'cubic' method requires the grid to have a uni-
form spacing.
Switching the method from 'cubic' to 'spline' because this con-
dition is not met.
> In interp1 (line 188)
vq =
   82.6626
```

To use 'makima' to calculate P80:

```
>> vq = interp1(y,x,80,'makima')
vq =
   82.6070
```

To use 'spline' to calculate P80:

```
>> vq = interp1(y,x,80,'spline')
vq =
   82.6626
```

It can be seen that all the above interpolation methods can give the accurate P80 as shown in Figure 3.7. The method of 'nearest' gives the result of 1 mm less than all others.

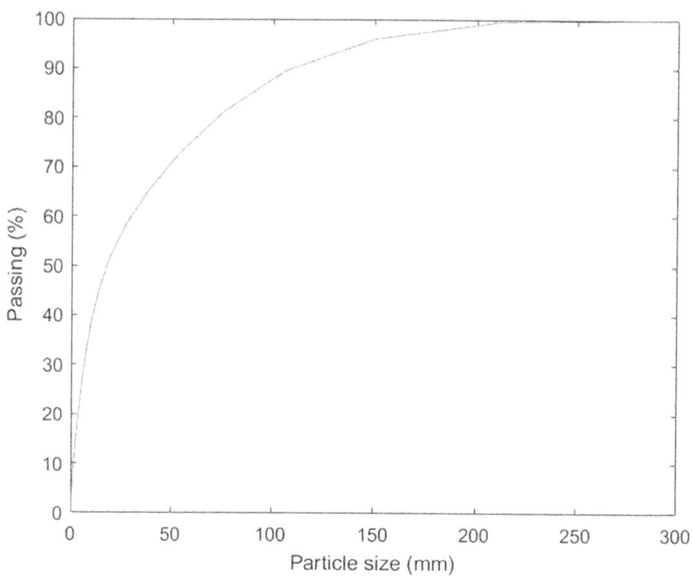

Figure 3.7 Particle size distribution curve of Example 3.2.

3.4.2 Spline Interpolation in MATLAB – *interp2*

The *interp2* is used for Interpolation for 2-D gridded data in a mesh grid format. The Help Center from MathWorks shows its syntax as follows:

```
Vq = interp2(X,Y,V,Xq,Yq)
Vq = interp2(V,Xq,Yq)
Vq = interp2(V)
Vq = interp2(V,k)
Vq = interp2(___,method)
Vq = interp2(___,method,extrapval)
```

The interpolation methods include 'inear', 'nearest', 'cubic', 'makima', or 'spline'. The default method is 'linear'.

Example 3.3

Determine the enthalpy of superheated steam at a particular condition. Table 3.3 lists the steam table's enthalpy [H, kJ/kg] data.

(1) To draw a surface plot (T, P, H);
(2) To determine the enthalpy of superheated steam at 420 K and 190 kPa.

Solution:

The code follows:

```
>> T = [300 350 400 450 500];
>> P = [150 200 250 300];
>> H = [3073.3 3174.7 3277.5 3381.7 3487.6
>> 3072.1 3173.8 3276.7 3381.1 3487.0
>> 3070.9 3172.8 3275.9 3380.4 3486.5
>> 3069.7 3171.9 3275.2 3379.8 3486.0];
>> figure;
>> surf(T,P,H);
>> title('The enthalpy of a superheated steam');
>> xlabel('Temperature (K)'); ylabel('Pressure (kPa)');
   zlabel('Enthalpy (kJ/kg)');
>> Enthalpy=interp2(T,P,H,420,190)

Enthalpy =

   3.3186e+03
```

Table 3.3 The enthalpy [H, kJ/kg] data of superheated steam.

P [kPa]*T*[*K*]	300	350	400	450	500
150	3073.3	3174.7	3277.5	3381.7	3487.6
200	3072.1	3173.8	3276.7	3381.1	3487.0
250	3070.9	3172.8	3275.9	3380.4	3486.5
300	3069.7	3171.9	3275.2	3379.8	3486.0

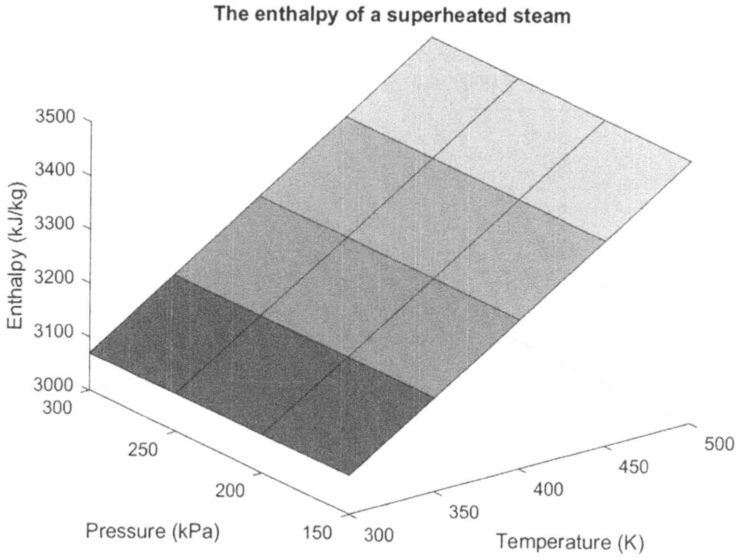

Figure 3.8 The surf plot of Example 3.3.

Figure 3.8 shows the surface plot (T, P, H). The result of the enthalpy of superheated steam at 420 K and 190 kPa is 3.3186e+03 kJ/kg.

3.5 Application in Chemical Engineering

The skills in model fitting and spline functions will be handy in the reader's career, especially their wide application in chemical engineering. This section demonstrates some examples of how to apply model fitting and spline functions.

3.5.1 Using Linear Equations

Example 3.4

Filtration is a physical separation process commonly used in chemical engineering to remove the suspended particles from a fluid, performed by a filter medium, septum, cloth, or bed of solids. A Buchner funnel with filter paper was used in the lab to filter limestone slurry. In Table 3.4, the filtrate volume was recorded at various times until no more filtrate drained from the cake. The ID of the Buchner funnel was 9.31 cm. The constant pressure drop was 6 inches Hg. The mass of limestone was 75.01 g. The mass of the filter paper was 0.90 g. The mass of the wet cake with filter paper after the filtration was 87.21 g. Allow the cake to dry overnight in an oven at 60 °C. Reweigh after cooling to obtain the mass of the dry cake with filter paper 73.08 g. The density of the limestone is 2740 kg/m^3. The viscosity of water is 1.0016 mPa.s.

(1) Calculate the specific resistance to filtration, α.
(2) Calculate the resistance of the medium, R_m.

Table 3.4 The volume of filtrate vs. time.

Time (s)	Volume (ml)
90	50
200	70
375	92
500	110
650	129
770	140
900	151
1080	170
1240	185
1380	200
1570	215
2000	245
2450	282
2700	301
2940	320
3130	330
3310	339
3500	340

Solution:

The modeling of cake filtration assumes that Darcy's law applied to both the filter medium (m) and the forming filter cake (c) (Holdich 2002):

$$\Delta P = \Delta P_c + \Delta P_m = \frac{\mu}{A}\frac{L}{k}\frac{dV}{dt} + \frac{\mu}{A}\frac{L_m}{k_m}\frac{dV}{dt} \tag{3.40}$$

where L represents the cake height at any instance in time. The medium height and permeability are essentially constant during the filtration. k is the permeability of the cake. The cake mass uniformly increases with the filtrate volume for incompressible cake filtration. To define the resistance of the medium, R_m. and the resistance of the cake, R_c.

$$R_m = \frac{L_m}{k_m} \tag{3.41}$$

$$R_c = \frac{L}{k} \tag{3.42}$$

The cake resistance increases proportionally with increasing cake mass or mass per unit filter area. The constant of proportionality is called specific filtration resistance (α). Hence, the cake resistance

$$R_c = \frac{\alpha W}{A} \tag{3.43}$$

where W is the total mass of cake deposited, which can be related to the volume of cake (LA) by the volume fraction of solids in the cake (C) and the density of the solids:

$$W = LAC\rho_s \tag{3.44}$$

The cake mass increases uniformly with the filtrate volume for incompressible cake filtration. This constant of proportionality is called dry cake mass per filtrate volume (c).

$$c = \frac{LAC\rho_s}{V} \tag{3.45}$$

Hence, Eq. 3.40 becomes

$$\frac{dt}{dV} = \frac{\mu\alpha c}{A^2\Delta P}V + \frac{\mu R_m}{A\Delta P} \tag{3.46}$$

Because in this example, the constant pressure drop was 6 inches Hg, then integrating from time zero to time t,

$$\int_0^t dt = \int_0^V \left(\frac{\mu\alpha c}{A^2\Delta P}V + \frac{\mu R_m}{A\Delta P}\right) dV \tag{3.47}$$

Then,

$$t = \frac{\mu\alpha c}{2A^2\Delta P}V^2 + \frac{\mu R_m}{A\Delta P}V \tag{3.48}$$

The above equation can be rearranged to assist data analysis to give

$$\frac{t}{V} = \frac{\mu\alpha c}{2A^2\Delta P}V + \frac{\mu R_m}{A\Delta P} \tag{3.49}$$

Therefore, there is a linear model between $\frac{t}{V}$ and V.

The model fitting in MATLAB for this example follows:

```
≫ t=[90 200 375 500 650 770 900 1080 1240...
 1380 1570 2000 2450 2700 2940 3130 3310 3500];
V=[50 70 92 110 129 140 151 170 185...
 200 215 245 282 301 320 330 339 340];
≫y=t./V;
≫ cftool
```

It can be seen from the above result (R-square: 0.9686) that there is a good linear relationship between $\frac{t}{V}$ and V, as shown in Figure 3.9.

(1) To calculate the specific resistance to filtration, α, the slope of the above straight line can be used (**Note: Make sure that all parameters need to have corresponding units**):

$$\frac{\mu\alpha c}{2A^2\Delta P} = 0.02494 \tag{3.50}$$

Except for α, the only unknown in the above is the dry cake mass per filtrate volume (c), which can be calculated from the mass of the dry cake (73.08–0.90 g) divided by the total filtrate volume (340 ml). Then, c is 212.29 kg/m^3. It is straightforward to obtain α value. We leave the calculation to readers and answer as follows:

$$\alpha = 2.21 \times 10^{11}\,\mathrm{m/kg}$$

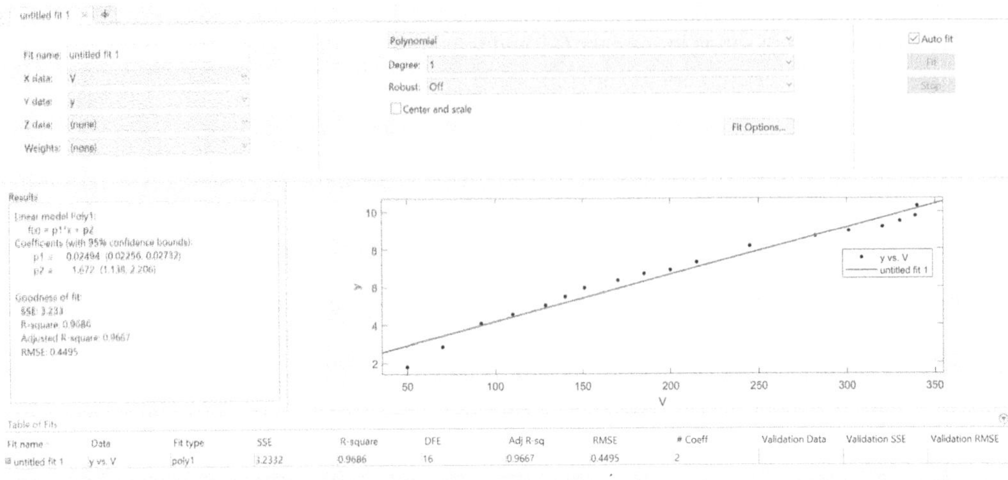

Figure 3.9 Model fitting using a linear relationship in MATLAB.

(2) To calculate the resistance of the medium, R_m. The intercept of the above straight line can be used (**Note: Make sure that all parameters need to have corresponding units**):

$$\frac{\mu R_m}{A\Delta P} = 1.672 \tag{3.51}$$

We leave the calculation to readers and answer as follows:

$$R_m = 2.31 \times 10^{11}/\text{m}$$

3.5.2 Using Nonlinear Equations

Example 3.5

Fluid flow through packed beds has important applications in chemical engineering, food processing, reactor design, and various other engineering fields. An example is the packed bed reactor (PBR). A PBR typically contains a porous solid-state catalyst used to increase the conversion rate in a chemical reaction. One factor that significantly affects the reaction rate is the pressure drop across the packed bed. The hollow cylindrical packing material produces significant frictional losses in the system. As shown in Figure 3.10, the Ergun correlation is used to calculate the friction factor for hollow objects, such as packing rings (Holdich 2002). The form of the Ergun equation has a correction term added to the laminar flow term to account for resistance due to turbulences. In Table 3.5, the experimental data were recorded, including the airflow rate and corresponding pressure drop. The ID of the tube containing the packing rings was 1.20 cm, and its length was 20 cm. The viscosity of air at 20 °C is 0.00001634 Pa.s. The hollow cylinder packing material has the following dimensions: $ID = 1.098$ mm, $OD = 1.885$ mm, and length = 1.311 mm. The porosity ε was 0.57. The specific surface area per unit volume, S_v, was 6608/m. The lab atmosphere pressure was 118977 Pa.

(1) To determine the coefficients of Ergun correlation via our experimental data.
(2) To use the spline function to obtain the pressure drop, ΔP, when the airflow rate is 1.0 m/s.

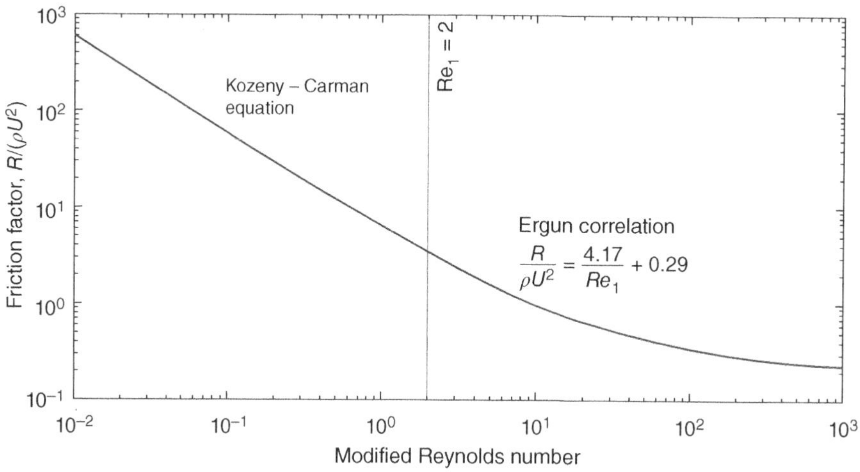

Figure 3.10 The Ergun correlation in a friction factor plot for fluid flow through porous media. (Holdich, 2002).

Table 3.5 The air volumetric flow rate vs. pressure drop.

Airflow rate (m/s)	ΔP (kPa)
0.70	1
0.83	1.2
1.11	2
1.32	3
1.67	4
1.74	4.5
1.81	5
1.98	6
2.09	6.1
2.12	6.5
2.16	7
2.36	8
2.47	9

Solution:

(1) The modified Reynolds number is

$$Re_1 = \frac{\varepsilon}{(1-\varepsilon)S_v}\frac{U_o}{\varepsilon}\frac{\rho}{\mu} = \frac{\rho U_o}{(1-\varepsilon)S_v\mu} \tag{3.52}$$

The pressure drop per unit length can be calculated by

$$\frac{\Delta P}{L} = \left(\frac{R}{\rho U^2}\right)\frac{S_v(1-\varepsilon)\rho U_o{}^2}{\varepsilon^3} \tag{3.53}$$

Ergun correlation has the form

$$\frac{R}{\rho U^2} = \frac{a}{Re_1} + b \tag{3.54}$$

To combine the above equations,

$$\Delta P = \frac{a \cdot S_v^2(1-\varepsilon)^2 \cdot \mu \cdot L}{\varepsilon^3} \cdot U_0 + \frac{b \cdot S_v(1-\varepsilon) \cdot \rho \cdot L}{\varepsilon^3} \cdot U_0^2 \tag{3.55}$$

The air density ρ can be calculated by ideal gas law using the lab atmosphere pressure: $\rho = 118977 * 29/(8.314 * 293 * 1000) = 1.416\ \text{kg/m}^3$
To substitute all known parameters into the equation,

$$\Delta P = a * 142.47 * U_0 + b * 4345.17 * U_0^2 \tag{3.56}$$

This example uses model fitting in MATLAB:

```
>> x=[0.70 0.83 1.11 1.32 1.67 1.74 1.81 1.98 2.09 2.12 2.16 2.36 2.47];
y=[1 1.2 2 3 4 4.5 5 6 6.1 6.5 7 8 9]*1000;
>> cftool
```

Then we can use Custom Equation $y = a * 142.47 * x + b * 4345.17 * x$ ^ 2, as shown in Figure 3.11.

It can be seen from the above result (R-square: 0.9954) that it is a good model between ΔP and U_0. The coefficients of the Ergun correlation are $a = 2.827$ and $b = 0.2946$. Compared to the Ergun correlation in Figure 3.10, b value is very close to, and a is quite different from the literature values.

(2) To use 'linear' to calculate the pressure drop, ΔP:

```
>> DP = interp1(x, y, 1.0,'linear')
DP =
    1.6857e+03
```

To use 'nearest' to calculate the pressure drop, ΔP:

```
>> DP= interp1(x, y, 1.0,'nearest')
DP =
          2000
```

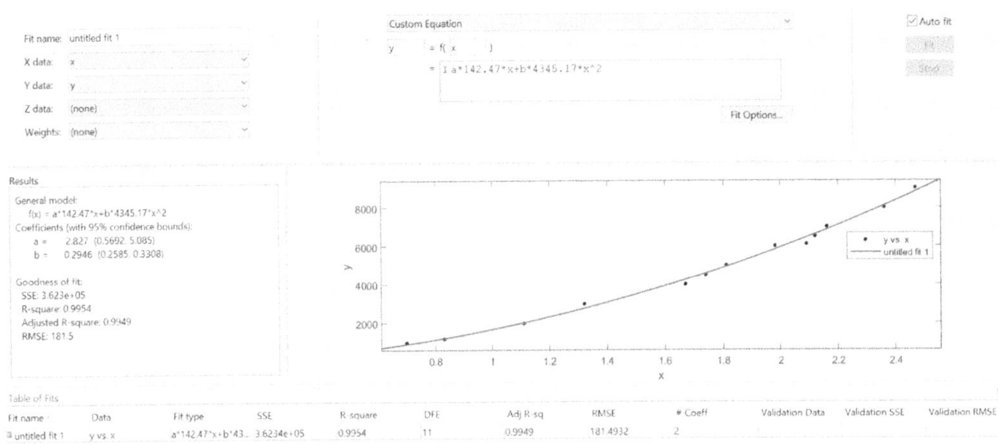

Figure 3.11 The model fitting using Custom Equation in MATLAB.

To use 'pchip' to calculate the pressure drop, ΔP:

```
≫ DP = interp1(x, y, 1.0,'pchip')
DP =
   1.6307e+03
```

Use 'cubic' to calculate the pressure drop, ΔP:

```
≫ DP= interp1(x, y, 1.0,'cubic')
Warning: The 'cubic' method requires the grid to have a
 uniform spacing.
Switching the method from 'cubic' to 'spline' because this
 condition is not met.
> In interp1 (line 188)
DP =
   1.5860e+03
```

Use 'makima' to calculate the pressure drop, ΔP:

```
≫ DP = interp1(x, y, 1.0,'makima')
DP =
   1.6313e+03
```

To use 'spline' to calculate the pressure drop, ΔP:

```
≫ DP = interp1(x, y, 1.0,'spline')
DP =
   1.5860e+03
```

All the above interpolation methods except 'nearest' can give quite accurate ΔP.

Learning Tips

- The linear equations are simplest and need to be tried first.
- The non-linear equations can be very complicated and need to have some experience to achieve better results.

3.6 Summary

This chapter presents model fitting and spline functions useful in chemical engineering design and simulation. Models in the form of algebraic equations are the focus of this chapter. Models with differential terms will be the focus of Chapter 9 of the Complex Equation Systems. Due to simplicity, linear models should be first considered, and model linearization can also be very useful for many complex systems.

To obtain an excellent model for the system of interest, the knowledge of the trend of certain model forms can be advantageous in the model fitting. In addition, some empirical models or models derived from theoretical backgrounds may better fit the experimental data.

Problems

3.1 Table 3.6 lists the particle size distribution.

(1) Plot the particle size distribution with passing (%) in terms of size (mm).
(2) Select models from Curve Fitting Toolbox in MATLAB and compare the goodness of the fitting.
(3) Use Custom Equation (Rosin–Rammler distribution model) for the model fitting and compare results with the above other models.
(4) Calculate P80 from the data.

Table 3.6 The particle size distribution shown with passing (%) in terms of size (mm).

Size (mm)	Passing (%)
19	100
13.2	96.59
9.5	83.35
6.7	65.08
4.75	46.90
3.35	32.90
2.36	24.90
1.7	18.42
1.18	14.28
0.85	11.09
0.6	9.05
0.425	7.01
0.3	5.67
0.212	4.42
0.15	3.42

3.2 The effect of specific comminution energy (E_{cs}) on the breakage index t_{10} (percentage passing an aperture of 1/10 of the original ore particle size, Napier-Munn et al. 1996). A and b: the ore breakage parameters determined from drop weight tester (DWT) laboratory tests. A*b is the slope of the curve of 'zero' energy input. It is the indicator of ore hardness. The greater the A*b value is, the softer the ore. Table 3.7 shows the lab data.

(1) To plot the breakage index t_{10} (%) vs. specific comminution energy (kWh/t).
(2) To select models from Curve Fitting Toolbox in MATLAB and compare the goodness of the fitting.
(3) To use Custom Equation $\left(t_{10} = A\left(1 - e^{-bE_{cs}}\right)\right)$ for the model fitting, and compare results with the above other models.
(4) To calculate t_{10} (%) value when $E_{cs} = 1$ kWh/t.

Table 3.7 The breakage index t_{10} (%) vs. specific comminution energy (kWh/t).

t_{10} (%)	E_{cs} (kWh/t)
37	1.83
31	1.16
18	0.6
18	0.55
15	0.42
10	0.27
7.6	0.24
7.4	0.23
6	0.17

3.3 Experimental data of reaction constant (k) for a first-order reaction as a function of temperature (T) are as shown in Table 3.8.
The Arrhenius equation could correlate the temperature dependence of the reaction constant:

$$k = Ae^{-\frac{E}{RT}}$$

A is the frequency factor, R is the gas constant ($= 8.314\,\text{J/mol·K}$), and E is the activation energy.
(1) To determine the activation energy for this reaction.
(2) What is the reaction constant at 295 K?
(3) Try to use the linearization method to obtain a more accurate value of the activation energy for this reaction and the reaction constant at 295 K.

Table 3.8 T vs. k data.

T [K]	287	290	293	296	299
k [s^{-1}]	7.9	10	12.6	15.8	20

3.4 Experimental density (ρ) data as a function of pressure (P) for N_2 at 200 K (T) are as shown in Table 3.9.

Table 3.9 Experimental density data as a function of pressure.

P [atm]	6.0	9.0	12.0	14.0	17.0	19.0	21.0
ρ [mol/l]	0.38	0.59	0.81	0.97	1.23	1.42	1.63

Using these data, estimate the second, third, and fourth virial coefficients.
The virial equation of state that relates pressure and density is:

$$Z = \frac{P}{\rho RT} = 1 + B\rho + C\rho^2 + D\rho^3$$

where B, C, and D are the second, third, and fourth virial coefficients, respectively, and R is the gas constant $=0.082\,\text{l} \cdot \text{atm/(mol} \cdot \text{K)}$.

4

Optimization

Optimization is an essential task for chemical engineers and engineering in general. Optimization is widely used in various areas, such as production planning, cost optimization, chemical economics, finance, and system design. This critical task calls for a deeper understanding of chemical engineering principles, as these principles are the foundation for all chemical phenomena occurring within chemical processes and operations. Therefore, various optimization methods must be used to determine the best possible solution. Other objective goals for chemical engineers are minimizing costs and maximizing production. Optimization is a beneficial tool to find optimal designs, accelerate design evaluations, and perform trade-off analyses for nonintuitive designs.

We might encounter several optimization problems in engineering, including extreme value problems, linear programming, quadratic programming, and nonlinear programming. Each of these types of problems is further discussed in this chapter. There are two main toolboxes for solving optimization problems: optimization toolbox (linear, quadratic, nonlinear, and least squares) and global optimization toolbox (nonlinear, non-smooth, stochastic, and noisy).

4.1 Optimization Overview and Solver Options

4.1.1 What Is Optimization?

Optimization is a mathematical technique with many practical applications whereby one can minimize or maximize a function. In other words, one may find the highest or lowest value belonging to a given function in a search region. This is immensely useful in engineering settings, providing a path to finding the most cost/energy/space efficient designs. The principle and technique of optimization can also be applied to other scenarios. This is done by taking in many variables as inputs and singling out the optimal balance of the system. The use of mathematical optimization of the system aids most decision-making processes.

Figure 4.1 is the optimization flow chart, which shows how optimization works in computing.

Chemical Engineering Analysis and Optimization Using MATLAB®, First Edition.
Weiguo Xie, Sam Toan, and Richard Davis.

Companion website: www.wiley.com/go/xie/chemicalengineering

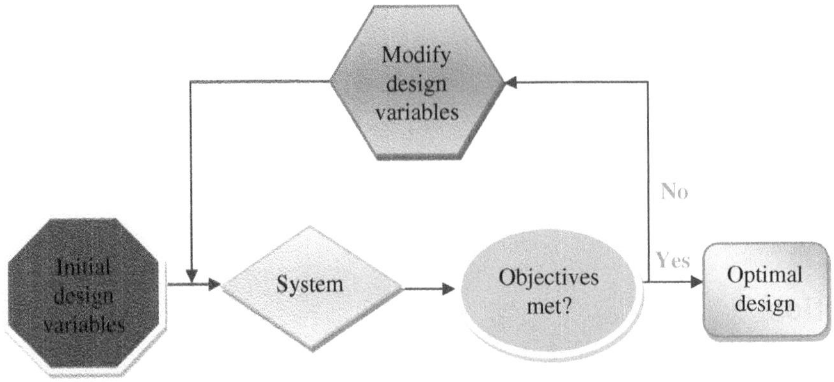

Figure 4.1 Optimization flowchart.

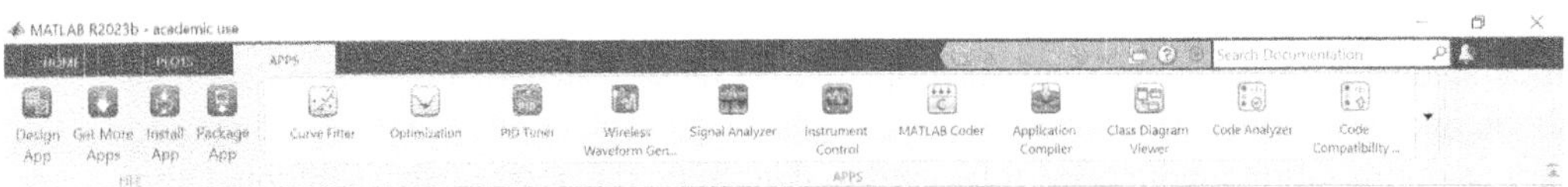

Figure 4.2 APPS tab options in MATLAB.

4.1.2 Built-In Function and MATLAB's Toolbox for Optimization

MATLAB has a very convenient and manageable toolbox for optimization problems (MathWorks Inc. 2022) which can be accessed by going to the APPS tab as shown in Figure 4.2, then choosing the optimization option.

An Optimization Tool Window appears. The optimization toolbox is very easy to use, and you may want to explore it by inserting some of the examples listed in this chapter to compare with the given answer.

The Optimization Tool Window can be separated into three sub-windows as shown in Figure 4.3: Problem Setup, Result/Options, and Quick Reference. In the Problem Setup Window, you may pick the solver, then enter the objective function and enter any additional information such as algorithm, start point, and constraints (if any). In the Options Window, you may or may not enter an input(s) as this section is optional. If you have a desired output or some value that is less than or greater than the default setting value, you may choose to enter your preferred options. The Quick Reference Window seen is not a working window but is used more as a reference section. If you need to know more about the solver or any additional information related to optimization, the Quick Reference Window is the best place to look for examples or answers. Clicking on the blue words below leads you to the information you want to access.

MATLAB has a whole built-in function map for different types of optimization solvers. To access this map, click the "Solver and Algorithm" tab under the Quick Reference Window, far right of the Optimization Tool Window. The "Optimization Decision Table" option appears, and click on it. Or you may want to access it from the help option under the HOME tabs → Optimization Toolbox → Optimization Problem Setup → Choose a Solver → Optimization Decision Table.

Depending on the type of problem you want to solve, as shown in Figure 4.4, the optimization solver figure guides you to pick the best possible solver for your problem.

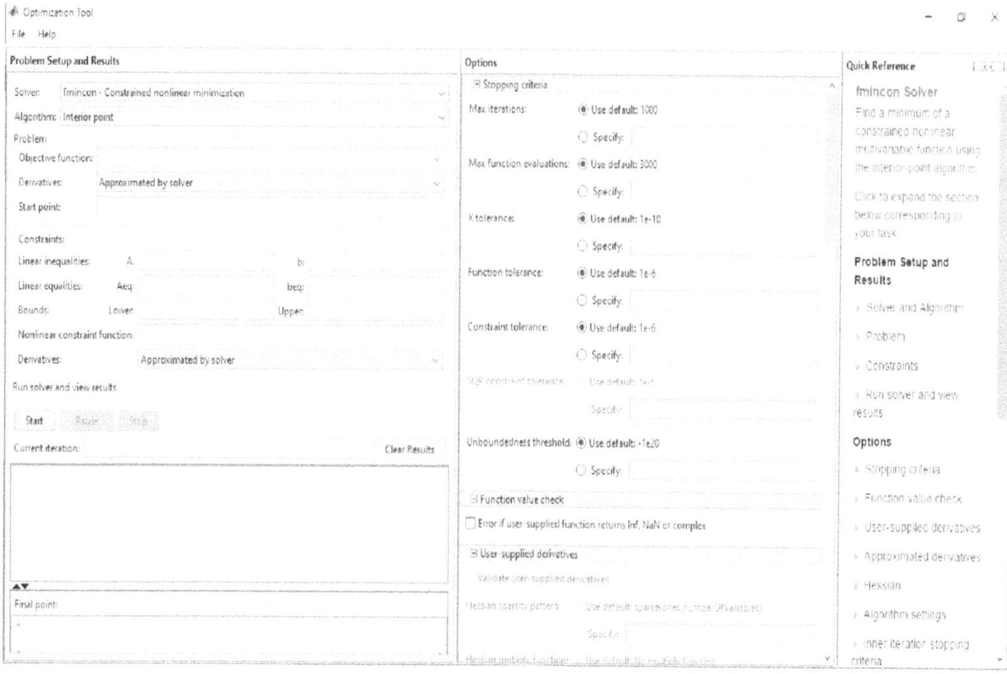

Figure 4.3 Optimization Window in MATLAB.

Constraint type	Objective type				
	Linear	**Quadratic**	**Least squares**	**Smooth nonlinear**	**Nonsmooth**
None	n/a (f= const, or min = $-\infty$)	`quadprog`, Information	`mldivide`, `lsqcurvefit`, `lsqnonlin`, Information	`fminsearch`, `fminunc`, Information	`fminsearch`, *
Bound	`linprog`, Information	`quadprog`, Information	`lsqcurvefit`, `lsqlin`, `lsqnonlin`, `lsqnonneg`, Information	`fminbnd`, `fmincon`, `fseminf`, Information	`fminbnd`, *
Linear	`linprog`, Information	`quadprog`, Information	`lsqlin`, Information	`fmincon`, `fseminf`, Information	*
General smooth	`fmincon`, Information	`fmincon`, Information	`fmincon`, Information	`fmincon`, `fseminf`, Information	*
Discrete, with bound or linear	`intlinprog`, Information	*	*	*	*

Figure 4.4 Optimization solver options table in MATLAB.

Table 4.1 MATLAB's optimization solvers description (https://www.mathworks.com/The MathWorks, Inc.).

Optimization solvers	Description
fmincon	Locates the minimum of a constrained nonlinear multivariable function
fminimax	Solves a minimax constraint problem
fminsearch	Locates the minimum of an unconstrained multivariable function using a derivative-free method
fminunc	Locates the minimum of an unconstrained multivariable function
fseminf	Locates the minimum of a semi-infinitely constrained multivariable function
fsolve	Solves a system of nonlinear equations
fzero	Locates the root of a nonlinear function
linprog	Solves linear programming problems
lsqcurvefit	Solves nonlinear curve-fitting (data-fitting) problems in the least square sense
lsqlin	Solves constrained linear least squares problems
lsqnonlin	Solves nonlinear least squares (nonlinear data-fitting) problems
lsqnonneg	Solves nonnegative least squares constraints problems
quadprog	Quadratic programming

Table 4.1 summarizes the built-in solvers for optimization problems. It lists the various types of solvers and briefly explains each solver. We do not review every solver in this table. We narrowly select only a specific solver to solve a specific problem. In the following sections, we introduce the *linprog* function to solve a linear equation with constraints, the *quadprog* function to solve a quadratic equation with constraints, and the *fmincon* function to solve a nonlinear equation with constraints.

4.2 Extreme Value Problem

An extreme value problem is an optimization problem of either minimization or maximization of a constrained or unconstrained objective function using a stationary point approach. Although MATLAB has very versatile functions for solving optimization problems, as shown later, we limit our discussion to conventional approaches using a stationary point. A conventional approach can give us a better understanding of the issues that arise when solving optimization problems.

4.2.1 Unconstrained Functions

For optimization of a function without any constraints, the solutions are conventionally obtained by finding the stationary points, i.e. equalizing the first derivatives of the functions (with respect to their independent variables) to zero and solving these equations.

To determine the types of stationary points, i.e. maxima, minima, or saddle points, we need to construct a matrix called the Hessian matrix at the stationary point. For optimizing a function of n independent variables, the Hessian matrix is an $n \times n$ matrix: $H = \lfloor h_{ij} \rfloor_{nxn}$

the element of which is defined as follows $h_{ij} = \frac{\partial^2 f}{\partial x_i \partial x_j}$ and evaluated at the stationary point where x_i and x_j are the independent variables. Then, the type of stationary point is a *minimum* if all of its eigenvalues are real and positive numbers (the Hessian matrix is positive definite), a *maximum* if all of its eigenvalues are real and negative numbers (the Hessian matrix is negative definite), and a *saddlepoint* otherwise (the Hessian matrix is indefinite). The eigenvalues of a matrix are calculated easily using the *eig* function. Example 4.1 shows this stationary point approach.

Example 4.1 shows that solving an extreme value problem using a stationary point approach is tedious. MATLAB offers easy ways to solve optimization problems, such as *fminsearch* in Table 4.1, which can be used to find a minimum or maximum stationary point, but not a saddle stationary point. Example 4.2 shows how to use these functions.

Example 4.1

Stationary point method for an unconstrained function.

Determine the values of the following function at the stationary points:

$$f(x,y) = (x^2 + y - 11)^2 + (x + y^2 - 7)^2 \tag{4.1}$$

Solution:

Step 1: Whenever possible, plot the function to get an idea of the solutions

```
clear all; clc; format compact
[x,y]=meshgrid(-5:0.5:5);
z=(x.^2+y-11).^2+(x+y.^2-7).^2;
figure
mesh(x,y,z)
title('Using mesh')
figure
surf(x,y,z)
title('Using surf')
figure
plot3(x,y,z)
title('Using plot3')
figure
contour3(x,y,z)
title('Using contour3')
figure
contour(x,y,z)
title('Using contour')
```

All the plots are shown in Figure 4.5. To create a 3D plot, the z coordinates must associate with a set of (x, y) coordinates. Thus we can consider $z = f(x,y) = (x^2 + y - 11)^2 + (x + y^2 - 7)^2$ and a set of (x, y) coordinates. Matrices must be created for the x and y coordinates, which can be done using the *meshgrid* function. Since x and y are not scalar, the z expression must have a period before *, /, or Ã. For a 3D line plot, you may use *plot3*. For a surface plot, you may use *mesh* or *surf*. For a 2D contour plot, you may use the *contour* command; for a 3D contour plot, you may use the *contour3* command. In the **Figure Window**, you can rotate the plot with the angle you want by clicking on the rotate 3D icon on the top of the **Figure Window.** This is a beneficial tool to help you look for a not-so-obvious stationary point. For nonlinear equations, initial guesses are essential because the output can be susceptible to the initial guess.

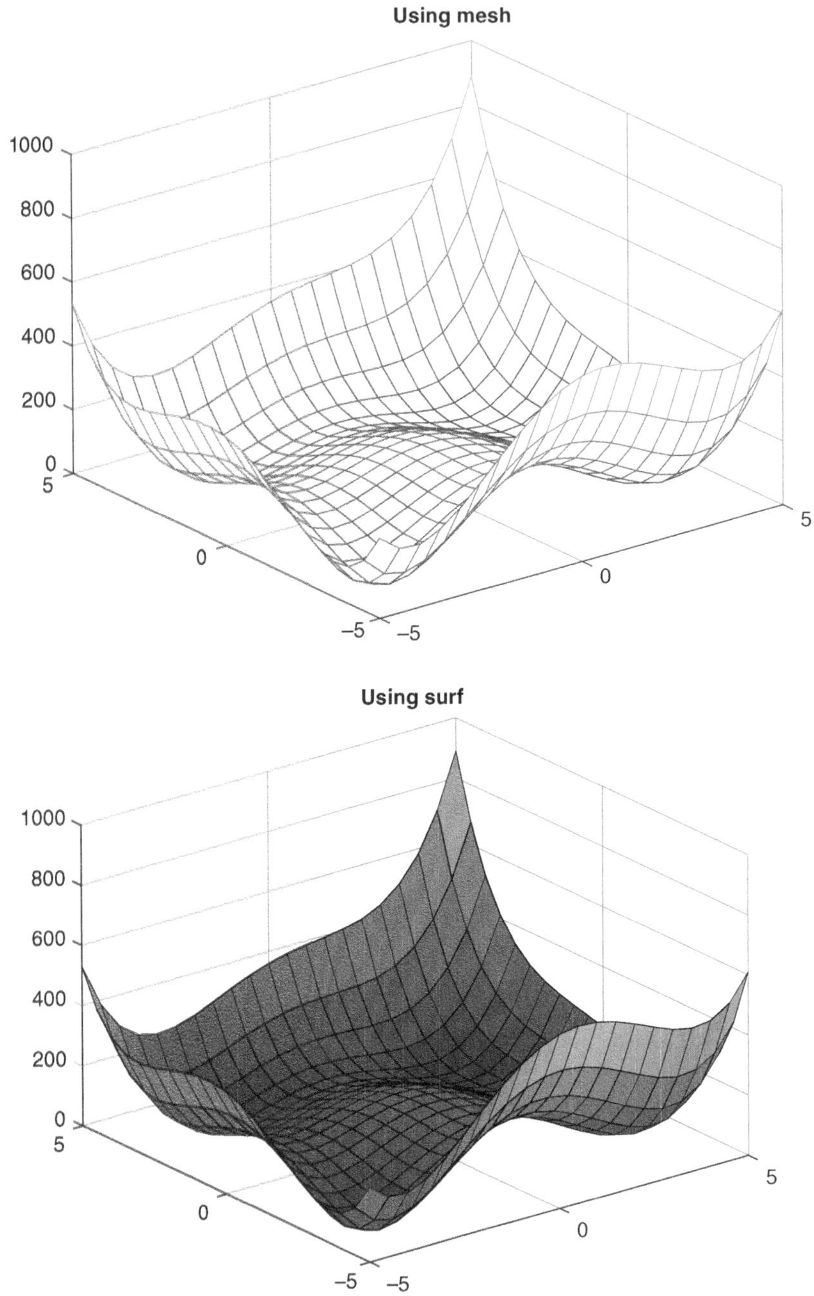

Figure 4.5 Plots of Example 4.1.

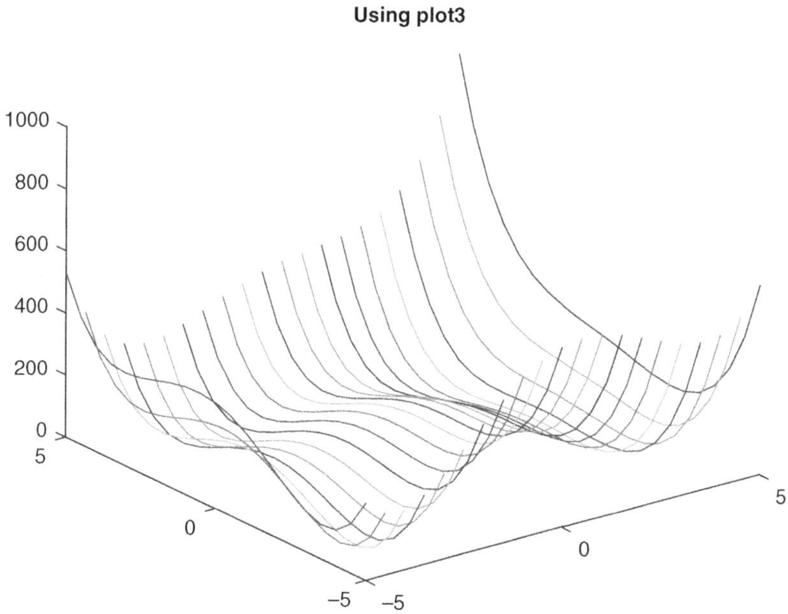

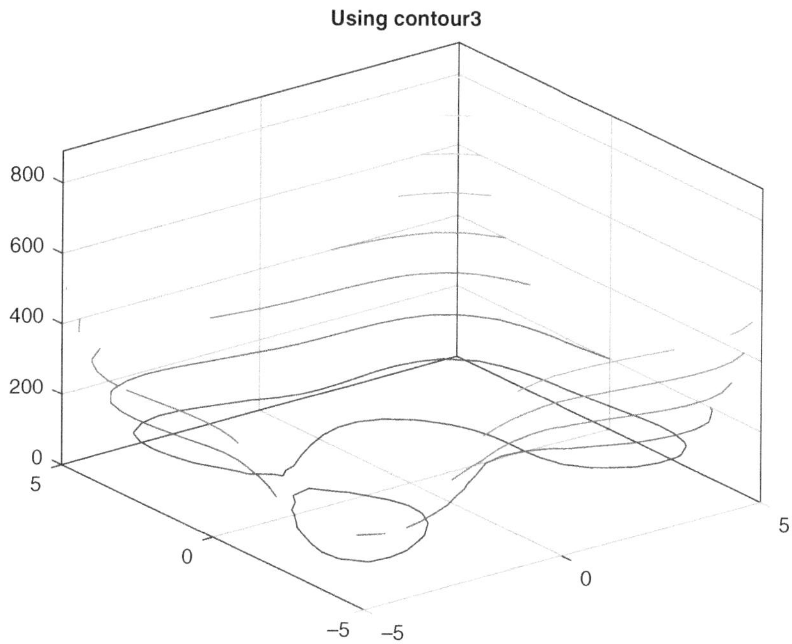

Figure 4.5 (*Continued*)

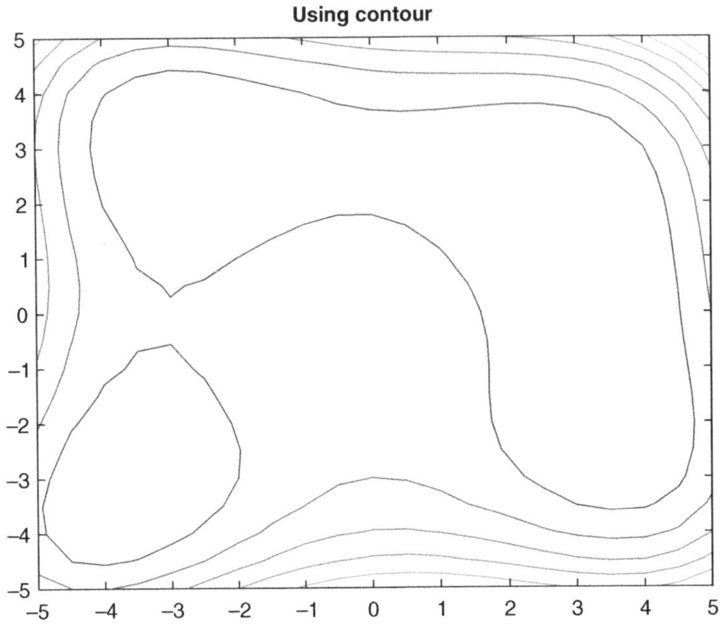

Figure 4.5 (*Continued*)

Step 2: Define functions for calculating all of the second derivatives needed in the Hessian matrix:

```
syms x y
f(x,y)=(x^2+y-11)^2+(x+y^2-7)^2;
fxx(x,y)=diff(f,x,x);
fxy(x,y)=diff(f,y,x);
fyy(x,y)=diff(f,y,y);
```

Step 3: Construct the Hessian matrix

```
H(x,y)=[fxx(x,y) fxy(x,y); fxy(x,y) fyy(x,y)];
```

Step 4: Create a function for calculating the eigenvalues of the Hessian matrix

```
Q=eig(H);
```

Step 5: Create a function to find all the stationary points using *vpasolve* or any other equations solver

```
eqn1=diff(f,x)==0;
eqn2=diff(f,y)==0;
[x,y]=vpasolve([eqn1,eqn2]);
x=eval(x)
y=eval(y)
```

x	y
3.0000	2.0000
0.0867	2.8843
−3.0730	−0.0814
3.3852	0.0739
−0.2708	−0.9230
−2.8051	3.1313
−0.1280	−1.9537
−3.7793	−3.2832
3.5844	−1.8481

As we can see from the output, there are nine stationary points, the next step is to determine the value of the function at each stationary point, and it is labeled using the definition of minimum, maximum, and saddle point.

Step 6: Determine the type and the value of the function at each stationary point

At (3.0000, 2.0000)

```
H1=H(x(1),y(1));
Q1=eval(eig(H1))
f1=eval(f(x(1),y(1)))
Q1 =
   25.7157
   82.2843
f1 =
     0
The Hessian matrix is indefinite →min
```

At (0.0867, 2.8843)

```
H2=H(x(2),y(2));
Q2=eval(eig(H2))
f2=eval(f(x(2),y(2)))
Q2 =
  -31.7066
   75.5076
f2 =
   67.7192
The Hessian matrix is indefinite →saddle
```

At (-3.0730, -0.0814)

```
H3=H(x(3),y(3));
Q3=eval(eig(H3))
f3=eval(f(x(3),y(3)))
```

```
Q3 =
  -39.6515
   72.4352
f3 =
  104.0152
The Hessian matrix is positive definite →saddle
At (3.3852, 0.0739)
H4=H(x(4),y(4));
Q4=eval(eig(H4))
f4=eval(f(x(4),y(4)))
Q4 =
  -14.1352
   97.5479
f4 =
   13.3119
The Hessian matrix is positive definite →saddle
At (-0.2708, -0.9230)
H5=H(x(5),y(5));
Q5=eval(eig(H5))
f5=eval(f(x(5),y(5)))
Q5 =
  -45.6052
  -16.0660
f5 =
  181.6165
The Hessian matrix is positive definite →max
At (-2.8051, 3.1313)
H6=H(x(6),y(6));
Q6=eval(eig(H6))
f6=eval(f(x(6),y(6)))
Q6 =
   64.8404
   80.5501
f6 =
    2.0089e-31
The Hessian matrix is indefinite →min
At (-0.1280, -1.9537)
H7=H(x(7),y(7));
Q7=eval(eig(H7))
f7=eval(f(x(7),y(7)))
Q7 =
  -50.6102
   20.2840
f7 =
  178.3372
The Hessian matrix is negative definite →saddle
At (-3.7793, -3.2832)
```

```
H8=H(x(8),y(8));
Q8=eval(eig(H8))
f8=eval(f(x(8),y(8)))
Q8 =
   70.7144
  133.7856
f8 =
   1.6613e-30
The Hessian matrix is indefinite →min
At (3.5844, -1.8481)
H9=H(x(9),y(9));
Q9=eval(eig(H9))
f9=eval(f(x(9),y(9)))
Q9 =
   28.6907
  105.4189
f9 =
   4.5403e-32
The Hessian matrix is positive definite →min
```

Example 4.1 is an overly length procedure to obtain all the available stationary points, whereas this process can be sped up by using built-in MATLAB functions to find the max and min points. Example 4.2 shows how to use the built-in function *fminsearch* for this purpose. Before using this built-in function, we can briefly explain it. *fminsearch* is a built-in function in MATLAB to obtain the minimum of a scalar unconstrained function of several variables (unconstrained nonlinear optimization), starting at an initial estimate using a derivative-free method.

The general format of ***fminsearch***

minimize $f^T x$

$[x, fval] = \text{fminsearch}(f, x0, options)$

It starts at the initial point, x0 (initial guess), and returns a local minimizer x of the function and the value *fval* of the objective function at the solution x.

By repeating Example 4.1, we can compare the stationary point method result with the built-in function *fminsearch* output.

Since this is a nonlinear equation, most built-in solvers are sensitive to initial points or guesses. A change in the initial guesses may give you a different output value. Thus plotting the function is always a good start and is also the first step to solving this type of problem. *fminsearch* can only return a minimum. However, there is a way around this to obtain the maximum of the function using *fminsearch*. To do so, you must flip your function upside down by multiplying the function by -1. In other words, you are now searching for the minimum of the flipped function. As a result, the negative of the output is the maximum of the given function in the problem statement. Performing maximization problems as a minimization of a negative function is the standard as it is computationally more efficient.

In MATLAB, you can also use *fminunc* (finds the minimum of an unconstrained multivariable function), which has a very similar format as *fminsearch*.

The general format of *fminunc*

minimize $f^T x$
$[x, fval] = \text{fminunc}(f, x0,)$

Example 4.2

Minimize/maximize for an unconstrained function

Determine the values of the following function at the stationary points:

$$f(x, y) = (x^2 + y - 11)^2 + (x + y^2 - 7)^2 \tag{4.2}$$

Solution:

Step 1: Whenever possible, plot the function to get an idea of the solutions

Step 2: Refine x = Y(1) and y = Y(2)

Step 3: Create a function for calculating the minimum stationary point using *fminsearch*

```
f=@(Y) (Y(1)^2+Y(2)-11)^2+(Y(1)+Y(2)^2-7)^2;
```

Step 4: Give an initial guess for min

```
Y0=[4;1];
```

Step 5: Calculating the minimum stationary point using *fminsearch*

```
[Ymin,fmin]=fminsearch(f,Y0,[])
```

Step 6: Create a function for calculating the maximum stationary point using *fminsearch*

```
f1=@(Y) -((Y(1)^2+Y(2)-11)^2+(Y(1)+Y(2)^2-7)^2);
```

Step 7: Give an initial guess for the maximum point

```
Y0=[0;0];
```

Step 8: Calculating the minimum stationary point using *fminsearch*

```
[Ymax,fmax]=fminsearch(f1,Y0,[])
Ymin =
     3.0000
     2.0000
fmin =
   1.4155e-08
Ymax =
   -0.2709
   -0.9230
fmax =
  -181.6165
The output is identical to the answers of Example 4.1. The
  Contour plot of the Example 4.2 is shown in Figure 4.6.
```

In the *fminsearch* command, *fminsearch*(f, Y0, []), the [] indicates optional arguments not required to be used in the function call. You may ignore it, such as *fminsearch*(f, Y0).

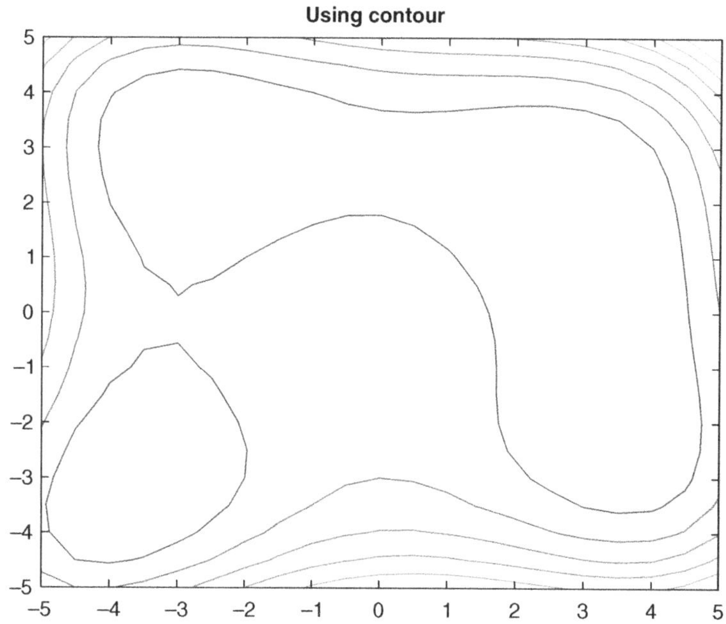

Figure 4.6 Plot of Example 4.2.

Note: In Example 4.2, the output of fmax = −181.6165. However, the actual maximum of the function is f(x,y) = 181.6165 because, as you recall, the function was "flipped" by multiplying by −1.

4.2.2 Constrained Functions

Consider the minimization of a function of *n* variables, referred to as the objective function, subject to *m* equality constraints:

$$\text{Minimize}\, f(x_1, x_2, \dots, x_n) \tag{4.3}$$

$$\text{Subject to} \begin{cases} g_1(x_1, x_2, \dots, x_n) = 0 \\ g_2(x_1, x_2, \dots, x_n) = 0 \\ \quad\quad . \\ \quad\quad . \\ g_m(x_1, x_2, \dots, x_n) = 0 \end{cases} \tag{4.4}$$

A more straightforward method is to use the built-in function, *fmincon*. Example 4.3 demonstrates the use of the *fmincon* function for this purpose.

Finding a min or max of a given function with constraint(s) is a lengthy procedure. Using MATLAB's built-in functions can be much faster. Since Example 4.3 is a nonlinear objective function with constraints, this example can be solved with a nonlinear multivariable objective functions solver with constraints. Thus, *fmincon* (finds the minimum of a constrained nonlinear multivariable function) is a good candidate in this case. If unsure, you may check the optimization solver table in MATLAB.

The general format of *fmincon*

$$\text{Minimize } f^T x \tag{4.5}$$

$$\text{Subject to } \begin{cases} c(x) \leq 0 \\ ceq(x) = 0 \\ A \cdot x \leq b \\ Aeq \cdot x = beq \\ lb \leq x \leq ub \end{cases} \tag{4.6}$$

$$[x, fval] = fmincon\,(f, x0, A, b, Aeq, beq, lb, ub, nonlcon) \tag{4.7}$$

where

- x0 is the initial guess
- f is the function $f(x)$ with the constraints
- b is a vector representing the constant vector in the constraints $A \cdot x \leq b$
- A is a matrix representing the linear coefficients in the constraints $A \cdot x \leq b$
- Aeq is a matrix representing the linear coefficients in the constraints $Aeq \cdot x = beq$
- beq is a vector representing the constant vector in the constraints $Aeq \cdot x = beq$
- ub is a vector representing the upper bounds in $lb \leq x \leq ub$
- lb is a vector representing the lower bounds in $lb \leq x \leq ub$
- $c(x)$ and $ceq(x)$ are functions that return vectors (c and ceq are used for nonlinear constraints)
- $f(x)$, $c(x)$, and $ceq(x)$ can be nonlinear functions, others must be linear
- x, lb, and ub can be passed as vectors or matrices
- $[x, fval]$ is the *fval*, the minimum of the function $f(x)$ that happens at location x

Again, *fmincon* is the command to find the minimum of a given function with constraints. Thus, to find a maximum of a given function, you must flip the function upside down and remember to change the sign of the fval at the end to obtain the correct maximum value of the function.

If no inequalities or equalities exist, you may set $A = []$, $b = []$ or $Aeq = []$, $beq = []$.

The order in the *fmincon*(fun, x0, *A*, *b*, *Aeq*, *beq*, *lb*, *ub*, *nonlcon*) command must be placed in the same order (http://www.mathworks.com).

Example 4.3

Solve the following optimization problem:

$$\text{Minimize } f(x, y) = (x - 2)^2 + y^2 \tag{4.8}$$

$$\text{Subject to } 2x + y - 2 = 0 \tag{4.9}$$

Solution:

Using *fmincon*

Step 1: Redefine x and y

```
x=x(1)  and y=x(2)
```

Step 2: Define the given function

```
fun=@(x) (x(1)-2)^2+x(2)^2;
```

Step 3: Define the constraints

```
Since the only constant in this problem is 2x+y=2, which is
 equivalent to Aeq*x=beq
Aeq=[2 1];
beq=2;
```

Since it must follow the order of fmincon(fun, x0, A, b, Aeq, beq), we have to define A and b in an empty square bracket

```
A=[];
b=[];
```

Step 4: Give an initial guess

```
x0=[4,2];
```

Step 5: Use the *fmincon* command to calculate the minimum of the function

```
x=fmincon(fun,x0,A,b,Aeq,beq);
f=fun(x)

f =
    0.8000
```

4.3 Linear Programming

In MATLAB, linear programming uses the *linprog* function. This built-in function can only optimize a linear objective function with linear inequalities and/or equality constraints. Use the fmincon built-in function to solve optimization for all nonlinear objective functions or constraints.

4.3.1 Standard Procedure of Linear Programming

In MATLAB, linear programs can be solved using the built-in function *linprog*.

The general format of *linprog*

$$\text{Minimize } f^T x \tag{4.10}$$

$$\text{Subject to} \begin{cases} Ax \leq b \\ A_{eq}x = b_{eq} \\ l_b \leq x \leq u_b \end{cases} \tag{4.11}$$

$$[x, fval] = linprog(f, A, b, Aeq, beq, lb, ub, x0) \tag{4.12}$$

where

- *f, x, b, beq, lb,* and *ub* are vectors, and *A* and *Aeq* are matrices
- *x0* is the initial guess. This can be ignored in linear programming
- f is the function $f(x)$ with the constraints
- b is a vector representing the constant vector in the constraints $A \cdot x \leq b$

- A is a matrix representing the linear coefficients in the constraints $A \cdot x \leq b$
- Aeq is a matrix representing the linear coefficients in the constraints $Aeq \cdot x = beq$
- beq is a vector representing the constant vector in the constraints $Aeq \cdot x = beq$
- ub is a vector representing the upper bounds in $lb \leq x \leq ub$
- lb is a vector representing the lower bounds in $lb \leq x \leq ub$
- $[x, fval]$ is the *fval*, the minimum of the function $f(x)$ that happens at location x

Note: All constraints with a different format than shown above must be converted to match. For example, if the given constraints have the following format, which is different than $Ax \leq b$.

$$\begin{cases} x_1 + x_2 - 5x_3 \leq 2 \\ x_1 + x_2 \geq 1 \end{cases} \tag{4.13}$$

Then, the first inequality constraint has the same format as $Ax \leq b$, but not the second constraint. Thus, we must convert it to the desired format by multiplying both sides by a negative 1 to change the sign. In other words, move one from right to left and $x_1 + x_2$ from left to right.

$$\begin{cases} x_1 + x_2 - 5x_3 \leq 2 \\ -x_1 - x_2 \leq -1 \end{cases} \tag{4.14}$$

If any of the constraints in $[x, fval] = linprog(f, A, b, Aeq, beq, lb, ub, x0)$ is not provided in the problem statement, you may use [] to present the lack of a constraint. Suppose a given variable only has an upper or lower bound while other variables have both upper and lower bounds. In that case, you may use *inf* (infinity) to present the absence of an upper or lower bound in the *lb* or *ub* vector. For example:

$$\begin{cases} x_1 + x_2 - x_3 \leq 20 \\ 2x_2 + 3x_1 \leq 15 \\ 0 \leq x_1 \leq 30 \\ 0 \leq x_2 \\ 0 \leq x_3 \leq 5 \end{cases} \tag{4.15}$$

Since no equalities constraints are given, we may write Aeq=[] and beq=[], and no upper bound constraint is given for x2. Thus, we can use the *inf* command to present it.

```
A=[1 1 -1; 3 2 0];
b=[20; 15];
Aeq=[];
beq=[];
lb=[0; 0; 0]; or lb=zeros(3,1);
ub=[30; inf; 5];
```

Again, *linprog* can only return a minimum value of a given function. You must flip the given function to find the maximum by placing a negative sign before the equation. The negative of the output is the maximum of the given function $f(x)$.

4.3.2 Solving Linear Programming in MATLAB

This section demonstrates the use of *linprog* in a linear optimization problem. We convert a linear optimization problem function and constraints into the given matrices or vectors listed in *linprog* format.

$$[x, fval] = linprog(f, A, b, Aeq, beq, lb, ub, x0) \tag{4.16}$$

Example 4.4
Minimization/maximization for linear programming.

$$\text{Maximize the following function}: f(x,y) = 3x + 2y \tag{4.17}$$

$$\text{Subject to}: -x + 2y \leq 4; \quad 0 \leq x \leq 3; \quad x - y \leq 3; \quad 3x + 2y \leq 14; \quad y \geq 0 \tag{4.18}$$

Solution:
Step 1: Define the objective function

In this step, the fundamental difference between fmincon and linprog is *that fmincon* can give an input objective function directly in a function file or using @(variable(s)). However, linprog and quadprog problems must also be placed in a matrix. Before placing it into a matrix format, you should be clear to yourself by implying

```
x=f(1) and y=f(2) or x=X(1) and y=X(2)
f=[-3,-2];
```

Since the problem asks for the maximum, not the minimum, the sign of both coefficients is flipped.

Step 2: Setup constraints

In this step, you must rearrange the format of the constraint to be the same as the *linprog* format. Now, let us regroup the given constraints into the desired format.

$$\begin{cases} -x + 2y \leq 4 \\ x - y \leq 3 \\ 3x + 2y \leq 14 \\ 0 \leq x \leq 3 \\ 0 \leq y \end{cases}$$

```
A=[-1 2; 1 -1; 3 2];
b=[4; 3; 14];
Aeq=[];
beq=[];
lb=[0;0];
ub=[3;inf];
```

Step 3: Using linprog to solve the problem

```
[x,fval]=linprog(f,A,b,Aeq,beq,lb,ub)
```

```
Optimal solution found.
x =
     3.0000
     2.5000
fval =-14
```

4.4 Quadratic Programming

When dealing with optimization problems, the objective function may not always be linear. If the objective function is a quadratic function with one or more variables, you may use quadratic programming's built-in function, *quadprog*, in MATLAB. One necessary condition to use *quadprog* is that the constraints must be linear. Otherwise, you may have to use other built-in functions in MATLAB, such as *fmincon*.

4.4.1 Standard Procedure of Quadratic Programming

quadprog has a similar format to *linprog* or *fmincon*.

The general format of *quadprog*

$$\text{Minimize } \frac{1}{2}x^T Hx + f^T x \tag{4.19}$$

$$\text{Subject to } \begin{cases} A \cdot x \leq b \\ Aeq \cdot x = beq \\ lb \leq x \leq ub \end{cases} \tag{4.20}$$

$$[x, fval] = quadprog(H, f, A, b, Aeq, beq, lb, ub, x0) \tag{4.21}$$

where

- *c*: Vector of doubles. Represents the linear in the expression $\frac{1}{2}x'Hx + f'x$
- *H*: Symmetric matrix of doubles. Represents the quadratic in the expression $\frac{1}{2}x'Hx + f'x$. Note: if $H = 0$, then the problem becomes a linear problem
- *A*: Matrix of doubles. Represents the linear coefficients in the constraints $A \cdot x \leq b$
- *b*: Vector of doubles. Represents the constant vector in the constraints $A \cdot x \leq b$
- *Aeq*: Matrix of doubles. Represents the linear coefficients in the constraints $Aeq \cdot x = beq$
- *beq*: Vector of doubles. Represents the constant vector in the constraints $Aeq \cdot x = beq$
- *lb*: Vector of doubles. Represents the lower bounds in $lb \leq x \leq ub$
- *ub*: Vector of doubles. Represents the upper bounds in $lb \leq x \leq ub$
- *x*0: Vector of doubles. Optional. The initial point for some *quadprog* algorithms:
  - o active-set
  - o trust-region-reflective when there are only bound constraints

If *x*0 is not provided, quadprog assigns all components of *x*0 to a point within the box's boundaries. The interior-point-convex and trust-region-reflective algorithms with equality constraints do not consider *x*0.

4.4.2 Solving Quadratic Programming in MATLAB

The general format is quite confusing. Thus, we use an example to explain it in more detail to clarify the use of quadprog.

Example 4.5

Quadratic programming.

Consider the following objective function:

$$f(x_i) = 0.025x_1^2 + 0.35x_2^2 + 0.0013x_3^2 + 0.0648x_1x_2 + 0.00032x_1x_3 - 0.00035x_2x_3 \tag{4.22}$$

Minimize this function subject to:

$$\begin{cases} 0.085x_1 + 0.068x_2 + 0.0425x_3 \geq 0.0153 \\ x_1 + x_2 + x_3 = 1 \\ 0 \leq x_1 \leq 1 \\ 0 \leq x_2 \leq 1 \\ 0 \leq x_3 \leq 1 \end{cases} \tag{4.23}$$

Solution:

Step 1: Convert the objective function into vector f and matrix H

Consider the following Table 4.2.
Place the coefficient in the highlighted part of Table 4.3.

$$H = \begin{bmatrix} 0.025 & 0.0648 & 0.00032 \\ 0.0648 & 0.35 & -0.00035 \\ 0.00032 & -0.00035 & 0.0013 \end{bmatrix}$$

Note: $x_1x_2 = x_2x_1$, and $x_3x_1 = x_1x_3$, and $x_3x_2 = x_2x_3.$
Thus they must be filled

Since there is no linear term, we can define $f = \begin{bmatrix} 0 \\ 0 \\ 0 \end{bmatrix}$

```
f=[0, 0, 0];
H=[0.025 0.0648 0.00032; 0.0648 0.35 -0.00035;...
 0.00032 -0.00035 0.0013];
```

Step 2: Setup constraints

```
A=[-0.085 -0.068 -0.0425];  % Remember to change the greater or
% equal sign to less than or equal b=[-0.00153];
Aeq=[];
```

Table 4.2 Convert the objective function into vector f and matrix H.

	x_1	x_2	x_3
x_1	x_1^2	x_1x_2	x_1x_3
x_2	x_2x_1	x_2^2	x_2x_3
x_3	x_3x_1	x_3x_2	x_3^2

Table 4.3 Place the coefficient into vector f and matrix H.

	x_1	x_2	x_3
x_1	$0.025x_1^2$	$0.0648x_1x_2$	$0.00032x_1x_3$
x_2	$0.0648x_2x_1$	$0.35x_2^2$	$-0.00035x_2x_3$
x_3	$0.00032x_3x_1$	$-0.00035x_3x_2$	$0.0013x_3^2$

```
beq=[];
lb=[0; 0; 0];
ub=[1; 1; 1];
```

Step 3: Solve using *quadprog*

```
[x,fval]=quadprog(H,f,A,b,Aeq,beq,lb,ub)

x =
     0.0023
     0.0003
     0.0310
fval =
   7.6439e-07
```

If creating the matrix H is cumbersome, you may use *fmincon* to solve this problem. However, it may be sensitive to the initial conditions.

4.5 Nonlinear Programming

If any equalities and inequalities in the constraints and/or the objective function are nonlinear, the optimization problem is called nonlinear programming, which includes any polynomial function. In the MATLAB optimization table, when you have both a nonlinear objective function and nonlinear constraints, you have fewer and fewer options for solving. Thus the most common solver for it is *fmincon*, which can solve any optimization problem. The general format of *fmincon* is discussed in Section 4.2.2; an example is also given. Therefore, in this section, we briefly go over another example with both a nonlinear objective function and nonlinear constraints.

Example 4.6
Minimize/maximize for nonlinear programming.
Solve the following optimization problem:

$$\text{Maximize}\, f(x) = 10x_1 + 4.4x_2^2 + 2x_3 \tag{4.24}$$

$$\text{Subject to} \begin{cases} x_1 \geq 0 \\ x_2 \geq 0 \\ x_3 \geq 0 \\ (x_2)^2 + 0.5(x_3)^2 \geq 3 \\ x_1 + 4x_2 + 5x_3 \leq 45 \\ x_1 + 3x_2 + 2x_3 \leq 40 \end{cases} \tag{4.25}$$

Solution:
Step 1: Set up the optimization objective function

```
fun=@(x)-(10*x(1)+4.4*x(2)^2+2*x(3));
```

Step 2: Set up all the linear constraints

```
A=[1 4 5;1 3 2];
b=[45; 40];
```

```
Aeq=[];
beq=[];
lb=[2;0;0];
ub=[];
```

In this case, we have five linear constraints and $(x_2)^2 + 0.5(x_3)^2 \geq 3$ is the nonlinear constraint. To deal with nonlinear constraints, we have to add an additional step.

Step 3: Set up all the nonlinear constraints

To set up the nonlinear constraint, let us review the general format of *fmincon*:

```
[x, fval]=fmincon(fun,x0,A,b,Aeq,beq,lb,ub,nonlcon);
```

Usually, the *nonlcon* term can be ignored because, thus far, we have not dealt with nonlinear constraints in the problem statement. However, in this specific problem, the nonlinear constraint is given, and it must be appropriately placed.

All nonlinear (in)equality constraints must be placed in a separate function file. Now let us create a nonlinear constraint function file and name this file as "unitdisk."

```
function [c, ceq]=unitdisk(x)
% 3-x2^2-0.5*x3^2<=0
c=3-x(2)^2-0.5*x(3)^2;
ceq=[];
end
```

In this file, c represents the nonlinear inequality constraints, and c is presented in the following format $c \leq 0$.

In this case, we have $(x_2)^2 + 0.5(x_3)^2 \geq 3 \rightarrow 3 - (x_2)^2 - 0.5(x_3)^2 \leq 0$.

ceq represents the nonlinear equalities which we do not have in this case.

In the main script file, we must define nonlcon.

nonlcon=@unitdisk;

Step 4: Give initial guesses

```
x0=[15,3,0];
```

Step 5: Solve

```
[x,fval]=fmincon(fun,x0,A,b,Aeq,beq,lb,ub,nonlcon)
fmax=-fval

x =
    2.0000   10.7500    0.0000
fval =
  -528.4750
fmax =
   528.4750
```

Note: with different initial guesses, you get a different solution because, for nonlinear programming, the *fmincon* function may find different local optima with different initial guesses. Thus it is better to try different initial guesses in different regions of the solvable area to confirm the actual maximum or minimum.

4.6 Application in Chemical Engineering

4.6.1 Reaction Kinetics Optimization

Substance D is produced by the decomposition reaction of A in a mixed reactor. Two other undesired side reactions accompany the reaction:

$$
\begin{aligned}
r_B &= 1 \\
r_C &= 0.2 \cdot C_A \\
r_D &= 0.5 \cdot C_A^2
\end{aligned}
\tag{4.26}
$$

C_A is the concentration of A in the exit stream, and r_B, r_C, and r_D are the reaction rates of B, C, and D production, respectively.

From the kinetic considerations, the concentration of D in the exit stream is given by

$$C_D = \frac{r_D}{r_B + r_c + r_D}(C_{A0} - C_A) \text{ and } \quad C_{A0} = 3 \text{ mol/l} \tag{4.27}$$

Determine the optimum A concentration in the exit stream that maximizes the desired product D. What is the maximum concentration of D at this condition?

Solution:

```
%% Application 4.6.1
clc; clear all; format compact;
%% Plot the function
CA0=3;
syms Ca
CD(Ca)=(0.5*Ca^2)*(CA0-Ca)/(1+0.2*Ca+0.5*Ca^2);
Ca=linspace(0,2);
plot(Ca,eval(CD(Ca)));
xlabel('CA');
ylabel('CD(CA)');
%% Using fminsearch or fminuncon
% In this case, we have to look for a maximum, thus need to flip
% the function upside
% down
fun=@(Ca) -((0.5*Ca^2)*(CA0-Ca)/(1+0.2*Ca+0.5*Ca^2));
X0=1.5;
[CA,CD]=fminsearch(fun,X0);
CA
CD=-CD % Remember to convert it back to positive

CA =
     1.4728
CD =
  0.6962
The plot of Application 4.6.1 is shown in Figure 4.7.
```

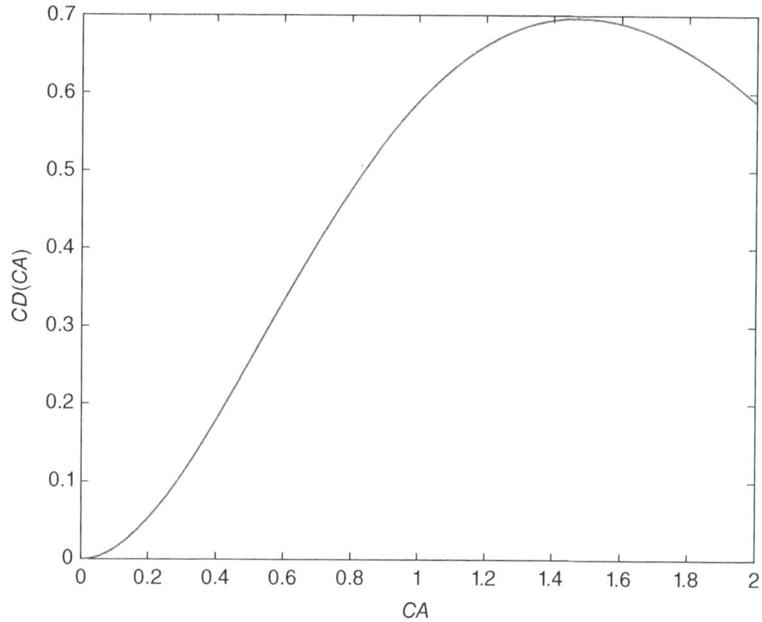

Figure 4.7 Plot of Application 4.6.1.

4.6.2 Reaction Productivity Optimization

Synthesis gas can be produced by catalytic reforming of methane (CH_4) with steam at high temperatures and atmospheric pressure:

$$CH_4(1) + 2H_2O(2) \rightarrow CO_2(4) + 4H_2(5) \tag{4.28}$$

The water-gas-shift reaction also occurs along with the above reaction:

$$CO_2 + H_2 \rightarrow CO(3) + H_2O \tag{4.29}$$

If the catalytic reforming is carried out at 1000 K and 1 bar, determine the mixture's equilibrium compositions (mol fractions). The reactants are supplied to the reactor in the ratio of 2 mol steam to 1 mol methane.

The equilibrium compositions can be calculated by minimizing the total Gibbs free energy of the system (G_t):

$$\text{Minimize } G_t(n_1, n_2, n_3, n_4, n_5) \tag{4.30}$$

$$\text{Subject to material balances: } n_1 + n_3 + n_4 = 14$$

$$4 \cdot n_1 + 2 \cdot n_2 + 2 \cdot n_5 = 8 \tag{4.31}$$

$$n_2 + n_3 + 2 \cdot n_4 = 2$$

where n_i is the number of moles of component i in the mixture

The mole fraction of component i and the total Gibbs free energy of the system can be calculated from:

$$y_i = \frac{n_i}{\sum_j n_j} \quad \text{and} \quad G_t = \sum_i (n_i \cdot \Delta G_i) + R \cdot T \cdot \sum_i \left(n_i \cdot \ln \left(\frac{n_i}{\sum_j n_j} \right) \right) \tag{4.32}$$

where the summation is over all components in the system, R is the gas constant, T is the absolute temperature, and ΔG_i is the standard Gibbs energy change of formation for component i. The standard Gibbs energy change of formation at 1000 K for each of the components is as follows:

$$\Delta G_1 = 19,720\,\mathrm{J \cdot mol^{-1}} \qquad \Delta G_2 = -192,420\,\mathrm{J \cdot mol^{-1}}$$

$$\Delta G_3 = -200,240\,\mathrm{J \cdot mol^{-1}} \quad \Delta G_4 = -395,790\,\mathrm{J \cdot mol^{-1}}$$

$$\Delta G_5 = 0\,\mathrm{J \cdot mol^{-1}}$$

Solution:

```
%% Application 4.6.2
clc; clear all; format compact;
%% Solving this using fmincon
% Since there are constraints in this problem.
% CH4 (1), H2O (2), CO (3), CO2 (4), H2 (5)
R=8.314472;
T=1000;
G=[19720 -192420 -200240 -395790 0];
fun=@(n) (n(1)*G(1)+n(2)*G(2)+n(3)*G(3)+...
n(4)*G(4)+n(5)*G(5))+R*T*...
    ((n(1)*log(n(1)/(n(1)+n(2)+n(3)+n(4)+n(5))))+...
    (n(2)*log(n(2)/(n(1)+n(2)+n(3)+n(4)+n(5))))+...
    (n(3)*log(n(3)/(n(1)+n(2)+n(3)+n(4)+n(5))))+...
    (n(4)*log(n(4)/(n(1)+n(2)+n(3)+n(4)+n(5))))+...
    (n(5)*log(n(5)/(n(1)+n(2)+n(3)+n(4)+n(5)))));
A=[];
b=[];
x0=[1; 1; 1; 1; 1];
Aeq=[1 0 1 1 0; 4 2 0 0 2; 0 1 1 2 0];
beq=[1; 8; 2];
[n,fminval]=fmincon(fun,x0,A,b,Aeq,beq)
y=n/(n(1)+n(2)+n(3)+n(4)+n(5))
```

```
n =
    0.0414
    0.7850
    0.7022
    0.2564
    3.1322
fminval =
  -4.3535e+05
y =
    0.0084
    0.1596
    0.1428
    0.0521
    0.6370
```

4.6.3 Refinery Economics

Five crude oils of different grades are processed in a refinery to produce four products: gasoline, heating oil, jet fuel, and lube oil. Table 4.4 shows the fractions of products that can be obtained from crude oil. Table 4.5 shows the cost, operating cost, and availability of crude oil. Table 4.6 shows the products, prices, and their maximum demand.

For maximum profit, determine the amount of each crude oil to be processed weekly. What is the profit per week?

Table 4.4 The fractions of products from crude oil.

Crude oil	Gasoline	Heating oil	Jet fuel	Lube oil
		bbl product per bbl of crude[a]		
Crude 1	0.6	0.2	0.1	0
Crude 2	0.5	0.2	0.2	0
Crude 3	0.3	0.3	0.3	0
Crude 4	0.4	0.3	0.2	0
Crude 5	0.4	0.1	0.2	0.2

a) The fractions do not add up to 1 due to losses in processing.

Table 4.5 The cost, operating cost, and availability of crude oil.

Crude oil	Cost ($/bbl)	Operating cost ($/bbl)	Availability (bbl/wk)
Crude 1	45.00	12.00	80,000
Crude 2	43.00	20.00	100,000
Crude 3	40.00	17.00	100,000
Crude 4	52.00	7.50	100,000
Crude 5	65.00	6.50	60,000

Table 4.6 The products, prices, and their maximum demand.

Product	Price ($/bbl)	Maximum demand (bbl/wk)
Gasoline	105.00	170,000
Heating oil	95.00	85,000
Jet fuel	61.00	75,000
Lube oil	140.00	30,000

This is an optimization (linear programming) problem:

$$\text{Maximize } P(x_1, x_2, x_3, x_4, x_5) \tag{4.33}$$

where P is the profit/week and x_i is the crude

$$\text{Subject to availability}: x_1 \leq 80000; \quad x_2 \leq 100000; \quad x_3 \leq 100000; \quad x_4 \leq 100000; \quad x_5 \leq 60000$$

Maximum product demand (what can be sold):

$$\begin{aligned} &0.6 \cdot x_1 + 0.5 \cdot x_2 + 0.3 \cdot x_3 + 0.4 \cdot x_4 + 0.4 \cdot x_5 \leq 170000 \\ &0.2 \cdot x_1 + 0.2 \cdot x_2 + 0.3 \cdot x_3 + 0.3 \cdot x_4 + 0.1 \cdot x_5 \leq 85000 \\ &0.1 \cdot x_1 + 0.2 \cdot x_2 + 0.3 \cdot x_3 + 0.2 \cdot x_4 + 0.2 \cdot x_5 \leq 75000 \\ &0.2 \cdot x_5 \leq 30000 \end{aligned} \tag{4.34}$$

$$\text{Positive variables } x_1 \geq 0; \; x_2 \geq 0; \; x_3 \geq 0; \; x_4 \geq 0; \; x_5 \geq 0$$

Solution:

```
% Application 4.6.3. Crude oil production optimization
% This is an optimization(Linear Programming) problem:
% Maximize P(x1,x2,x3,x4,x5) where P is the profit/week and xi
% is the crude
clc; clear all; format compact
syms x1 x2 x3 x4 x5
P=(0.6*x1+0.5*x2+0.3*x3+0.4*x4+0.4*x5)*105+(0.2*x1+0.2*x2+...
0.3*x3+0.3*x4+0.1*x5)*95+(0.1*x1+0.2*x2+0.3*x3+0.2*x4+...
0.2*x5)*61+0.2*x5*140-(45*x1+43*x2+...
40*x3+52*x4+65*x5)-...
(12*x1+20*x2+17*x3+...
7.5*x4+6.5*x3);
expand(P) ;
% After expanding the equation, P(x)=:
(311*x1)/10 + (207*x2)/10 + (74*x3)/5 + (116*x4)/5 +...
 (267*x5)/10;
% Maximize this function
f=[-311/10,-207/10,-74/5,-116/5,-267/10];
% Maximum product demand(what can be sold)
A=[0.6,0.5,0.3,0.4,0.4;0.2,0.2,0.3,0.3,0.1;...
0.1,0.2,0.3,0.2,0.2;0,0,0,0,0.2];
b=[170000;85000;75000;20000];
Aeq=[];
beq=[];
% All x needs to be a positive value and limited to the avail-
ability
lb=[0;0;0;0;0];
ub=[80000;100000;100000;100000;60000];
% Linear program function:
```

```
[x,fval]=linprog(f,A,b,Aeq,beq,lb,ub)
P_max=-(fval)

x =1.0e+05 *
     0.8000
     0.8333
     0.5444
     1.0000
     0.6000
P_max =
   8.9408e+06
```

4.6.4 Chemical Production Optimization

A company manufactures three chemical products: product 1, product 2, and product 3. The required resources for each product are shown in Table 4.7.

There are 100 hours of engineering, 600 hours of labor, and 300 lb of material available daily. The unit profits on these products are US$10, US$6, and US$4, respectively.

a. Determine the units of each product to be produced for maximum profit per day. What is the maximum profit?
b. What unit profit of product 3 is required to become economical to produce?
c. If it is possible to increase the direct labor hours by 10% by scheduling overtime that requires an additional labor cost of US$60/day, analyze whether this increase is economical.

This is a linear programming problem:

$$\text{Maximize} \, f(x_1, x_2, x_3) \text{ (profit per day)} \tag{4.35}$$

$$\begin{aligned} \text{Subject to } x_1 + 2 \cdot x_2 + x_3 &\leq 100 \text{ (engineering service)} \\ 10 \cdot x_1 + 4 \cdot x_2 + 5 \cdot x_3 &\leq 600 \text{ (direct labor)} \\ 3 \cdot x_1 + 2 \cdot x_2 + x_3 &\leq 300 \text{ (raw material)} \end{aligned} \tag{4.36}$$

$$x_1 \geq 0 \quad x_2 \geq 0 \quad x_3 \geq 0$$

Solution:

```
%% Application 4.6.4. Part a
clc; clear all; format compact;
f=[-10,-6,-4];
```

Table 4.7 The required resources for each product.

Product	Engineering service (h)	Direct labor (h)	Raw material (lb)
1	1	10	3
2	2	4	2
3	1	5	1

```
A=[1,2,1;10,4,5;3,2,1];
b=[100;600;300];
Aeq=[];
beq=[];
lb=[0;0;0];
[x1,fval]=linprog(f,A,b,Aeq,beq,lb); % PR1
x1
f1val = fval

x1 =
    50
    25
     0
f1val =
     650.0000
```

It means that the maximum profit is US$650.0000/day

Part b: As we see from part (a), product 3 is not economical when the unit profit is US$4. To analyze this problem, it is better to write the objective function as a function of the unit profits so that we can plot x3 as a function of the unit profit of product 3:

Part c: To analyze this, we need to rewrite the objective function and change the constraints

```
%% Part b
F=linprog(f,A,b,Aeq,beq,lb);
n=1;
while F(3)<0.3;
    n=n+1;
    f=[-10 -6 (-4-.01*n)];
    F=linprog(f,A,b,Aeq,beq,lb);
end
Unitprofit=(n/100)+4

Unitprofit =
    5.6300
```

Thus the unit profit of product 3 should be US$5.63 before it becomes economical to produce

```
%% Part c
f=[-10,-6,-4];
A=[1,2,1;10,4,5;3,2,1];
b=[100;600*1.1;300];
Aeq=[];
beq=[];
lb=[0;0;0];
[x3,fval]=linprog(f,A,b,Aeq,beq,lb); % PR2
x3
f3val=-fval
```

```
PR2-PR1= fval-f1val=52.5 It is not economical to schedule over-
time because the increase in profit is less than the addi-
tional cost needed (US$60)
x3 =
   57.5000
   21.2500
         0
f3val =
 702.5000
It means that the maximum profit is US$702.5000/day under the
 new constraints.
```

4.6.5 Productivity Profit Optimization

A chemical company sells three products: *A*, *B*, and *C*. Its revenue function can be represented as:

$$R(x) = 10 \cdot x_1 + 4 \cdot x_2 + 5.5 \cdot x_3 \ \ \$10^4/\text{month} \tag{4.37}$$

where x_1, x_2, and x_3 are the monthly production rates of *A*, *B*, and *C*. From the breakeven analysis, the minimum production rates of *A*, *B*, and *C* must be 5, 2, and 1 ton/month, respectively. The production rates of *B* and *C* must also follow the following restriction:

$$2 \cdot (x_2)^2 + 3 \cdot (x_3)^2 \geq 15 \tag{4.38}$$

Additional restrictions must be considered due to the raw material availability:

$$x_1 + 4 \cdot x_2 + 8 \cdot x_3 \leq 40 \quad x_1 + 3 \cdot x_2 + x_3 \leq 35 \tag{4.39}$$

For maximum revenue, determine the production rate per month for each chemical. What is the revenue at this optimum point?

Solution:

```
%% Application 4.6.5
clc; clear all; format compact;
fun=@(x) -(10*x(1)+4*(x(2))+5.5*x(3));
nonlcon=@noncon;
A=[1 4 8; 1 3 1];
b=[40; 35];
Aeq=[];
beq=[];
lb=[5;2;1];
ub=[];
x0=[2;2;0];
[x,fval]=fmincon(fun,x0,A,b,Aeq,beq,lb,ub,nonlcon)
fmax=-fval
function [c,ceq]=noncon(x)
c=15-2*(x(2))^2-3*(x(3))^2;
ceq=[];
end
```

```
x =
   22.2020
    2.4495
    1.0000
fval =
 -237.3184
fmax =
  237.3184
```

4.6.6 Catalyst System Design Optimization

Thygeson and Grossmann (1970) studied the optimal circulation system design for drying catalyst pellets. To maximize the production rate, the following optimization problem was obtained:

$$\text{Minimize } f(x) := 0.0064x_1 \cdot \left[\exp\left[-0.184(x_1)^{0.3} \cdot x_2\right] - 1\right] \tag{4.40}$$

$$\text{Subject to } 1.2 \cdot 10^{13} - (3000 + x_1) \cdot (x_1)^2 \cdot x_2 \geq 0$$

$$4.1 - \exp\left[0.184 \cdot (x_1)^{0.3} \cdot x_2\right] \geq 0 \tag{4.41}$$

$x_1 \geq 0; \quad x_2 \geq 0$
where x_1 is the fluid velocity, and x_2 is the bed depth. Determine the design fluid velocity and the bed depth that gives the maximum production rate.

Solution:

```
%% Application 4.6.6
clc; clear all; format compact;
fun=@(x) 0.0064*x(1)*1e4*(exp(-0.184*((1e4*x(1))^0.3)*x(2))-1);
nonlcon=@noncon966;
A=[];
b=[];
Aeq=[];
beq=[];
lb=[0;0];
ub=[];
x0=[0.5;5];
[x,fval]=fmincon(fun,x0,A,b,Aeq,beq,lb,ub,nonlcon)
fmax1=-fval
function [c,ceq]=noncon966(x)
c=[-(1.2e13-(3000+1e4*x(1))*(1e4*x(1))^2*x(2));...
 -(4.1-exp(0.184*((1e4*x(1))^0.3)*x(2)))];
ceq=[];
end

x =
    3.1766
    0.3421
fval =
 -153.7143
```

```
fmax1 =
  153.7143
```

4.6.7 Reaction Kinetic

Two simultaneous reactions occur at 900 K and 1 atm:

$$\text{Reaction A: } 2CO + 2H_2 \leftrightarrow CO_2 + CH_4 \tag{4.42}$$

$$\text{Reaction B: } CO + 3H_2 \leftrightarrow H_2O + CH_4 \tag{4.43}$$

At this condition, the equilibrium constants of these two reactions are:

$$K_A = 1.8283; \quad K_B = 0.7902 \tag{4.44}$$

Determine the equilibrium composition of the gas mixture if the initial numbers of CO (n_1) and H_2 (n_2) moles are 2 and 1, respectively.

The following two equations can be set up from the definition of equilibrium constant:

$$K_A = \frac{\xi_A \cdot (\xi_A + \xi_B) \cdot (n_1 + n_2 - 2 \cdot \xi_A - 2 \cdot \xi_B)^2}{(n_1 - 2 \cdot \xi_A - \xi_B)^2 \cdot (n_2 - 2 \cdot \xi_A - 3 \cdot \xi_B)^2} \tag{4.45}$$

$$K_B = \frac{\xi_B \cdot (\xi_A + \xi_B) \cdot (n_1 + n_2 - 2 \cdot \xi_A - 2 \cdot \xi_B)^2}{(n_1 - 2 \cdot \xi_A - \xi_B) \cdot (n_2 - 2 \cdot \xi_A - 3 \cdot \xi_B)^3} \tag{4.46}$$

where ξ_A and ξ_B are the extents of reactions A and B, respectively.

The extent of reaction should be a positive number: $\xi_A \geq 0 \quad \xi_B \geq 0$

Since the amount of reacted CO and H_2 cannot exceed their initial amount, two additional inequalities can be obtained:

$$2 \cdot \xi_A + \xi_B \leq 2; \quad 2 \cdot \xi_A + 3 \cdot \xi_B \leq 1 \tag{4.47}$$

The equilibrium composition can be calculated using the following equations:

$$\begin{aligned} Y_{CO} &= \frac{n_1 - 2 \cdot \xi_A - \xi_B}{n_1 + n_2 - 2 \cdot \xi_A - 2 \cdot \xi_B}; \quad Y_{H_2} = \frac{n_2 - 2 \cdot \xi_A - 3 \cdot \xi_B}{n_1 + n_2 - 2 \cdot \xi_A - 2 \cdot \xi_B}; \\ Y_{CO_2} &= \frac{\xi_A}{n_1 + n_2 - 2 \cdot \xi_A - 2 \cdot \xi_B}; \quad Y_{CH_4} = \frac{\xi_A + \xi_B}{n_1 + n_2 - 2 \cdot \xi_A - 2 \cdot \xi_B}; \\ Y_{H_2O} &= \frac{\xi_B}{n_1 + n_2 - 2 \cdot \xi_A - 2 \cdot \xi_B} \end{aligned} \tag{4.48}$$

Solution

```
% Application 4.6.7
clc; clear all;
n1=2;
n2=1;
syms eA eB
eqn1=(eA*(eA+eB)*(n1+n2-2*eA-2*eB)^2)/...
((n1-2*eA-eB)^2*(n2-2*eA-3*eB)^2)==1.8283;
eqn2=(eB*(eA+eB)*(n1+n2-2*eA-2*eB)^2)/...
((n1-2*eA-eB)*(n2-2*eA-3*eB)^3)==0.7902;
[eA, eB]=vpasolve([eqn1, eqn2],[eA,eB]);
% There is more than one answer for each extent of the reaction
```

```
% Now try with fmincon
lb=[0,0];
A=[2 1; 2 3];
b=[2;1];
e0=[0.2;0.01]; % initial guess
x=fmincon(@(x)0,e0,A,b,[],[],lb,[],@fminconstr)

function F=EoR(e)
F(1)=((e(1)*(e(1)+e(2))*(2+1-2*e(1)-2*e(2))^2)/...
((2-2*e(1)-e(2))^2*(1-2*e(1)-3*e(2))^2))-1.8283;
F(2)=((e(2)*(e(1)+e(2))*(2+1-2*e(1)-2*e(2))^2)/...
((2-2*e(1)-e(2))*(1-2*e(1)-3*e(2))^3))-0.7902;
end

function [c,ceq]=fminconstr(e)
c=[]; % no nonlinear inequality
ceq=EoR(e);
end

x =
     0.2747
     0.0301
```

4.7 Summary

Optimization and its applications in chemical engineering have been presented in this chapter. Optimization techniques can be used to minimize or maximize a given function in a search region. MATLAB has a very useful optimization toolbox including solvers such as fmincon, fminsearch, fminunc, fsolve, fzero, linprog, quadprog, and so on. It is a common practice for a given problem to use different solvers to crosscheck the optimization results. Furthermore, since complexity of optimization problems, most built-in solvers are sensitive to initial points or guesses, it is worthwhile to change the initial guesses to check the output values.

The optimization toolbox from MATLAB has its limitations and we still need to learn the enhanced optimization methods in Chapter 5 and global optimization in Chapter 6.

Problems

4.1 Substance A reacts according to an elementary autocatalytic reaction:

$$A + B \rightarrow B + B \text{ and } -r_A = kC_{A0}^2 X_A (1 - X_A) \tag{4.49}$$

where $-r_A$ is the rate of reaction, k is the rate constant [= 1 l/(mol.min)], C_{A0} is the concentration of A in the feed [= 1 mol/l], and X_A is the conversion of A.

The process is carried out in a recycle reactor, which has the following design equation:

$$\tau = C_{A0}(R+1)\int_{\frac{RX_{Af}}{R+1}}^{X_{Af}} \frac{1}{-r_A} dX_A \tag{4.50}$$

where τ is the space time (residence time), R is the recycle ratio, and X_{Af} is the final conversion of A [= 0.9]. Determine the optimum recycle ratio that gives minimum reactor volume (minimum space time). What is the residence time of the reacting mixture in the reactor at this optimum recycle ratio?

4.2 Consider a gas pipeline transmission system where compressor stations are placed L miles apart (Sherwood 1963). The total annual cost of this transmission system and its operation is

$$C(D, P_1, L, r) = 7.84D^2P_1 + 450{,}000 + 36{,}900D + \frac{6.57 \times 10^6}{L} + \frac{772 \times 10^6}{L}(r^{0.219} - 1) \tag{4.51}$$

where D is the pipe inside diameter (in), P_1 is the compressor discharge pressure (psia), L is the length between compressor stations (miles), r is the compression ratio (P_1/P_2), and P_2 is the compressor inlet pressure (psia). If the flow rate is 100×10^6 scf/day, determine the design variables, i.e. D, P_1, L, and r, for a minimum annual cost.
The flow rate (Q) is related to the design variables:

$$Q = 3.39\sqrt{\frac{\left(P_1^2 - P_2^2\right) D^5}{fL}} \text{(scf/h)} \tag{4.52}$$

where f is the friction factor ($= 0.008D^{-1/3}$).

4.3 Product R is produced at 80 mol/h (F_R) in a mixed flow reactor, in which the following first-order reaction takes place:

$$A \to R \text{ and } r_R = kC_A \tag{4.53}$$

where r_R is the reaction rate of R formation, k is the rate constant ($= 0.5\,\text{h}^{-1}$), and C_A is the concentration of A in the reactor. The concentration of reactant A in the feed (C_{A0}) is 0.2 mol/l. The reactant (Ps) cost at this concentration is US\$0.80/mol A, whereas the reactor cost includes installation, instrumentation, and depreciation. (P_r), is US\$0.02/(h · l). Determine the reactor size, feed rate, and conversion that should be used for optimum operations, i.e. minimum cost. Also, determine the unit cost of R for these conditions. Assume that the unreacted A is discarded.
On an hourly basis, the total cost is

$$P_t = VP_r + F_{A0}P_s \tag{4.54}$$

where V is the volume of the reactor, and F_{A0} is the molar flow rate of A. For a first-order reaction, the volume of the reactor, which is derived from mole balance, is given by

$$V = \frac{F_{A0}X_A}{kC_{A0}(1 - X_A)} \tag{4.55}$$

where
X_A is the conversion of A, from which we can also relate the product rate to the feed rate:

$$F_R = F_{A0}X_A \tag{4.56}$$

Note that if the conversion is low, the reactor volume needed is small, but the amount of reactant A needed for a specific product rate becomes large. On the other hand, if the conversion is high, the amount of reactant A is small, but the rector volume needed is large.

4.4 A company produces alcohol in two grades through a single process stream. The unit profit on product A is US\$1.5/gal, and on product B is US\$0.9/gal. Since the purity of product A is higher than that of product B, product A needs 1.5 times the processing times of product B. Product B alone can be produced at 1500 gal/day. Contract sales require producing at least 300 gal/day of product B. For optimum profit, determine the amount of products A and B that should be produced daily. What is the profit?

4.5 In a three-stage compressor, which is adiabatically reversible, one needs to optimize the discharge pressures from the first and second stages to minimize the total work needed. For a given diatomic gas and final pressure ($P_3 = 78$ bar), the optimal pressures are obtained as the solution to the following problem:

$$\text{Minimize}\, f(P_1, P_2) = P_1^{\frac{2}{7}} + \left(\frac{P_2}{P_1}\right)^{\frac{2}{7}} + \left(\frac{P_3}{P_2}\right)^{\frac{2}{7}} \tag{4.57}$$

Subject to: $P_1 \geq 1$ (the gas enters the three-stage compressor at 1 bar)

$$P_1 \leq P_2 \leq P_3 \text{ (the gas pressures are monotonically increasing from inlet to outlet)} \tag{4.58}$$

where P_1 and P_2 are the discharge pressures from the first and the second stages, respectively. Determine the optimal discharge pressures from the first and the second stages.

4.6 As shown in Figure 4.8, the total heat transfer area of three cross-flow heat exchangers in series is to be minimized. Cold fluid is heated in each heat exchanger by hot fluid. A steady-state lumped model can model each heat exchanger:

$$WC_p(T_i - T_{i-1}) = U_i A_i (t_i - T_i) \tag{4.59}$$

where

WC_p = heat capacity of the cold fluid [$= 10^5$ Btu/(h°F)]
T_i = temperature of cold fluid leaving heat exchanger i
T_{i-1} = temperature of the cold fluid entering heat exchanger i
U_i = overall heat transfer coefficient in heat exchanger i
A_i = heat transfer area of heat exchanger i
t_i = temperature of the hot fluid entering heat exchanger i

a. What design variables can be adjusted to minimize the total heat transfer area of this three-stage heat exchanger?
b. Write the objective function, i.e. the total heat transfer area as a function of the design variables.
c. Write the constraints considering that the temperatures of the cold fluid are monotonically increasing from inlet to outlet, and the temperature of the hot fluid entering a heat exchanger must be higher than that of the cold fluid leaving the heat exchanger.
d. Determine the optimum design variables and the system's minimum total heat transfer area.

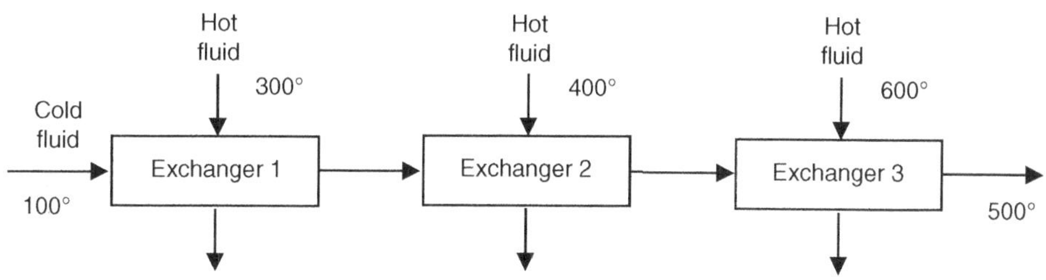

Figure 4.8 Three cross-flow heat exchangers in series.

Table 4.8 The data of the three products.

	Product *A*	**Product *B***	**Product *C***
Profit ($/kg)	40	60	80
Time needed to process one kg of product in every step, hours			
Step 1	3	2	4
Step 2	1	1	7
Step 3	1	3	3

Data: $U_1 = 120$ Btu/(h ft^2·°F)
$U_2 = 80$ Btu/(h ·ft^2·°F)
$U_3 = 40$ Btu/(h ·ft^2·°F)

4.7 In the pharmaceutical industry, three highly demanded products (A, B, and C) are produced in a single process stream consisting of three consecutive steps as shown in Table 4.8. Running the equipment in each step is possible 24 hours a day. Contract sales require producing at least 2 kg/day of product A.

a. How many kilograms per day should each product be produced to give maximum profit? What is the maximum profit?
b. Is unused time available on any of the steps with the optimal solution? If yes, which step does not run 24 hours a day?
c. Unfortunately, there is an energy crisis leading to widespread blackouts. A few power generators with limited capacity are available in the production site, but one cannot run for more than 12 hours/day. Consequently, one of the production steps must not run more than 12 hours a day. As a production manager, decide which step should operate for no more than 12 hours. Do some calculations to support your reasoning.

5

Enhanced Optimization

This chapter develops the reader's capability of enhanced optimization, which cannot be easily obtained using MATLAB's Optimization Toolbox. After learning from this chapter, readers will have more confidence to work on broader optimization problems.

As in Chapter 4, MATLAB has a built-in Optimization Toolbox with many optimization solvers, such as fmincon, fminimax, fminsearch, fminunc, fseminf, fsolve, and fzero. There are many books or articles in which researchers used specific solvers from this list to tackle their problems. However, some readers may wonder whether they have never tried to compare all those solvers to see whether their results are correct and their methods are robust, or overcome the limitations of MATLAB software (Xie and Davis 2022). It is a valid point to try to find a way to compare the above solvers. For example, fmincon finds the minimum of a constrained nonlinear multivariable function. In contrast, fminunc and fminsearch find the minimum of an unconstrained multivariable function, and fseminf finds the minimum of a semi-infinitely constrained, multivariable nonlinear function.

In this chapter, we introduce enhanced optimization by adding constraints on the search region, including the Lagrange method, inequality constraints converted to equality constraints, the penalty function method, and the enhanced constraints method. After the treatment of constraints, different optimization solvers can be applied and compared to optimization problems. It can expand readers' skills in optimization and not be restricted by the limitation of specific solvers in MATLAB. It can also help readers to build their confidence in their optimization results. In addition, the methods from this chapter can also be used for Chapter 6 on global optimizations.

5.1 Introduction

5.1.1 Differences Among Optimization Solvers in MATLAB

It is clearly shown in the following table the differences from the syntax of all different optimization solvers in MATLAB. The fmincon, fminimax, and fseminf can work on the constrained nonlinear multivariable function, while fminsearch, fminunc, fsolve, and fzero can only work on the unconstrained multivariable function. Table 5.1 shows the main syntax of different optimization solvers in MATLAB. There is no way to directly use fminsearch or fminunc to solve optimization problems with constraints because you did not follow their

Chemical Engineering Analysis and Optimization Using MATLAB®, First Edition.
Weiguo Xie, Sam Toan, and Richard Davis.

Companion website: www.wiley.com/go/xie/chemicalengineering

Table 5.1 Syntax of optimization solvers (https://www.mathworks.com/The MathWorks, Inc.).

Solver with input arguments	Description
fmincon(fun,x0,A,b,Aeq,beq,lb,ub,nonlcon,options)	Find a minimum of a constrained nonlinear multivariable function
fminimax(fun,x0,A,b,Aeq,beq,lb,ub,nonlcon,options)	Solve a minimax constraint problem. Seeks a point that minimizes the maximum of a set of objective functions
fminsearch(fun,x0,options)	Find a minimum of an unconstrained multivariable function using a direct, derivative-free method
fminunc(fun,x0,options)	Find a minimum of an unconstrained multivariable function
fminbnd(fun,x1,x2,options)	Find a minimum of a single-variable function on a fixed interval $x_1 < x < x_2$
fseminf(fun,x0,ntheta,seminfcon,A,b,Aeq,beq,lb,ub,options)	Find a minimum of a semi-infinitely constrained multivariable nonlinear function
fsolve(fun,x0,options)	Solve a system of nonlinear equations
fzero(fun,x0,options)	Solve for a root of a nonlinear function

syntax and would get error messages. In this chapter, we discuss how to treat the constraints to make them unconstrained and then use fminsearch or fminunc and compare results to those of fmincon.

5.1.2 The Differences in Algorithms

Many people may not know or think about what kinds of algorithms have been used for all different optimization solvers in MATLAB. Table 5.2 shows algorithm options for different optimization solvers in MATLAB.

Example 5.1
The function is

$$f(x) = (\cos(e^x) + x^2 - 1)^2 \tag{5.1}$$

with x range (0.01, 2).

(1) To plot the curve of $f(x)$ vs. X

Solution:

```
>> x=linspace(0.01,2,200);
y=(cos(exp(x))+x.^2-1).^2; plot(x,y);
```

Table 5.2 Options for optimization solver algorithms in MATLAB (https://www.mathworks.com/The MathWorks, Inc.).

Solver	Options
fmincon	'interior-point' (default) 'trust-region-reflective' 'sqp' 'sqp-legacy' 'active-set'
fminimax	'interior-point'
fminsearch	The Nelder–Mead simplex algorithm
fminunc	'quasi-Newton' (default) 'trust-region'
fminbnd	Based on golden section search and parabolic interpolation
fseminf	'sqp'
fsolve	'trust-region-dogleg' (default) 'trust-region' 'Levenberg-Marquardt'
fzero	A combination of bisection, secant, and inverse quadratic interpolation methods

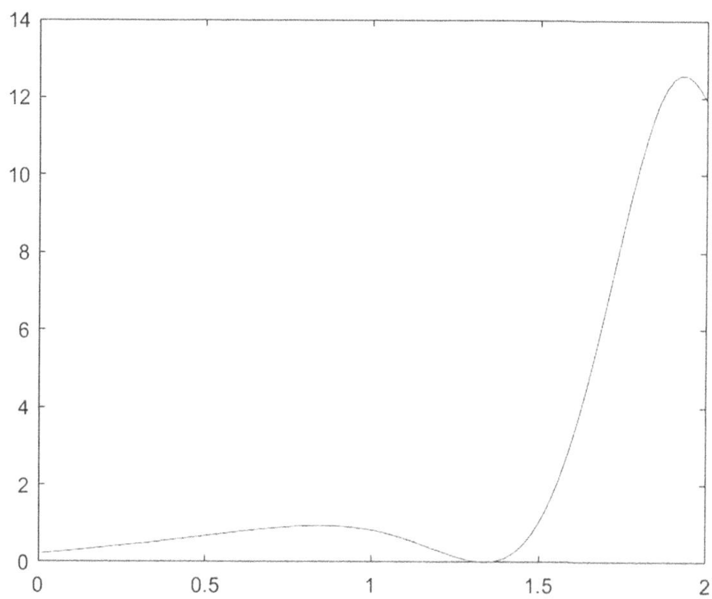

Figure 5.1 The plot of Example 5.1.

Figure 5.1 shows the plot of the curve.

(2) To use fminsearch to find the minimum point with the initial x0=1.

```
>> [x1 fval] = fminsearch(@(x)(cos(exp(x))+x.^2-1).^2, 1)
x1 =
  1.3367
fval =
  1.8686e-08
```

(3) To use fminunc to find the minimum point with the initial x0=1.

```
>> [x1 fval] = fminunc(@(x)(cos(exp(x))+x.^2-1).^2, 1)
Local minimum found.
```

```
Optimization completed because the size of the gradient is less than the
value of the optimality tolerance.
<stopping criteria details>
x1 =
  1.3367
fval =
  2.2105e-15
```

(4) To use fminbnd to find the minimum point with the interval [0.01 2].

```
>> [x1 fval] = fminbnd(@(x)(cos(exp(x))+x.^2-1).^2, 0.01, 2)
x1 =
    1.3367
fval =
  3.2668e-10
```

(5) To use fmincon to find the minimum point with the initial x0=1, lower bound lb = 0.01, and upper bound ub = 2.

```
>> [x1 fval] = fmincon(@(x)(cos(exp(x))+x.^2-1).^2, 1, [], [], [], [],...
 0.01, 2)
Local minimum found that satisfies the constraints.
Optimization completed because the objective function is non decreasing in
feasible directions, to within the value of the optimality tolerance, and
constraints are satisfied to within the value of the constraint tolerance.
<stopping criteria details>
x1 =
  1.3367
fval =
  2.6546e-15
```

(6) To use fsolve to find the minimum point with the interval [0.01 2].

```
>> [x1 fval] = fsolve(@(x)(cos(exp(x))+x.^2-1).^2, 1)
Equation solved.
fsolve completed because the vector of function values is near zero as
measured by the value of the function tolerance, and the problem appears
regular as measured by the gradient.
<stopping criteria details>
x1 =
  1.3362
fval =
  6.2538e-06
```

(7) To use fzero to find the minimum point with the interval [0.01 2].

```
>> [x1 fval] = fzero(@(x)(cos(exp(x))+x.^2-1).^2, 1)
Exiting fzero: aborting search for an interval containing a sign change
because NaN or Inf function value encountered during search.
(Function value at 927.819 is NaN.)
Check the function or try again with a different starting value.
x1 =
  NaN
fval =
  NaN
```

We find from the above example that other than fzero and fsolve, all other optimization solvers can give very close optimal results. However, the fval, the function value, is different from all solvers, indicating that different solvers use different algorithms.

5.2 Enhanced Optimization Through Treatment of Constraints on the Search Region

5.2.1 Lagrange Method

Lagrange multiplier theorem: The Lagrange multiplier theorem is a mathematical concept that explains how to find local maxima or minima of a function while considering equality constraints (Lagrange 1811). To apply this theorem, constraint qualification must be met first. The theorem states that the function's gradient at a point can be expressed as a linear combination of the constraints' gradients at that same point, with the Lagrange multipliers serving as coefficients. In simpler terms, it means that any direction perpendicular to all constraint gradients is also perpendicular to the function's gradient. Additionally, for every possible direction, the function's directional derivative is zero.

$$min f(\overline{x}) \tag{5.2}$$

$$\text{s.t. } h_j(\overline{x}) = 0 \; j = 1, 2, \ldots l \tag{5.3}$$

If $\overline{x}^*$ is a local optimal solution and satisfies the constraint; f and h_j have continuous first derivatives around $\overline{x}^*$; $\nabla h_1(\overline{x}^*)$, $\nabla h_2(\overline{x}^*)$, …, $\nabla h_l(\overline{x}^*)$ are linearly independent, then $\nabla f(\overline{x}^*) - \sum_{j=1}^{l} \lambda_j^* \nabla h_j(\overline{x}^*) = \overline{0}$ has a set of solution $\lambda_1^*, \lambda_2^*, \ldots \lambda_l^*$.

$$\nabla f(\overline{x}^*) = \left[\frac{\partial f(\overline{x}^*)}{\partial x_1}, \frac{\partial f(\overline{x}^*)}{\partial x_2}, \ldots \frac{\partial f(\overline{x}^*)}{\partial x_n}\right]^T, \text{ and } \nabla h_j(\overline{x}^*) = \left[\frac{\partial h_j(\overline{x}^*)}{\partial x_1}, \frac{\partial h_j(\overline{x}^*)}{\partial x_2}, \ldots \frac{\partial h_j(\overline{x}^*)}{\partial x_l}\right]^T \tag{5.4}$$

- Lagrange function:

$$L(\overline{x}, \lambda_1, \lambda_2, \ldots, \lambda_l) = f(\overline{x}) - \sum_{j=1}^{l} \lambda_j h_j(\overline{x}) \tag{5.5}$$

- The Lagrange multiplier method changes the equality-constrained optimization to an unconstrained optimization with $n + l$ dimensions.
- There is a computational method to construct a new S function:

$$s = (\nabla_{\overline{x}} L)^2 + (\nabla_{\overline{\lambda}} L)^2 = \sum_{i=1}^{n} \left(\frac{\partial L}{\partial x_i}\right)^2 + \sum_{j=1}^{l} (h_j(\overline{x}))^2 \tag{5.6}$$

Example 5.2 shows how to optimize a constrained function using the method of Lagrange multipliers. Example 4.3 can be transformed into an unconstrained optimization problem by solving for y (in terms of x) from the constraint and then substituting it into the objective function. Thus, the objective function will have only one independent variable, i.e. x. However, this cannot always be done, although trying to reduce the number of independent variables by substitution is always a good idea.

5.2.1.1 Method of Lagrange Multiplier

Example 5.2

Lagrange multiplier

Solve the following optimization problem:

$$\text{Minimize } f(x, y) = (x - 2)^2 + y^2 \tag{5.7}$$

$$\text{Subject to } 2x + y - 2 = 0 \tag{5.8}$$

Solution:

Step 1: Create the function to be optimized and a function representing the constraint

$$f(x, y) = (x-2)^2 + y^2 \quad g(x, y) = 2x + y - 2$$

```
clc; clear all;
syms x y lambda
f(x,y)=(x-2)^2+y^2;
g(x,y)=2*x+y-2;
```

Step 2: Whenever possible, plot the objective function and constraint, as shown in Figure 5.2, to get the idea of the solution:

```
clear all; clc; format compact
[x,y]=meshgrid(-5:0.5:5);
f=(x-2).^2+y.^2;
g=2*x+y-2;
figure
mesh(x,y,f)
hold on
mesh(x,y,g)
title('Using mesh')
hold off
figure
plot3(x,y,f)
hold on
plot3(x,y,g)
title('Using plot3')
hold off
figure
surf(x,y,f)
hold on
surf(x,y,g)
title('Using surf')
hold off
figure
contour3(x,y,f)
hold on
contour3(x,y,g)
title('Using contour3')
hold off
figure
contour(x,y,f)
hold on
contour(x,y,g)
title('Using contour')
hold off
legend('f', 'g')
```

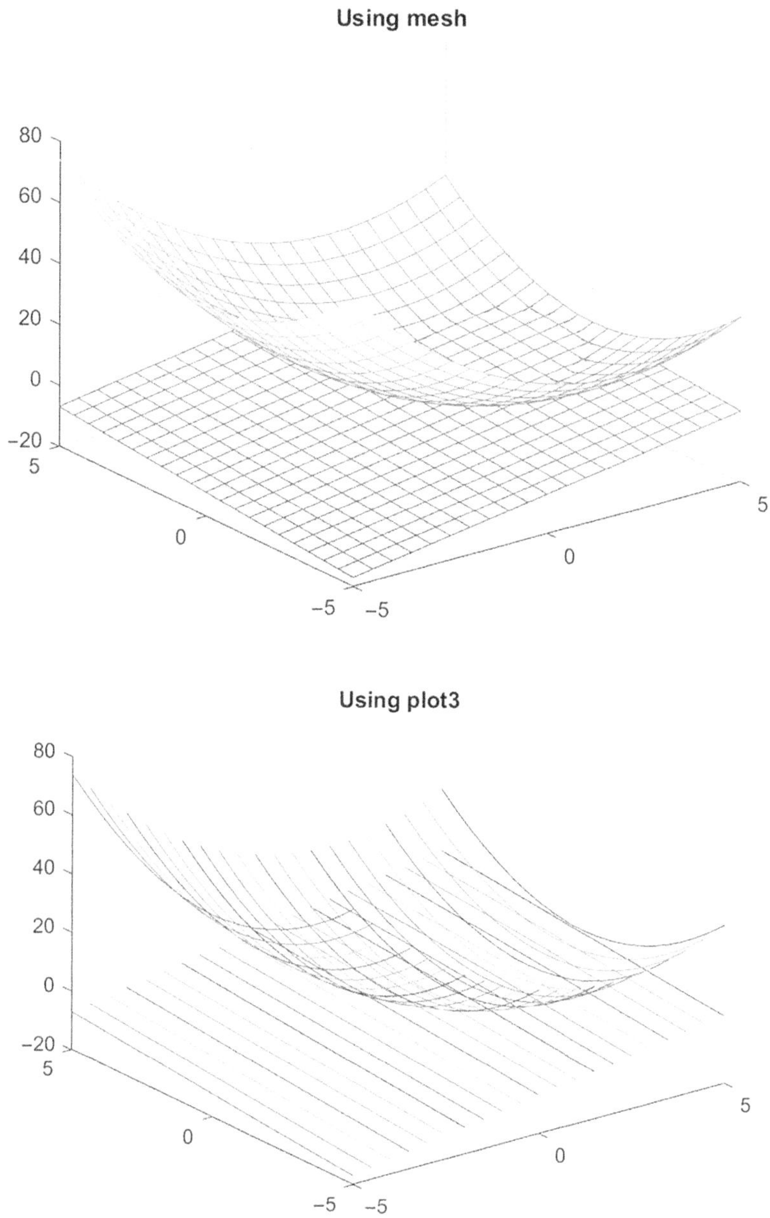

Figure 5.2 The plots of Example 5.2.

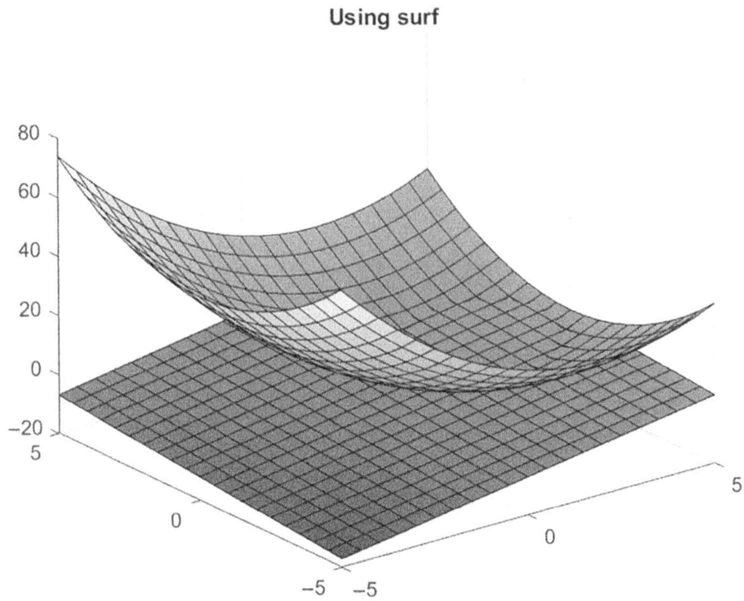

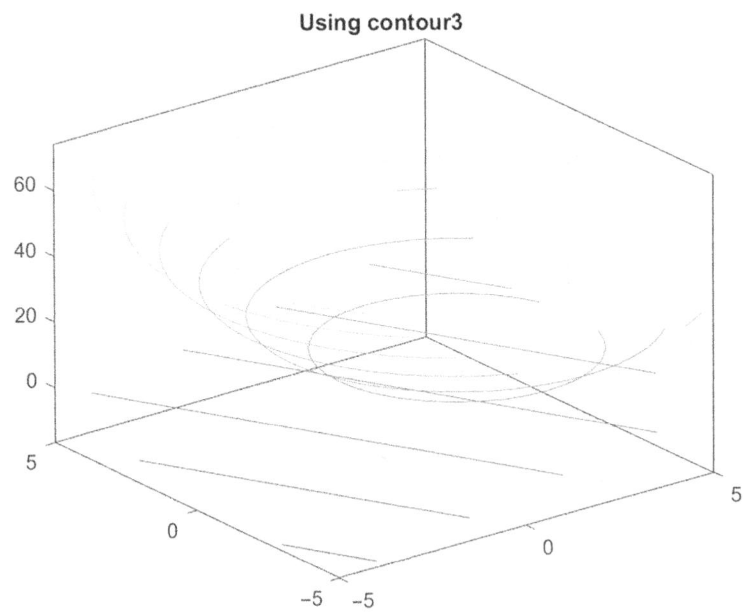

Figure 5.2 (*Continued*)

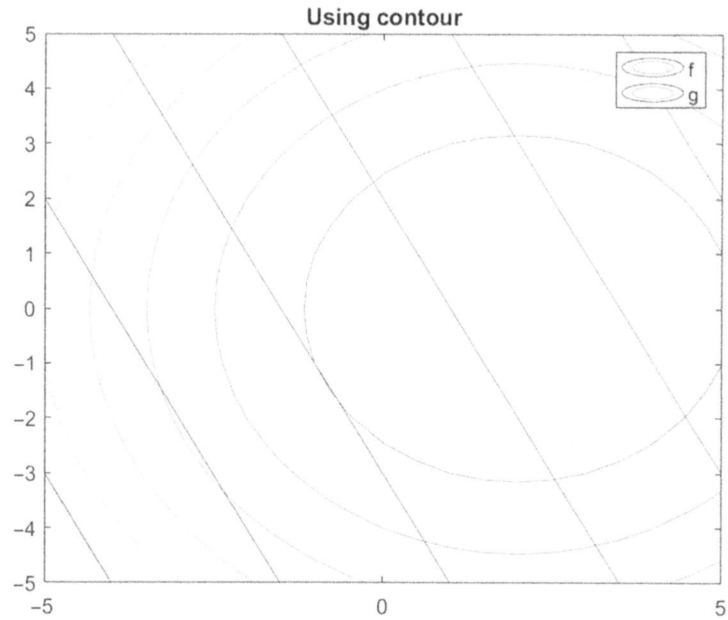

Figure 5.2 (*Continued*)

Step 3: Construct the unconstrained function, i.e. the Lagrangian function L

```
clc; clear all;
syms x y lambda
f(x,y)=(x-2)^2+y^2;
g(x,y)=2*x+y-2;
L(x,y,lambda)=f(x,y)-lambda*g(x,y);
```

Step 4: Define functions for calculating all of the second derivatives needed in the Hessian matrix

```
Lxx(x,y,lambda)=diff(L,x,x);
Lxy(x,y,lambda)=diff(L,y,x);
Lyy(x,y,lambda)=diff(L,y,y);
```

Step 5: Construct the Hessian matrix

```
H(x,y,lambda)=[Lxx(x,y,lambda) Lxy(x,y,lambda);...
Lxy(x,y,lambda) Lyy(x,y,lambda)];
```

Step 6: Create a function for calculating the eigenvalues of the Hessian matrix

```
Q=eig(H);
```

Step 7: Find a stationary point using the *vpasolve* function

```
eqn1=diff(L,x)==0;
eqn2=diff(L,y)==0;
eqn3=diff(L,lambda)==0;
[x,y,lambda]=vpasolve([eqn1,eqn2,eqn3], [x,y,lambda])
```

Step 8: Determine the type of the stationary point and the value of the function

```
H1=H(x,y,lambda);
Q1=eval(eig(H1))
f(x,y)
vpa(g(x,y), 15)
```

```
x =
1.2
y =
-0.4
lambda =
-0.8
Q1 =
  2
  2
F =
0.8
G =
0.0
```

In Example 5.2, from Step 7, we obtained the coordinate of x and y at the minimum of the function $f(x,y)$ with the given constraint of $g(x,y)$. We can use the x and y values (1.2, −0.4) to obtain the eigenvalues of the Hessian matrix, $Q = [2; 2]$. Since all the eigenvalues are real and positive numbers, this type of stationary point is a minimum stationary point. $f(1.2, -0.4) = 0.8$ is the minimum of the given function $f(x,y)$.

This is a lengthy procedure to find a min or max of a given function with constraint(s). Using MATLAB's built-in functions can be much faster. Since Example 5.2 is a nonlinear equation with constraints. This example can be solved with a nonlinear multivariable objective functions solver with constraints. Thus, *fmincon* (finds the minimum of a constrained nonlinear multivariable function) is a good candidate in this case. If unsure, you may check the optimization solver table in MATLAB. The reader can use *fmincon* to solve this optimization problem, as shown in Example 4.3, and compare the results with the Lagrange multipliers method output.

5.2.1.2 Method of Optimizing S Function

To use Example 5.2, construct the S function and perform optimization on it.

Solve the following optimization problem:

$$\text{Minimize } f(x,y) = (x-2)^2 + y^2 \tag{5.9}$$

$$\text{Subject to } 2x + y - 2 = 0 \tag{5.10}$$

Solution:

Step 1: Use vector form. Replace x using x_1, and replace y using x_2

$$f(\overline{x}) = (x_1 - 2)^2 + x_2^2 \text{ Subject to } h(\overline{x}) = 2x_1 + x_2 - 2 = 0 \tag{5.11}$$

Step 2: Create a Lagrange function

$$L(\overline{x}, \lambda_1) = f(\overline{x}) - \lambda_1 h(\overline{x}) = (x_1 - 2)^2 + x_2^2 - \lambda_1(2x_1 + x_2 - 2\ ) \tag{5.12}$$

Step 3: Construct S function

$$s = (\nabla_{\overline{x}} L)^2 + (\nabla_{\overline{\lambda}} L)^2 = \sum_{i=1}^{n} \left(\frac{\partial L}{\partial x_i} \right)^2 + \sum_{j=1}^{l} (h_j(\overline{x}))^2 = (2(x_1 - 2) - 2\lambda_1)^2 + (2x_2 - \lambda_1)^2 + (2x_1 + x_2 - 2)^2 \quad (5.13)$$

(1) To treat λ_1 as x_3 and use fminsearch to find the minimum point of S function with the initial x0=[1 1 1]. Then, to calculate $f(x, y)$ and $h(x, y)$ using x_1 and x_2.

```
>> [x] = fminsearch (@(x) ((2*(x(1)-2)-2*x(3))^2+(2*x(2)-...
x(3))^2+(2*x(1)+x(2)-2)^2), [1 1 1])
x =
   1.2000   -0.4000   -0.8000
>> f=(x(1)-2)^2+x(2)^2
f =
   0.8000
>> h=2*x(1)+x(2)-2
h =
   8.7816e-06
```

(2) To treat λ_1 as x_3 and use fminunc to find the minimum point of the S function with the initial x0=[1 1 1]. Then, to calculate $f(x, y)$ and $h(x, y)$ using x_1 and x_2.

```
>> [x] = fminunc (@(x) ((2*(x(1)-2)-2*x(3))^2+(2*x(2)-x(3))^2+(2*x(1)+...
x(2)-2)^2), [1 1 1])
Local minimum found.
Optimization completed because the size of the gradient is less than
the value of the optimality tolerance.
<stopping criteria details>
x =
   1.2000   -0.4000   -0.8000
>> f=(x(1)-2)^2+x(2)^2
f =
   0.8000
>> h=2*x(1)+x(2)-2
h =
   1.8915e-07
```

(3) To treat λ_1 as x_3 and use fsolve to find the minimum point of S function with the initial x0=[1 1 1]. Then, to calculate $f(x, y)$ and $h(x, y)$ using x_1 and x_2.

```
>> [x] = fsolve (@(x) ((2*(x(1)-2)-2*x(3))^2+(2*x(2)-x(3))^2+(2*x(1)+...
x(2)-2)^2), [1 1 1])
Warning: Trust-region-dogleg algorithm of FSOLVE cannot handle non-square
systems; using Levenberg-Marquardt algorithm instead.
> In fsolve (line 336)
Equation solved.
fsolve completed because the vector of function values is near zero as
measured by the value of the function tolerance and the problem appears
regular as measured by the gradient.
<stopping criteria details>
x =
   1.2006   -0.3997   -0.7993
>> f=(x(1)-2)^2+x(2)^2
f =
   0.7989
```

```
>> h=2*x(1)+x(2)-2
h =
  0.0014
```

The unconstrained optimization solvers (such as fminsearch, fminunc, and fsolve) apply to the equality constraints optimization via the above method of optimizing the S function. The result from fsolve is worse than fminsearch and fminunc.

The fmincon solver also applies to the unconstrained optimization:

x = fmincon(fun,x0). For the above example,

```
>> [x] = fmincon (@(x) ((2*(x(1)-2)-2*x(3))^2+(2*x(2)-x(3))^2+(2*x(1)+...
x(2)-2)^2), [1 1 1])
Local minimum found that satisfies the constraints.
Optimization completed because the objective function is non decreasing
in feasible directions, to within the value of the optimality tolerance, and
constraints are satisfied to within the value of the constraint tolerance.
<stopping criteria details>
x =
 1.2000    -0.4000    -0.8000
>> f=(x(1)-2)^2+x(2)^2
f =
  0.8000
>> h=2*x(1)+x(2)-2
h =
  -7.0467e-09
```

If we compare h values, fmincon provides better results than other solvers for this case, even just using it for unconstrained optimization.

5.2.2 Convert Inequality Constraints to Equality Constraints

There is a way to add nonnegative variable items (such as squares of variables) to convert inequality constraints to equality constraints. After this conversion, we can extend the Lagrange method to cover inequality constraints.

For example:

$$\text{Inequality constraints: } g_i(\overline{x}) \le 0 \quad i = 1, 2, \ \dots m \tag{5.14}$$

$$\text{They can be converted to } g_i(\overline{x}) + (x_{m+i})^2 = 0 \quad i = 1, 2, \ \dots m \tag{5.15}$$

Example 5.3

$$min f(\overline{x}) = 2x_1^2 - 2x_1x_2 + 2x_2^2 - 6x_1 \tag{5.16}$$

$$\text{s.t. } g_1(\overline{x}) = 3x_1 + 4x_2 - 6 \le 0 \tag{5.17}$$

$$g_2(\overline{x}) = -x_1 + 4x_2 - 2 \le 0 \tag{5.18}$$

(1) To use fmincon in MATLAB to solve the problem; x0=[0 0];

Solution:

```
>> fun = @(x)2*x(1)^2-2*x(1)*x(2)+2*x(2)^2-6*x(1);
>> x0 = [0,0];
>> [x,fval]  = fmincon(fun,x0,[3 4;-1 4],[6;2],[],[],[],[])
Local minimum found that satisfies the constraints.
```

```
Optimization completed because the objective function is non
decreasing in feasible directions, to within the value of the
optimality tolerance, and constraints are satisfied to within
the value of the constraint tolerance.
<stopping criteria details>
x =
  1.4595   0.4054
fval =
  -5.3514
```

(2) To construct the S function and then use fminunc in MATLAB to solve the problem.

Step 1: Convert inequality constraints to equality constraints

$$h_1(\bar{x}) = 3x_1 + 4x_2 - 6 + x_3^2 = 0 \tag{5.19}$$

$$h_2(\bar{x}) = -x_1 + 4x_2 - 2 + x_4^2 = 0 \tag{5.20}$$

Step 2: Create a Lagrange function

$$L = 2x_1^2 - 2x_1x_2 + 2x_2^2 - 6x_1 - \lambda_1\left(3x_1 + 4x_2 - 6 + x_3^2\right) - \lambda_2\left(-x_1 + 4x_2 - 2 + x_4^2\right) \tag{5.21}$$

Step 3: Construct S function

$$\begin{aligned} s = (\nabla_{\bar{x}}L)^2 + (\nabla_{\bar{\lambda}}L)^2 = \sum_{i=1}^{n}\left(\frac{\partial L}{\partial x_i}\right)^2 + \sum_{j=1}^{l}(h_j(\bar{x}))^2 = (4x_1 - 2x_2 - 6 - 3\lambda_1 + \lambda_2)^2 \\ + (-2x_1 + 4x_2 - 4\lambda_1 - 4\lambda_2)^2 + (2\lambda_1 x_3)^2 + (2\lambda_2 x_4)^2 + \left(3x_1 + 4x_2 - 6 + x_3^2\right)^2 \\ + \left(-x_1 + 4x_2 - 2 + x_4^2\right)^2 \end{aligned} \tag{5.22}$$

To treat λ_1 as x_5 and λ_2 as x_6, then the S function will have six variables.

Note: The initial x_3, x_4, x_5, and x_6 cannot be zero. Otherwise, inequality constraints will become equality constraints, and the constraints can be ignored. We can give the initial values as x0 = [0 0 1 1 1 1].

```
>> fun = @(x) ((4*x(1)-2*x(2)-6-3*x(5)+x(6))^2+(-2*x(1)+...
4*x(2)-4*x(5)-4*x(6))^2+(2*x(5)*x(3))^2+(2*x(6)*x(4))^2+...
(3*x(1)+4*x(2)-6+x(3)^2)^2+(-x(1)+4*x(2)-2+x(4)^2)^2);
>> x0 = [0 0 1 1 1 1];
>> x  = fminunc(fun,x0)

Local minimum found.
Optimization completed because the size of the gradient is less
than the value of the optimality tolerance.
<stopping criteria details>
x =
  1.4595    0.4054    0.0000    1.3557   -0.3243   -0.0000

>> f =2*x(1)^2-2*x(1)*x(2)+2*x(2)^2-6*x(1)
f =
  -5.3514
```

(3) To use fminsearch in MATLAB to solve the problem based on the S function.

```
>> fun = @(x)  ((4*x(1)-2*x(2)-6-3*x(5)+x(6))^2+(-2*x(1)+...
4*x(2)-4*x(5)-4*x(6))^2+(2*x(5)*x(3))^2+(2*x(6)*x(4))^2+...
(3*x(1)+4*x(2)-6+x(3)^2)^2+(-x(1)+4*x(2)-2+x(4)^2)^2);
>> x0 = [0 0 1 1 1 1];
>> x  = fminsearch(fun,x0)

x =
   1.2202    0.6280   -0.0032    0.3558   -0.5736    0.5773

>> f =2*x(1)^2-2*x(1)*x(2)+2*x(2)^2-6*x(1)
f =
   -5.0873
```

(4) To use fsolve in MATLAB to solve the problem based on the S function.

```
>> fun = @(x)  ((4*x(1)-2*x(2)-6-3*x(5)+x(6))^2+(-2*x(1)+...
4*x(2)-4*x(5)-4*x(6))^2+(2*x(5)*x(3))^2+(2*x(6)*x(4))^2+...
(3*x(1)+4*x(2)-6+x(3)^2)^2+(-x(1)+4*x(2)-2+x(4)^2)^2);
>> x0 = [0 0 1 1 1 1];
>> x  = fsolve(fun,x0)

Warning: Trust-region-dogleg algorithm of FSOLVE cannot handle
non-square systems; using Levenberg-Marquardt algorithm instead.
> In fsolve (line 336)
Solver stopped prematurely.
fsolve stopped because it exceeded the function evaluation
limit, options.MaxFunctionEvaluations = 1.200000e+03.
x =
   1.4593    0.4048    0.0534    1.3564   -0.3242   -0.0003

>> f =2*x(1)^2-2*x(1)*x(2)+2*x(2)^2-6*x(1)
f =
   -5.3504
```

The unconstrained optimization solvers (such as fminsearch, fminunc, and fsolve) apply to the inequality constraints optimization via the above method of optimizing the S function. However, the newly generated S function has six variables, making the optimization much more difficult than the original two variables. The result from fminunc is similar to that from fmincon, while both fminsearch and fsolve could not give good results. In this case, the 'quasi-Newton' algorithm for fminunc is robust. The Levenberg–Marquardt algorithm for fsolve is also acceptable.

5.2.3 Penalty Function Method

Penalty methods solve constrained optimization problems by a series of unconstrained problems whose solutions can converge to the solution of the original problem. The new objective, or penalty function, is formed from the original objective function by adding a term consisting of a penalty parameter multiplied by some measure of violating the constraints. The violation measure is nonzero when the constraints are violated and zero in the region where constraints are satisfied.

$$\text{The original objective function: } minf(\overline{x}) \tag{5.23}$$

$$\text{With constraints : } \quad g_i(\overline{x}) \leq 0 \quad i = 1, 2, \ \ldots m \tag{5.24}$$

$$\text{To construct a penalty function } \theta(\overline{x}, u) = f(\overline{x}) + u * \left(\sum_{1}^{m} ((max(0, g_i(\overline{x})))^2) \right) \tag{5.25}$$

where u can start from u_0. Then multiply with a constant each time, such as ($\beta = 10$).

Now, we change constrained optimization for the original objective function $minf(\overline{x})$ to unconstrained optimization for $min\theta(\overline{x}, u)$.

Example 5.4

$$minf(\overline{x}) = (x_1 - 3)^2 + (x_2 - 4)^2 \tag{5.26}$$

$$\text{s.t. } g_1(\overline{x}) = x_1^2 - x_2 \leq 0 \tag{5.27}$$

$$g_2(\overline{x}) = e^{-x_1} - x_2 \leq 0 \tag{5.28}$$

$$g_3(\overline{x}) = -x_1 + 2x_2 - 2 \leq 0 \tag{5.29}$$

(1) To use fmincon in MATLAB to solve the problem; x0=[1 1].

Solution:

The following code can be saved as a .m file and run to solve the problem.

```
fun = @(x)(x(1)-3)^2+(x(2)-4)^2;
x0 = [1 1];
nonlcon = @c_ineq;
[x,fval] = fmincon(fun,x0,[],[],[],[],[],[],nonlcon)
function [c,ceq] = c_ineq(x)
c(1) = x(1)^2-x(2);
c(2)=exp(-x(1))-x(2);
c(3)=-x(1)+2*x(2)-2;
ceq = [];
end
```

```
The results are as follows:
Local minimum found that satisfies the constraints.
Optimization completed because the objective function is non
decreasing in feasible directions, to within the value of the
optimality tolerance, and constraints are satisfied to within
the value of the constraint tolerance.
<stopping criteria details>
x =
  1.2808    1.6404
fval =
  8.5235
```

(2) To use the penalty function and fminunc in MATLAB to solve the problem.

Step 1: Construct a penalty function

$$\theta(\overline{x}, u) = f(\overline{x}) + u * \left(\sum_{1}^{m} ((max(0, g_i(\overline{x})))^2) \right) \tag{5.30}$$

Step 2: Give the initial value to u

$$u = u_0 = 1 \tag{5.31}$$

Step 3: Optimize the penalty function

$$\begin{aligned} \min\ \theta(\overline{x}, u) = \min\ &\big((x_1 - 3)^2 + (x_2 - 4)^2 + u_0 * \big((max(0, x_1^2 - x_2))^2 \\ &+ (max(0, e^{-x_1} - x_2))^2 + (max(0, -x_1 + 2x_2 - 2))^2\big)\big) \end{aligned} \tag{5.32}$$

Step 4: Increase value u by multiplying a constant β such as $\beta = 10$ and return to Step 2 each iteration (i.e. $m+1 \rightarrow m$) and run optimization until convergence

$$u = u_0 * \beta^m \tag{5.33}$$

The code for the first iteration is as follows:

```
>> u0=1; beta=10; m=0; u=u0*beta^m;
>> fun = @(x) (x(1)-3)^2+(x(2)-4)^2+u*((max(0,x(1)^2-x(2)))^...
2+(max(0,exp(-x(1))-x(2)))^2+(max(0,-x(1)+2*x(2)-2))^2);
>> x0 = [1 1];
>> [x,ThetaVal] = fminunc(fun,x0)
>> f=(x(1)-3)^2+(x(2)-4)^2
The results for the first iteration are as follows:
Local minimum found.
Optimization completed because the size of the gradient is less
than the value of the optimality tolerance.
<stopping criteria details>
x =
  1.7623    2.4384
ThetaVal =
  5.6579
f =
  3.9705
```

The code for the second iteration is as follows:

```
>> u0=1; beta=10; m=1; u=u0*beta^m;
>> fun = @(x) (x(1)-3)^2+(x(2)-4)^2+u*((max(0,x(1)^2-x(2)))^...
2+(max(0,exp(-x(1))-x(2)))^2+(max(0,-x(1)+2*x(2)-2))^2);
>> x0 = [1 1];
>> [x,ThetaVal] = fminunc(fun,x0)
>> f=(x(1)-3)^2+(x(2)-4)^2
The results for the second iteration are as follows:
Local minimum found.
Optimization completed because the size of the gradient is less
than the value of the optimality tolerance.
<stopping criteria details>
```

```
x =
  1.3769    1.7744
ThetaVal =
  8.0308
f =
  7.5876
```

Then, we can carry on for more iterations using fminunc. The results from six iterations are shown in Table 5.3.

(3) To use the penalty function and fminsearch in MATLAB to solve the problem. We can change fminunc to fminsearch based on the above question (2).

The code for the first iteration is as follows:

```
≫ u0=1; beta=10; m=0; u=u0*beta^m;
≫ fun = @(x) (x(1)-3)^2+(x(2)-4)^2+u*((max(0,x(1)^2-x(2)))^...
2+(max(0,exp(-x(1))-x(2)))^2+(max(0,-x(1)+2*x(2)-2))^2);
≫ x0 = [1 1];
≫ [x,ThetaVal]  = fminsearch(fun,x0)
≫ f=(x(1)-3)^2+(x(2)-4)^2
The results for the first iteration are as follows:
x =
  1.7623    2.4384
ThetaVal =
  5.6579
f =
  3.9704
```

The code for the second iteration is as follows:

```
≫ u0=1; beta=10; m=1; u=u0*beta^m;
≫ fun = @(x) (x(1)-3)^2+(x(2)-4)^2+u*((max(0,x(1)^2-x(2)))^...
2+(max(0,exp(-x(1))-x(2)))^2+(max(0,-x(1)+2*x(2)-2))^2);
≫ x0 = [1 1];
≫ [x,ThetaVal]  = fminsearch(fun,x0)
≫ f=(x(1)-3)^2+(x(2)-4)^2
The results for the second iteration are as follows:
```

Table 5.3 To use the penalty function and fminunc in MATLAB to solve Example 5.4.

u	x_{min}	f_{min}	θ_{min}
1	(1.7623, 2.4384)	3.9705	5.6579
10	(1.3769, 1.7744)	7.5876	8.0308
100	(1.2919, 1.6553)	8.4151	8.4690
1000	(1.2819, 1.6419)	8.5125	8.5180
10,000	(1.2809, 1.6405)	8.5224	8.5229
100,000	(1.2808, 1.6404)	8.5234	8.5234

```
x =
   1.3768    1.7744
ThetaVal =
   8.0308
f =
   7.5880
```

Then, we can carry on for more iterations using fminsearch. The results from six iterations are shown in Table 5.4.

(4) To use the penalty function and fsolve in MATLAB to solve the problem. We can change fminunc to fsolve based on the above question (2).

The code for the first iteration is as follows:

```
>> u0=1; beta=10; m=0; u=u0*beta^m;
>> fun = @(x)(x(1)-3)^2+(x(2)-4)^2+u*((max(0,x(1)^2-x(2)))^...
2+(max(0,exp(-x(1))-x(2)))^2+(max(0,-x(1)+2*x(2)-2))^2);
>> x0 = [1 1];
>> [x,ThetaVal] = fsolve(fun,x0)
>> f=(x(1)-3)^2+(x(2)-4)^2
The results for the first iteration are as follows:
Warning: Trust-region-dogleg algorithm of FSOLVE cannot handle
non-square systems; using Levenberg-Marquardt algorithm instead.
> In fsolve (line 336)
No solution found.
fsolve stopped because the last step was ineffective. However,
the vector of function values is not near zero, as measured by
the value of the function tolerance.
<stopping criteria details>
x =
   1.7611    2.4401
ThetaVal =
   5.6580
f =
   3.9680
```

The code for the second iteration is as follows:

```
>> u0=1; beta=10; m=1; u=u0*beta^m;
```

Table 5.4 To use the penalty function, then fminsearch in MATLAB to solve Example 5.4.

u	x_{min}	f_{min}	θ_{min}
1	(1.7623, 2.4384)	3.9704	5.6579
10	(1.3769, 1.7744)	7.5880	8.0308
100	(1.2920, 1.6553)	8.4150	8.4690
1000	(1.2819, 1.6419)	8.5123	8.5180
10,000	(1.2809, 1.6406)	8.5222	8.5230
100,000	(1.2808, 1.6404)	8.5234	8.5235

```
≫ fun = @(x)(x(1)-3)^2+(x(2)-4)^2+u*((max(0,x(1)^2-x(2)))^...
2+(max(0,exp(-x(1))-x(2)))^2+(max(0,-x(1)+2*x(2)-2))^2);
≫ x0 = [1 1];
≫ [x,ThetaVal] = fsolve(fun,x0)
≫ f=(x(1)-3)^2+(x(2)-4)^2
The results for the second iteration are as follows:
Warning: Trust-region-dogleg algorithm of FSOLVE cannot handle
non-square systems; using Levenberg-Marquardt algorithm instead.
> In fsolve (line 336)
No solution found.
fsolve stopped because the last step was ineffective. However,
the vector of function values is not near zero, as measured by
the value of the function tolerance.
<stopping criteria details>
x =
  1.3765    1.7742
ThetaVal =
  8.0308
f =
  7.5898
```

Then, we can carry on for more iterations using fsolve. The results from six iterations are shown in Table 5.5.

It can be seen that the unconstrained optimization solvers (such as fminsearch, fminunc, and fsolve) can be applied to the constraints optimization via the penalty function method. Compared to the results from fmincon in Example 5.3, the results from both fminunc and fminsearch give excellent results after six iterations, i.e. when $u = 100{,}000$. The first four iterations from fsolve are also close to fminunc and fminsearch. Please note that the penalty function method may not give excellent results for the cases when it reaches the limit of computer capability, e.g. the fifth and sixth iterations using fsolve.

5.2.4 Enhanced Constraints Method

There may be a problem with an unwanted "zero" local optimal point, which sometimes physically means doing nothing or not starting it. However, it needs to be avoided most of the time. It is only a local optimal point; we must move away from it. We will introduce a method to add a square value of an additional variable to avoid the zero result for the calculated variable.

Table 5.5 To use the penalty function, then fsolve in MATLAB to solve Example 5.4.

u	x_{min}	f_{min}	θ_{min}
1	(1.7611, 2.4401)	3.9680	5.6580
10	(1.3765, 1.7742)	7.5898	8.0308
100	(1.2920, 1.6552)	8.4152	8.4690
1000	(1.2819, 1.6419)	8.5127	8.5180
10,000	(1.1808, 1.5906)	9.1150	9.1160
100,000	(1.2142, 1.4743)	9.5680	9.5681

Example 5.5

$$minf(\bar{x}) = \frac{1}{3}(x_1+1)^3 - 2x_2^2 + x_2 \quad (5.34)$$

$$\text{s.t. } g_1(\bar{x}) = 3x_1^2 + x_2 - 6 \le 0 \quad (5.35)$$

$$g_2(\bar{x}) = -x_2 \le 0 \quad (5.36)$$

$$g_3(\bar{x}) = x_2 - 3 \le 0 \quad (5.37)$$

(1) To use fmincon in MATLAB to solve the problem; x0=[0 0].

Solution:

The following code can be saved as a .m file and run to solve the problem.

```
fun = @(x)1/3*(x(1)+1)^3-2*x(2)^2+x(2);
x0 = [0 0];
nonlcon = @c_ineq;
[x,fval]  = fmincon(fun,x0,[],[],[],[],[],[],nonlcon)
function [c,ceq] = c_ineq(x)
c(1) = 3*x(1)^2+x(2)-6;
c(2)=-x(2);
c(3)=x(2)-3;
ceq = [];
end
```

```
The results are as follows:
Local minimum found that satisfies the constraints.
Optimization completed because the objective function is non
decreasing in feasible directions, to within the value of the
optimality tolerance, and constraints are satisfied to within
the value of the constraint tolerance.
<stopping criteria details>
x =
   -0.9993    0.0000
fval =
   2.5613e-07
```

It can be seen that the result is the unwanted "zero" local optimal point ($x_2 = 0$ and fval closed to zero), although we used the fmincon solver.

(2) To move away from the "zero" optimal point, change to $g_2(\bar{x}) = -x_2 + x_3^2 \le 0$, with x0=[0 0 1], use fmincon in MATLAB to solve the problem.

The following code can be saved as a .m file and run to solve the problem.

```
fun = @(x)1/3*(x(1)+1)^3-2*x(2)^2+x(2);
x0 = [0 0 1];
nonlcon = @c_ineq;
[x,fval]  = fmincon(fun,x0,[],[],[],[],[],[],nonlcon)
function [c,ceq] = c_ineq(x)
c(1) = 3*x(1)^2+x(2)-6;
```

```
c(2)=-x(2)+x(3)^2;
c(3)=x(2)-3;
ceq = [];
end

The results are as follows:
Local minimum found that satisfies the constraints.
Optimization completed because the objective function is non
decreasing in feasible directions, to within the value of the
optimality tolerance, and constraints are satisfied to within
the value of the constraint tolerance.
<stopping criteria details>
x =
  -0.9924    3.0000    0.5588
fval =
  -15.0000
```

It can be seen that the result is moved away from the unwanted "zero" local optimal point and is a much better result with a function value equal to −15.0000.

Learning Tips

- You can use enhanced optimization to expand the existing MATLAB optimization solvers' capability.
- Comparison of results from fmincon, fminunc, fminsearch and fsolve can build up our confidence in the results.
- The penalty function method can be tried first because it does not increase number of variables and can be easily implemented.

5.3 Application in Chemical Engineering

5.3.1 Optimization of a Three-Stage Compressor

Example 5.6

In a three-stage compressor, which is adiabatically reversible, one needs to optimize the discharge pressures from the first and second stages to minimize the total work needed. For a given diatomic gas and final pressure ($P_3 = 80$ bar), the optimal pressures are obtained as the solution to the following problem with a constraint $P_1 + P_2 \geq 50$:

$$min f(P_1, P_2) = P_1^{\frac{2}{7}} + \left(\frac{P_2}{P_1}\right)^{\frac{2}{7}} + \left(\frac{P_3}{P_2}\right)^{\frac{2}{7}} \tag{5.38}$$

$$\text{s.t. } g(P_1, P_2) = P_1 + P_2 \geq 50 \tag{5.39}$$

where P_1 and P_2 are the discharge pressures from the first and the second stages, respectively. Determine the optimal discharge pressures from the first and the second stages. The initial values $P_1 = 1$; $P_2 = 1$.

(1) To use fmincon to solve the problem

Solution:

```
% Example 5.6
clc; clear all; format compact;
P=80;
fun=@(x)x(1)^(2/7)+(x(2)/x(1))^(2/7)+(P/x(2))^(2/7);
A=[-1 -1];
B=[-50];
x0=[1;1];
[x,fval]=fmincon(fun,x0,A,B)
```

```
The results are as follows:
Local minimum found that satisfies the constraints.
Optimization completed because the objective function is non
decreasing in feasible directions, to within the value of the
optimality tolerance, and constraints are satisfied to within
the value of the constraint tolerance.
<stopping criteria details>
x =
  7.1005
  42.8995
fval =
  4.6175
```

(2) To construct the S function and perform optimization using fminsearch.

Step 1: Use vector form, replace P_1 using x_1, and replace P_2 using x_2

$$f(\bar{x}) = x_1^{\frac{2}{7}} + \left(\frac{x_2}{x_1}\right)^{\frac{2}{7}} + \left(\frac{80}{x_2}\right)^{\frac{2}{7}} \quad \text{Subject to } g(\bar{x}) = 50 - x_1 - x_2 \leq 0 \tag{5.40}$$

Step 2: Convert inequality constraints to equality constraints

$$h_1(\bar{x}) = 50 - x_1 - x_2 + x_3^2 = 0 \tag{5.41}$$

Step 3: Create a Lagrange function

$$L(\bar{x}, \lambda_1) = f(\bar{x}) - \lambda_1 h(\bar{x}) = x_1^{\frac{2}{7}} + \left(\frac{x_2}{x_1}\right)^{\frac{2}{7}} + \left(\frac{80}{x_2}\right)^{\frac{2}{7}} - \lambda_1\left(50 - x_1 - x_2 + x_3^2\right) \tag{5.42}$$

Step 4: Construct S function

$$s = (\nabla_{\bar{x}}L)^2 + (\nabla_{\bar{\lambda}}L)^2 = \sum_{i=1}^{n}\left(\frac{\partial L}{\partial x_i}\right)^2 + \sum_{j=1}^{l}(h_j(\bar{x}))^2 = \left(\frac{2}{7}(x_1)^{-\frac{5}{7}} - \frac{2}{7}x_2^{\frac{2}{7}}x_1^{-\frac{9}{7}} + \lambda_1\right)^2$$
$$+ \left(\frac{2}{7}(x_2)^{-\frac{5}{7}}(x_1)^{-\frac{2}{7}} - \frac{2}{7}\cdot 80^{\frac{2}{7}}\cdot x_2^{-\frac{9}{7}} + \lambda_1\right)^2 + (2\lambda_1 x_3)^2 + \left(50 - x_1 - x_2 + x_3^2\right)^2 \tag{5.43}$$

To treat λ_1 as x_4 and use fminsearch to find the minimum point of the S function with the initial x0=[1 1 0 0]. Then, to calculate $f(\bar{x})$ using x_1 and x_2.

```
>> [x] = fminsearch(@(x) (2/7*x(1)^(-5/7)-2/7*x(2)^(2/7)*...
x(1)^(-9/7)+x(4))^2+(2/7*x(2)^(-5/7)* x(1)^(-2/7) -2/7*...
```

```
80^(2/7)*x(2)^(-9/7)+x(4))^2+(2*x(4)*x(3))^2+(50-x(1)-...
x(2)+x(3)^2)^2, [1 1 0 0])
≫ f= x(1)^(2/7)+(x(2)/x(1))^(2/7)+(80/x(2))^(2/7)
The results are as follows:
x =
  7.0989    42.9022    -0.0324    -0.0032
f =
  4.6175
```

(3) To construct the *S* function and perform optimization using fminunc.

Based on question (2), use fminunc to find the minimum point of the *S* function with the initial x0=[1 1 0 0]. Then, to calculate $f(\overline{x})$ using x_1 and x_2.

```
≫ [x] = fminunc(@(x) (2/7*x(1)^(-5/7)-2/7*x(2)^(2/7)*...
 x(1)^(-9/7)+x(4))^2+(2/7*x(2)^(-5/7)* x(1)^(-2/...
7)-2/7*80^(2/7)*x(2)^(-9/7)+x(4))^2+(2*x(4)*x(3))^2+...
(50-x(1)-x(2)+x(3)^2)^2, [1 1 0 0])
≫ f= x(1)^(2/7)+(x(2)/x(1))^(2/7)+(80/x(2))^(2/7)
The results are as follows:
Local minimum found.
Optimization completed because the size of the gradient is less
than the value of the optimality tolerance.
<stopping criteria details>
x =
  25.0832    24.9169    -0.0000    -0.0064
f =
  4.9045
```

(4) To construct the *S* function and perform optimization using fmincon(fun,x0).

Based on question (2), use fmincon to find the minimum point of the *S* function with the initial x0=[1 1 0 0]. Then, to calculate $f(\overline{x})$ using x_1 and x_2.

```
≫ [x] = fmincon(@(x) (2/7*x(1)^(-5/7)-2/7*x(2)^(2/7)*...
 x(1)^(-9/7)+x(4))^2+(2/7*x(2)^(-5/7)* x(1)^(-2/7) -2/7*...
 80^(2/7)*x(2)^(-9/7)+x(4))^2+(2*x(4)*x(3))^2+...
 (50-x(1)-x(2)+x(3)^2)^2, [1 1 0 0])
≫ f= x(1)^(2/7)+(x(2)/x(1))^(2/7)+(80/x(2))^(2/7)
Local minimum found.
Local minimum found that satisfies the constraints.
Optimization completed because the objective function is non
decreasing in feasible directions, to within the value of the
optimality tolerance, and constraints are satisfied to within
the value of the constraint tolerance.
<stopping criteria details>
x =
  7.1152    42.8848    -0.0017    -0.0032
f =
  4.6175
```

It can be seen that the *S* function of the Lagrange method with both fminsearch and fmincon(fun, x0) can give excellent results, but the fminunc give result far away from optimal.

5.3.2 Productivity Profit Optimization

Example 5.7

A chemical company sells three products: *A*, *B*, and *C*. Its revenue function can be represented as:

$$R(x) = 10 \cdot x_1 + 15 \cdot x_2 + 5.5 \cdot x_3 \ \$10^4/\text{month} \tag{5.44}$$

where x_1, x_2, and x_3 are the monthly production rates of *A*, *B*, and *C*. The production rates of *A*, *B*, and *C* must also satisfy the following restriction:

$$(x_1)^2 + 2 \cdot (x_2)^2 + 2 \cdot x_3 \leq 30 \tag{5.45}$$

An additional restriction must be considered due to the raw material availability:

$$x_1 + 3 \cdot x_2 + x_3 \leq 10 \tag{5.46}$$

For maximum revenue, determine the production rate per month for each chemical. What is the revenue at this optimum point?

(1) To use fmincon to solve the problem

Solution:

```
% Example 5.7
function [c,ceq]=noncon5_7(x)
c=x(1)^2+x(2)^2+2*x(3)-30;
ceq=[];
end
clc; clear all; format compact;
fun=@(x) -(10*x(1)+15*x(2)+5.5*x(3));
nonlcon=@noncon5_7;
A=[1 3 1];
b=[10];
Aeq=[];
beq=[];
lb=[0;0;0];
ub=[];
x0=[1;1;1];
[x,fval]=fmincon(fun,x0,A,b,Aeq,beq,lb,ub,nonlcon)
fmax=-fval
```

```
The results are as follows:
Local minimum found that satisfies the constraints.
Optimization completed because the objective function is non
decreasing in feasible directions, to within the value of the
optimality tolerance, and constraints are satisfied to within
the value of the constraint tolerance.
```

```
<stopping criteria details>
x =
  5.2424
  1.5850
  0.0028
fval =
 -76.2132
fmax =
  76.2132
```

(2) To use the penalty function and fminunc in MATLAB to solve the problem.

Step 1: Construct a penalty function

$$\theta(\overline{x}, u) = f(\overline{x}) + u * \left(\sum_{1}^{m} ((max(0, g_i(\overline{x})))^2) \right) \tag{5.47}$$

Step 2: Give the initial value to u

$$u = u_0 = 1 \tag{5.48}$$

Step 3: Optimize the penalty function

$$\begin{aligned} \min\ \theta(\overline{x}, u) = \min\ & \big(-(10 \cdot x_1 + 15 \cdot x_2 + 5.5 \cdot x_3) + u_0 \\ & * \Big(\big(max\left(0, 30 - (x_1)^2 + 2 \cdot (x_2)^2 + 2 \cdot x_3\right)\big)^2 \\ & + (max\ (0, x_1 + 3 \cdot x_2 + x_3 - 10))^2\Big)\Big) \end{aligned} \tag{5.49}$$

Step 4: Increase value u by multiplying a constant β such as $\beta = 10$ and return to Step 2 each iteration (i.e. $m + 1 \rightarrow m$) and run optimization until convergence

$$u = u_0 * \beta^m \tag{5.50}$$

The code for the first iteration is as follows:

```
≫ u0=1; beta=10; m=0; u=u0*beta^m;
≫ fun = @(x)-(10*x(1)+15*x(2)+5.5*x(3))+u*((max(0,x(1)^...
2+x(2)^2+2*x(3)-30))^2+(max(0,x(1)+3*x(2)+x(3)-10))^2);
≫ x0 = [1 1 1];
≫ [x,Val] = fminunc(fun,x0)
≫ ThetaVal=-Val
≫ f=10*x(1)+15*x(2)+5.5*x(3)
The results for the first iteration are as follows:
Local minimum found that satisfies the constraints.
Optimization completed because the objective function is non
decreasing in feasible directions, to within the value of the
optimality tolerance, and constraints are satisfied to within
the value of the constraint tolerance.
<stopping criteria details>
x =
  4.7927    1.7358    2.1568
Val =
 -81.0862
```

```
ThetaVal =
  81.0862
f =
  85.8258
```

It can be seen that the result is close to that of fmincon of question (1) when $u = 1000$. Then, the result is worse than fmincon of question (1) when $u = 10{,}000$.

(3) To use penalty function and then use fmincon(fun,x0) without constraints in MATLAB to solve the problem.

The code is modified from question (2) by replacing fminunc(fun,x0) with fmincon(fun,x0) as follows:

```
≫ u0=1; beta=10; m=0; u=u0*beta^m;
≫ fun = @(x)-(10*x(1)+15*x(2)+5.5*x(3))+u*((max(0,x(1)^...
2+x(2)^2+2*x(3)-30))^2+(max(0,x(1)+3*x(2)+x(3)-10))^2);
≫ x0 = [1 1 1];
≫ [x,Val] = fmincon(fun,x0)
≫ ThetaVal=-Val
≫ f=10*x(1)+15*x(2)+5.5*x(3)
The results for the first iteration are as follows:
Local minimum found that satisfies the constraints.
Optimization completed because the objective function is non
decreasing in feasible directions, to within the value of the
optimality tolerance, and constraints are satisfied to within
the value of the constraint tolerance.
<stopping criteria details>
x =
    4.7927    1.7358    2.1568
Val =
  -81.0862
ThetaVal =
   81.0862
f =
   85.8258
```

By comparing Table 5.7 with Table 5.6, we see that the first four rows are the same. The fifth row of Table 5.7 still works close to the result of question (1). The final results of Table 5.7 are better than that of Table 5.6. It means that the default algorithm of fmincon's 'interior-point' is better than the default algorithm of fminunc 'quasi-Newton' for this example.

5.3.3 Optimization for the Operation of a Continuous Stirred Tank Reactor (CSTR)

Example 5.8

A CSTR reactor, as shown in Figure 5.3, is used to convert liquid chemical A to product B with the following reaction:

$$2A \rightarrow B \tag{5.51}$$

Table 5.6 Results of using a penalty function with fminunc in MATLAB to solve Example 5.7.

u	x_{min}	f_{min}	θ_{min}
1	(4.7927,1.7358,2.1568)	85.8258	81.0862
10	(5.1983,1.6005,0.2215)	77.2092	76.7118
100	(5.2381,1.5870,0.0230)	76.3131	76.2632
1000	(5.2415,1.5837,0.0097)	76.2232	76.2182
10,000	(5.0746,1.1567,1.4550)	76.1000	76.1000

Table 5.7 Use penalty function and fmincon(fun,x0) in MATLAB to solve Example 5.7.

u	x_{min}	f_{min}	θ_{min}
1	(4.7927,1.7358,2.1568)	85.8258	81.0862
10	(5.1983,1.6005,0.2215)	77.2092	76.7118
100	(5.2381,1.5870,0.0230)	76.3131	76.2632
1000	(5.2415,1.5837,0.0097)	76.2232	76.2182
10000	(5.2337,1.5594,0.0883)	76.2138	76.2133
100000	(5.1397,1.3079,0.9367)	76.1667	76.1667

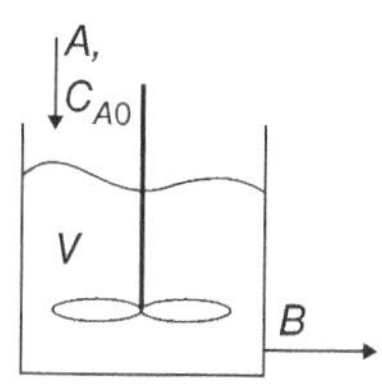

Figure 5.3 A CSTR reactor is used to convert liquid chemical A to product B.

The reaction rate is

$$-r_A = \frac{-dC_A}{dt} = 2.0 \cdot C_A^2 = 2.0 \cdot C_{A_0}^2 \cdot (1 - x_A)^2 \tag{5.52}$$

where x_A is the conversion rate, which is no more than 100%.

The price of chemical A is provided as

$$S_A = 4.0 \cdot C_{A_0}^{1.4} (\$/L) \tag{5.53}$$

Other costs, such as labor costs and depreciation of equipment,

$$S_{ld} = 0.4 \cdot V^{0.6} (\$) \tag{5.54}$$

where V is the load volume of liquid inside the reactor.

To perform the minimization of the costs for the operation of the CSTR reactor with the requirement of a production rate of B greater than 50 kg/h.

(1) To use fmincon with constraint to solve the problem

Solution:
Step 1: To formulate the optimization problem
Mass balance inside the CSTR reactor:

$$F_{A_0} \cdot C_{A_0} = F_{A_0} \cdot C_{A_0}(1 - x_A) - r_A V = F_{A_0} \cdot C_{A_0}(1 - x_A) + 2.0 \cdot C_{A_0}^2 \cdot (1 - x_A)^2 \cdot V \tag{5.55}$$

where F_{A_0} is the feed rate of chemical A.
Then,

$$V = \frac{F_{A_0} \cdot C_{A_0} \cdot x_A}{2.0 \cdot C_{A_0}^2 \cdot (1 - x_A)^2} = \frac{F_{A_0} \cdot x_A}{2.0 \cdot C_{A_0} \cdot (1 - x_A)^2} \tag{5.56}$$

The production rate of B is:

$$P_B = \frac{1}{2} F_{A_0} \cdot C_{A_0} \cdot x_A \tag{5.57}$$

Therefore, the optimization problem is:

$$minf = 4.0 \cdot C_{A_0}^{1.4} \cdot F_{A_0} + 0.4 \cdot \left(\frac{F_{A_0} \cdot x_A}{2.0 \cdot C_{A_0} \cdot (1 - x_A)^2} \right)^{0.6} \tag{5.58}$$

$$\text{s.t.} \frac{1}{2} F_{A_0} \cdot C_{A_0} \cdot x_A \geq 50 \tag{5.59}$$

To treat F_{A_0} as x_1, C_{A_0} as x_2, and x_A as x_3, then it becomes:

$$minf(\overline{x}) = 4.0 \cdot x_2^{1.4} \cdot x_1 + 0.4 \cdot \left(\frac{x_1 \cdot x_3}{2.0 \cdot x_2 \cdot (1 - x_3)^2} \right)^{0.6} \tag{5.60}$$

$$\text{s.t.} \ \frac{1}{2} x_1 \cdot x_2 \cdot x_3 \geq 50 \tag{5.61}$$

$$x_3 \leq 1 \tag{5.62}$$

Step 2: To use fmincon to solve the problem
The code is as follows:

```
function [c,ceq]=noncon(x)
c(1)=50-1/2* x(1)*x(2)*x(3);
c(2)=x(3)-1;
ceq=[];
end
clc; clear all; format compact;
fun=@(x)4.0*x(2)^1.4*x(1)+0.4*(((x(1)*x(3))/(2.0*x(2)*...
  (1-x(3))^2)))^0.6;
nonlcon=@noncon;
A=[];
b=[];
Aeq=[];
beq=[];
lb=[0;0;0];
ub=[];
x0=[100;1;0.5];
[x,fval]=fmincon(fun,x0,A,b,Aeq,beq,lb,ub,nonlcon)
```

```
The results are as follows:
Local minimum found that satisfies the constraints.
Optimization completed because the objective function is non
decreasing in feasible directions, to within the value of the
optimality tolerance, and constraints are satisfied to within
the value of the constraint tolerance.
<stopping criteria details>
x =
  587.6146
    0.2383
    0.7143
fval =
  420.6695
```

(2) To use fminsearch with penalty function method

```
u0=1; beta=10; m=0; u=u0*beta^m;
fun = @(x)4.0*x(2)^1.4*x(1)+0.4*(((x(1)*x(3))/(2.0*x(2)*(1-x(3))^2)))^...
 0.6+u*((max(0,50-1/2*x(1)*x(2)*x(3)))^2+(max(0, x(3)-1))^2);
x0 = [100 1 0.5];
[x,ThetaVal] = fminsearch(fun,x0)
f=4.0*x(2)^1.4*x(1)+0.4*(((x(1)*x(3))/(2.0*x(2)*(1-x(3))^2)))^0.6
```

```
The results are as follows:
x =
  532.0985    0.2430    0.7143
ThetaVal =
  406.2229
f =
  391.6595
```

It can be seen in Table 5.8 that the first three simulations were approaching the minimization of the costs for the operation of the CSTR reactor. However, due to the limit of the fminsearch solver and the limitation of computer calculation, the simulations were moving away from the target (the result shown in question 1) and falling into another local optimal.

(3) To use fminunc with penalty function method

```
u0=1; beta=10; m=0; u=u0*beta^m;
```

Table 5.8 Results from using a penalty function with fminsearch(fun,x0) in MATLAB to solve Example 5.8.

u	x_{min}	f_{min}	θ_{min}
1	(532.0985,0.2430,0.7143)	391.6595	406.2229
10	(582.0537,0.2387,0.7143)	417.7994	419.2350
100	(571.1212,0.2436,0.7181)	420.4070	420.5497
1000	(340.1653,0.3749,0.7839)	429.8356	430.1645
10,000	(221.4195,0.5493,0.8222)	450.9968	451.1013
100,000	(163.3346,0.7196,0.8508)	473.0381	473.0381

```
fun = @(x)4.0*x(2)^1.4*x(1)+0.4*(((x(1)*x(3))/(2.0*x(2)*(1-x(3))^2)))^...
0.6+u*((max(0,50-1/2*x(1)*x(2)*x(3)))^2+(max(0, x(3)-1))^2);...
x0 = [100 1 0.5];
[x,ThetaVal] = fminunc(fun,x0)
f=4.0*x(2)^1.4*x(1)+0.4*(((x(1)*x(3))/(2.0*x(2)*(1-x(3))^2)))^0.6
```

```
The results are as follows:
Local minimum possible.
fminunc stopped because it cannot decrease the objective function along the
current search direction.
<stopping criteria details>
x =
   170.8848     0.1092     5.3189
ThetaVal =
   59.8332
f =
   41.0415
```

It can be seen in Table 5.9 that all simulations missed the target (the result shown in question 1) and cannot reach the state of $f_{min} = \theta_{min}$. It means that fminunc does not work at all for this example.

(4) To use fmincon(fun,x0) with penalty function method

```
u0=1; beta=10; m=0; u=u0*beta^m;
fun = @(x)4.0*x(2)^1.4*x(1)+0.4*(((x(1)*x(3))/(2.0*x(2)*(1-x(3))^2)))^...
0.6+u*((max(0,50-1/2*x(1)*x(2)*x(3)))^2+(max(0, x(3)-1))^2);
x0 = [100 1 0.5];
[x,ThetaVal] = fmincon(fun,x0)
f=4.0*x(2)^1.4*x(1)+0.4*(((x(1)*x(3))/(2.0*x(2)*(1-x(3))^2)))^0.6
```

```
The results are as follows:
Local minimum found that satisfies the constraints.
Optimization completed because the objective function is non decreas-
ing in feasible directions, to within the value of the optimality tol-
erance, and constraints are satisfied to within the value of the con-
straint tolerance.
<stopping criteria details>
x =
   170.8849     0.1092     5.3189
ThetaVal =
   59.8332
f =
   41.0414
```

Table 5.9 Results of using a penalty function with fminunc(fun,x0) in MATLAB to solve Example 5.8.

u	x_{min}	f_{min}	θ_{min}
1	(170.8848,0.1092,5.3189)	41.0415	59.8332
10	(204.8748,0.1708,2.8459)	85.7225	120.1689
100	(151.2122,0.3918,1.6875)	183.1117	230.3816
1000	(99.9979,0.8107,1.2334)	328.9867	383.5012
10,000	(99.9982,0.9230,1.0834)	448.3297	517.8578
100,000	(99.9974,0.9640,1.0374)	605.6271	745.3231

Table 5.10 Results of using a penalty function with fmincon(fun,x0) in MATLAB to solve Example 5.8.

u	x_{min}	f_{min}	θ_{min}
1	(170.8848,0.1092,5.3189)	41.0415	59.8332
10	(251.9528,0.1399,2.8331)	85.5799	119.2414
100	(281.5407,0.2073,1.7126)	165.9230	216.7273
1000	1.0e + 11*(−5.2321,0.1259,−0.0381)	−2.8898e + 26	−2.8898e + 26
10,000	1.0e + 14 *(−1.5929,0.0291,−0.0066)	−1.7921e + 32	−1.7921e + 32
100,000	(62.0079,1.5633,1.0316)	618.1677	717.9929

Table 5.10 shows that all simulations missed the target (the result shown in question 1) terribly and cannot reach the state of $f_{min} = \theta_{min}$. Thus, fmincon(fun,x0) with the penalty function method does not work for this example.

Learning Tips

- It will help to build your confidence to use more solvers to your optimization problem to crosscheck your results because any one MATLAB optimization solver can have failures or give wrong results.
- The advantage of converting constrained optimization to unconstrained optimization will be able to use more solvers, even including other global optimization solvers in next chapter.

5.4 Summary

In this chapter, we have presented enhanced optimization methods useful in chemical engineering design and simulation. It is apparent that any one of the solvers in the Optimization Toolbox from MATLAB has its limitations and may not give correct answers for the case of interest. Therefore, developing our readers' capability to tackle more challenging problems is necessary.

The enhanced optimization methods in this chapter include the Lagrange method, inequality constraints converted to equality constraints, the penalty function method, and the enhanced constraints method. These methods can help us use more solvers to our optimization problem to crosscheck our results to avoid failing or wrong results because of relying on any single solver.

The methods learned from this chapter can also be applied to global optimization solvers in Chapter 6.

Problems

5.1

$$min f(\overline{x}) = x_1^3 + 2x_1^2 - 2x_1x_2 + 2x_2^2 - 6x_1 \quad (5.63)$$

$$\text{s.t. } g_1(\overline{x}) = 3x_1^2 + 4x_2 - 6 \le 0 \quad (5.64)$$

$$g_2(\overline{x}) = -x_1^2 + 4x_2 - 2 \le 0 \quad (5.65)$$

(1) To use fmincon in MATLAB to solve the problem; x0=[1 1].
(2) To construct the S function, and then use fminunc in MATLAB to solve the problem.
(3) To construct the S function, and then use fminsearch in MATLAB to solve the problem.

5.2

$$minf(\overline{x}) = (x_1 - 2)^2 - 6ln\left((x_2 - 1)^2 + 1\right) \tag{5.66}$$

$$\text{s.t.}\, g_1(\overline{x}) = x_1^2 + x_2 - 3 \leq 0 \tag{5.67}$$

$$g_2(\overline{x}) = e^{-x_1} - x_2 \leq 0 \tag{5.68}$$

(1) Use fmincon in MATLAB to solve the problem; x0=[2 3].
(2) Use a penalty function with fminunc in MATLAB to solve the problem.
(3) Use a penalty function with fminsearch in MATLAB to solve the problem.

5.3

$$minf(\overline{x}) = x_1^3 - x_2^3 - x_1^2 + x_2^2 \tag{5.69}$$

$$\text{s.t. } g_1(\overline{x}) = x_1^2 + x_2^2 - 9 \leq 0 \tag{5.70}$$

$$g_2(\overline{x}) = -x_2 \leq 0 \tag{5.71}$$

$$g_3(\overline{x}) = x_2 - 3 \leq 0 \tag{5.72}$$

(1) Use fmincon in MATLAB to solve the problem; x0=[0 0].
(2) To move away from the "zero" optimal point, change to

$$g_2(\overline{x}) = -x_2 + x_3^2 \leq 0 \tag{5.73}$$

with x0=[0 0 1], then use fmincon in MATLAB to solve the problem.
(3) Use a penalty function with fminsearch in MATLAB to solve the problem.

6

Global Optimization

Global optimization in this chapter further develops the readers' capability to deal with many challenging optimization problems in which multiple local minima (or maxima) exist to make optimization solvers fail to find the global minima (or maxima). Finding the global optima of a function is far more complicated: analytical methods are frequently not applicable, and using numerical solution strategies often leads to tough challenges.

MATLAB has a built-in Global Optimization Toolbox, with many optimization solvers, such as GlobalSearch, MultiStart, patternsearch, particleswarm, ga, simulannealbnd, and surrogateopt. It differs from previous Chapter 4, the MATLAB Built-in Optimization Toolbox with many optimization solvers, such as fmincon, fminimax, fminsearch, fminunc, fseminf, fsolve, and fzero. Those solvers in Chapter 4 can only provide results of local optima. Even the enhanced optimization methods of Chapter 5 are based on the solvers in Chapter 4. Therefore, their results are limited to local optima. In real chemical engineering problems, multiple local minima (or maxima) exist due to complex chemical and physical phenomena in the system of interest. New methods are needed to avoid falling into local optima and find global optima.

In this chapter, we introduce the Global Optimization Toolbox in MATLAB and the selection of global optimization methods from GlobalSearch, MultiStart, patternsearch, particleswarm, ga, simulannealbnd, surrogateopt, and tabuSearch. The knowledge and skills of the enhanced optimization developed from Chapter 5 are applied to this chapter to expand readers' optimization skills. Doing these can help us use more solvers for our optimization problem to crosscheck our results to avoid failing or getting wrong results because of relying on any single solver.

6.1 Introduction

6.1.1 Local vs. Global Optima

Let us revisit Example 2.6 in Chapter 2 and the plot shown in Figure 6.1. There are three local optima for the function $f(x) = \cos(x) - 2\ln(x)$, with x range (0.01, 20), while there is a global optimum. A local minimum of the function is to compare its nearby points to find a point's function value is minimum. However, it may be greater than at a distant point. A global minimum is to find a point's function value compared to all other feasible points.

As shown in Example 2.6 in Chapter 2, the "fminsearch," "fminunc," and "fmincon" solvers give the local optimal point (3.7108, −3.4648), and the "fminbnd" solver gives a

Chemical Engineering Analysis and Optimization Using MATLAB®, First Edition.
Weiguo Xie, Sam Toan, and Richard Davis.

Companion website: www.wiley.com/go/xie/chemicalengineering

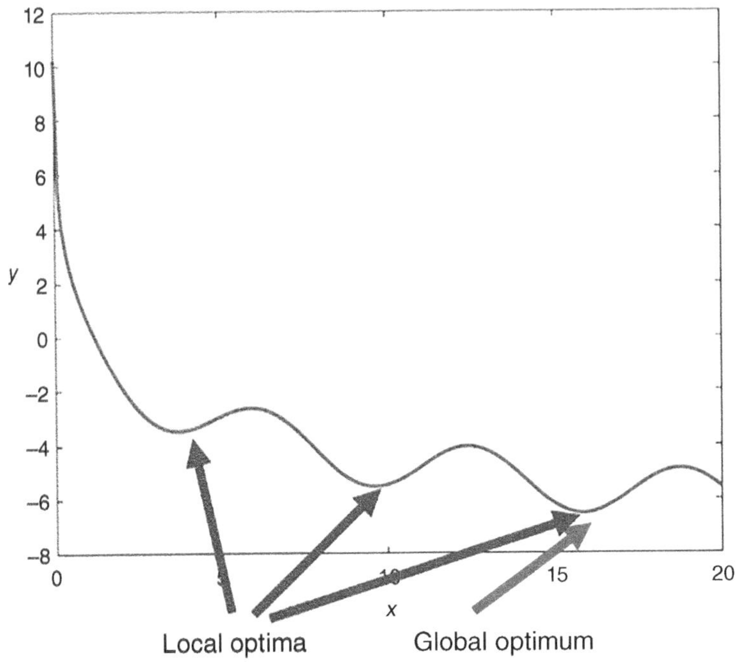

Figure 6.1 There are three local optima and a global optima for the function $f(x) = \cos(x) - 2\ \ln(x)$, with x range (0.01, 20).

global optimal point (15.8346, −6.5164) for this example. However, the "fminbnd" solver can only find the minimum of a single-variable function on a fixed interval, which cannot be used for multivariable optimization. Therefore, it is necessary to explore new methods to find global optima.

Using classical local optimization methods to find an arbitrary local minimum is relatively straightforward. However, finding the global minimum of a function is very challenging because analytical methods are frequently unavailable.

6.1.2 Four Standard Heuristic Techniques (Often Incorporate Randomization)

Since there are no deterministic methods for global optimization, heuristic techniques incorporating randomization will be employed for practical approaches to provide sufficient accurate solutions for the optimization problems. Four standard heuristic techniques are as follows:

(1) Simulated annealing (SA) algorithm

It is an analogy to the annealing process in metallurgy, a technique involving heating and controlled cooling of a material to increase the size of its crystals and reduce their defects. The temperature progressively decreases from an initial positive value to zero. It approximates global optimization in a large search space for an optimization problem. It is handy when the search space is discrete (e.g. the traveling salesman problem).

The algorithm is hill-climbing, except it picks a random move instead of the best one. If the selected move improves the solution, then it is always accepted. Otherwise, the algorithm makes the move anyway with some probability less than 1. The probability decreases exponentially with the "badness" of the move, which is the amount ΔE by which the solution is worsened (i.e. energy is increased.)

Restarting from a significantly better solution is sometimes better than moving from the current state. This process is called restarting of SA. The decision to restart could be based on several criteria:

- based on a fixed number of steps;
- based on whether the current energy is too high compared to the best energy obtained so far;
- restarting randomly.

The simulannealbnd solver implemented in MATLAB finds a minimum of a function using stochastic iterates with bound constraints from a user-supplied starting point. There are no gradients needed for the simulannealbnd solver in MATLAB.

(2) Genetic algorithm (GA)

A GA is a metaheuristic inspired by the process of natural selection that belongs to the larger class of evolutionary algorithms (EA). GAs are commonly used to generate high-quality solutions to optimization and search problems by relying on biologically inspired operators such as mutation, crossover, and selection. GAs are randomized search methods that repeatedly modify a population of individual solutions. At each step, the GA randomly selects individuals from the current population to be parents and uses them to produce the children for the next generation. Over successive generations, the population "evolves" toward an optimal solution.

The GA in MATLAB solves both constrained and unconstrained optimization problems based on natural selection, which drives biological evolution. You can apply the GA to solve various optimization problems that are not well suited for standard optimization algorithms, including problems in which the objective function is discontinuous, nondifferentiable, stochastic, or highly nonlinear. The GA can address mixed integer programming problems, where some variable components are restricted to integers.

The GA uses three main types of rules at each step to create the next generation from the current population:

- Selection rules for the individuals, called parents, who contribute to the next generation's population.
- Crossover rules combine two parents to form children for the next generation.
- Mutation rules apply random changes to individual parents to form children.

The GA differs from a classical, derivative-based optimization algorithm in two main ways, as summarized in the following table.

Classical algorithm	**Genetic algorithm**
Generates a single point at each iteration. The sequence of points approaches an optimal solution.	Generates a population of points at each iteration. The best point in the population approaches an optimal solution.
Selects the next point in the sequence by a deterministic computation.	Selects the next population by computation, which uses random number generators.

(3) Particle swarm optimization (PSO)

PSO is a computational method that optimizes a problem by iteratively trying to improve a candidate solution with regard to a given measure or quality. It solves a problem by having a population of candidate solutions, here dubbed particles, and moving them around in the search space according to simple mathematical formulae over the particle's position and velocity. Each particle's movement is influenced by its local best-known position and guided toward the best-known positions in the search space, updated as other particles find better positions. This is expected to move the swarm toward the best solutions.

PSO is a metaheuristic as it makes few or no assumptions about the problem being optimized and can search vast spaces of candidate solutions. However, metaheuristics such as PSO do not guarantee that an optimal solution is ever found.

PSO does not use the gradient of the problem being optimized, which means PSO does not require a differentiable optimization problem as required by classic optimization methods such as gradient descent and quasi-Newton methods.

The particleswarm solver has been implemented in MATLAB, which can also be applied to multi-objective problems, in which the objective function comparison takes Pareto dominance into account when moving the PSO particles, and non-dominated solutions are stored to approximate the Pareto front.

(4) Tabu search

Tabu search, created by Fred W. Glover in 1986 and formalized in 1989, is a metaheuristic search method employing local (neighborhood) search methods used for mathematical optimization. It enhances the performance of local search by relaxing its fundamental rule. At each step, worsening moves can be accepted if no improving move is available (like when the search is stuck at a strict local minimum).

Prohibitions (henceforth, the term tabu) are introduced to discourage the search from returning to previously visited solutions. Tabu search implementation uses memory structures describing the visited solutions or user-provided rules. If a potential solution has been previously visited within a certain short-term period or violated a rule, it is marked as "tabu" (forbidden) so that the algorithm does not consider that possibility repeatedly.

The memory structures used in Tabu search can roughly be divided into three categories:

- Short term: The list of solutions recently considered. If a potential solution appears on the tabu list, it cannot be revisited until it expires.
- Intermediate term: Intensification rules intended to bias the search towards promising areas of the search space.
- Long term: Diversification rules that drive the search into new regions (i.e. regarding resets when the search becomes stuck in a plateau or a suboptimal dead-end).

The tabuSearch algorithm in MATLAB, implemented since version R2023a, solves a quadratic unconstrained binary optimization (QUBO) problem. It can be used to solve quantum computing problems. It is newly implemented and has narrow application areas currently. Therefore, it will not be used in examples in our book.

6.2 Global Optimization Toolbox in MATLAB

6.2.1 Global Optimization Solvers in MATLAB

Table 6.1 shows the leading global optimization solvers in MATLAB. It is shown in the following table that some of them (such as GlobalSearch, MultiStart, and patternsearch) are based on deterministic iterates. In contrast, others (surrogateopt, particleswarm, ga, and simulannealbnd) are based on stochastic iterates. The solvers such as GlobalSearch and MultiStart need a gradient, but other solvers do not need a gradient.

Table 6.1 Global optimization solvers in MATLAB (https://www.mathworks.com/The MathWorks, Inc.).

Solver	Description	Syntax
GlobalSearch	Find a global minimum using deterministic iterates, gradient-based, automatic stochastic start points, and remove many start points heuristically	• gs = GlobalSearch • gs = GlobalSearch(Name,Value) • gs = GlobalSearch(oldGS,Name,Value) • gs = GlobalSearch(ms)
MultiStart	Find multiple local minima using deterministic iterates, gradient-based, stochastic or deterministic start points, or a combination of both	• ms = MultiStart • ms = MultiStart(Name,Value) • ms = MultiStart(oldMS,Name,Value) • ms = MultiStart(gs)
patternsearch	Find a minimum of a function using deterministic iterates. No gradient is needed, user-supplied start point	• x = patternsearch(fun,x0) • x = patternsearch(fun,x0,A,b) • x = patternsearch(fun,x0,A,b,Aeq,beq) • x = patternsearch(fun,x0,A,b,Aeq,beq,lb,ub) • x = patternsearch(fun,x0,A,b,Aeq,beq,lb, ub,nonlcon) • x = patternsearch(fun,x0,A,b,Aeq,beq, lb,ub,nonlcon,options) • x = patternsearch(problem) • [x,fval] = patternsearch(___) • [x,fval,exitflag,output] = patternsearch(___)
surrogateopt	Find global minimization of time-consuming objective functions using stochastic iterates, requires bound constraints, accepts nonlinear inequality constraints, allows integer constraints; no gradients needed, automatic start points or user-supplied points, or a combination of both	• x = surrogateopt(objconstr,lb,ub) • x = surrogateopt(objconstr,lb,ub,intcon) • x = surroga-teopt(objconstr,lb,ub,intcon,A,b,Aeq,beq) • x = surrogateopt(___,options) • x = surrogateopt(problem) • x = surrogateopt(checkpointFile) • x = surrogateopt(checkpointFile,opts) • [x,fval] = surrogateopt(___) • [x,fval,exitflag,output] = surrogateopt(___) • [x,fval,exitflag,output,trials] = surrogateopt(___)

(Continued)

Table 6.1 (Continued)

Solver	Description	Syntax
particleswarm	Find a vector x that achieves a local minimum of the function using stochastic iterates, population-based, no gradients needed, automatic start population or user-supplied population, or a combination of both only bound constraints	• x = particleswarm(fun,nvars) • x = particleswarm(fun,nvars,lb,ub) • x = particleswarm(fun,nvars,lb,ub,options) • x = particleswarm(problem) • [x,fval,exitflag,output,points] = particleswarm(___)
ga	Find a minimum of a function using stochastic iterates, population-based, no gradients needed, allows integer constraints, automatic start population or user-supplied population, or a combination of both	• x = ga(fun,nvars) • x = ga(fun,nvars,A,b) • x = ga(fun,nvars,A,b,Aeq,beq) • x = ga(fun,nvars,A,b,Aeq,beq,lb,ub) • x = ga(fun,nvars,A,b,Aeq,beq,lb,ub,nonlcon) • x = ga(fun,nvars,A,b,Aeq,beq,lb,ub,nonlcon, options) • x = ga(fun,nvars,A,b,Aeq,beq,lb,ub,nonlcon, intcon) • x = ga(fun,nvars,A,b,Aeq,beq,lb,ub,nonlcon, intcon,options) • x = ga(problem) • [x,fval] = ga(___) • [x,fval,exitflag,output] = ga(___) • [x,fval,exitflag,output,population,scores] = ga(___)
simulannealbnd	Find the minimum of function using stochastic iterates, no gradients needed, user-supplied start point, only bound constraints	• x = simulannealbnd(fun,x0) • x = simulannealbnd(fun,x0,lb,ub) • x = simulannealbnd(fun,x0,lb,ub,options) • x = simulannealbnd(problem) • [x,fval] = simulannealbnd(___) • [x,fval,exitflag,output] = simulannealbnd(___)
tabuSearch	Tabu search algorithm for QUBO (Quadratic Unconstrained Binary Optimization) solve	• ts = tabuSearch • ts = tabuSearch(Name=Value)

More details in MATLAB online materials for global optimization: https://www.mathworks.com/products/global-optimization.html

6.2.2 The Differences in the Syntax of the Solvers

Table 6.1 also shows the syntax of the global optimization solvers in MATLAB. It can be seen that particleswarm and simulannealbnd can be used for optimization with upper and lower bound constraints. The patternsearch, surrogateopt, and ga allow linear/nonlinear and upper and lower bound constraints. GlobalSearch and MultiStart can also be used for optimization with constraints, but they must be specified in the createOptimProblem.

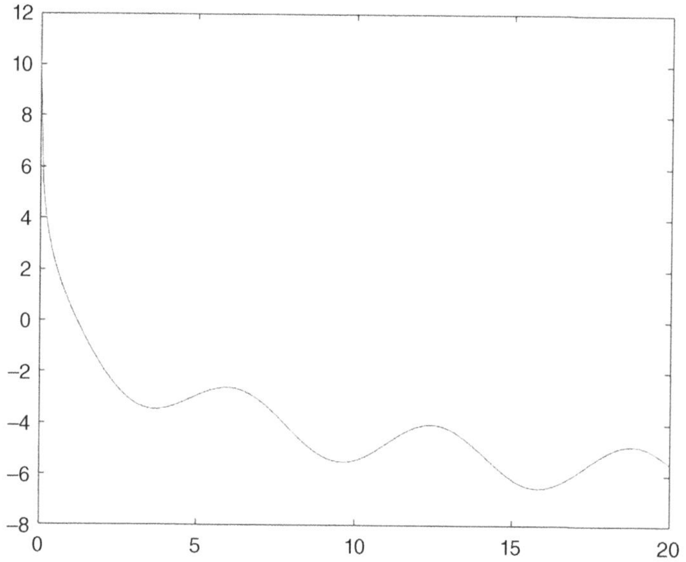

Figure 6.2 The plot of Example 6.1.

Example 6.1

(1)

$$f(x) = \cos(x) - 2\ \ln(x) \tag{6.1}$$

with x range (0.01, 20). Plot the curve of f(x) vs. x

```
Solution:
≫ x=linspace(0.01,20,200);
y=cos(x)-2*log(x);plot(x,y);
```

The plot of Example 6.1 is shown in Figure 6.2.

(2) Use simulannealbnd to find the minimum point with the initial x0=1 and lb and ub.

```
≫ [x, y]=simulannealbnd(@(x)cos(x)-2*log(x), 1, 0.01,20)
The results are as follows:
Optimization terminated: change in best function value less
 than options.FunctionTolerance.
x =
   15.8331
y =
   -6.5164
```

(3) Use ga to find the minimum point with the initial x0=1 and lb and ub.

```
≫ [x, y]=ga(@(x)cos(x)-2*log(x), 1,[],[],[],[], 0.01,20)
The results are as follows:
Optimization terminated: average change in the fitness value
 less than options.FunctionTolerance.
```

```
x =
   15.8346
y =
   -6.5164
```

(4) Use particleswarm to find the minimum point with the initial x0=1 and lb and ub.

```
>> [x, y]= particleswarm(@(x)cos(x)-2*log(x), 1, 0.01,20)
The results are as follows:
Optimization ended: relative change in the objective value over
 the last OPTIONS.MaxStallIterations iterations is less than
 OPTIONS.FunctionTolerance.
x =
   15.8346
y =
   -6.5164
```

(5) Use surrogateopt to find the minimum point with the initial x0=1 and lb and ub.

```
>> [x, y]= surrogateopt(@(x)cos(x)-2*log(x), 0.01,20)
The results are as follows:
Surrogateopt stopped because it exceeded the function evalua-
tion limit set by 'options.MaxFunctionEvaluations'.
x =
   15.8393
y =
   -6.5164
```

The plot of surrogateopt results of Example 6.1 is shown in Figure 6.3.

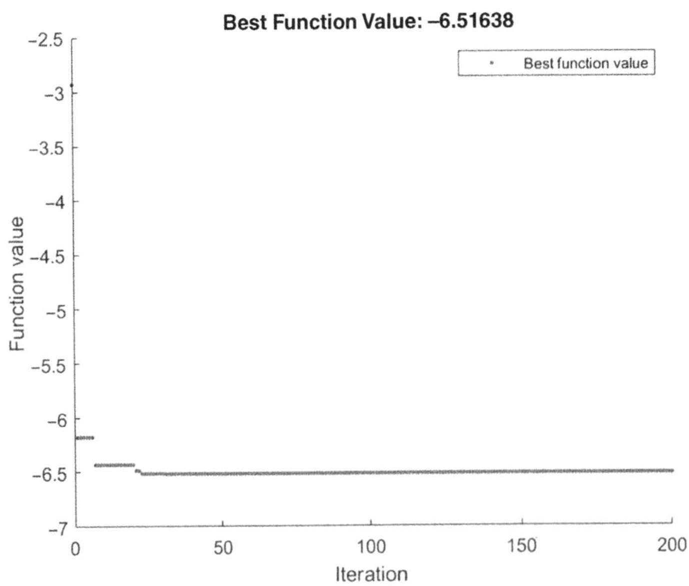

Figure 6.3 The plot of surrogateopt results of Example 6.1.

(6) Use patternsearch to find the minimum point with the initial x0=1.

```
≫ [x, y]= patternsearch(@(x)cos(x)-2*log(x), 1)
The results are as follows:
Optimization terminated: mesh size less than
 options.MeshTolerance.
x =
   15.8346
y =
   -6.5164
```

(7) Use MultiStart to find the minimum point with the initial x0=1 and lb and ub.

```
≫ fun = @(x) cos(x)-2*log(x);
rng default % For reproducibility
opts = optimoptions(@fmincon,'Algorithm','sqp');
problem = createOptimProblem('fmincon','objective',...
    fun,'x0',1,'lb',0.01,'ub',20,'options', opts);
ms = MultiStart;
[x,y] = run(ms,problem,20)

The results are as follows:
MultiStart completed the runs from all start points.
All 20 local solver runs converged with a positive local solver
 exit flag.
x =
   15.8346
y =
   -6.5164
```

(8) Use GlobalSearch to find the minimum point with the initial x0=1 and lb and ub.

```
≫ fun = @(x) cos(x)-2*log(x);
rng default % For reproducibility
gs = GlobalSearch;
problem = createOptimProblem('fmincon','x0',1,...
    'objective',fun,'lb',0.01,'ub',20);
[x,y] = run(gs,problem)

The results are as follows:
GlobalSearch stopped because it analyzed all the trial points.
All 28 local solver runs converged with a positive local solver
 exit flag.
x =
   15.8346
y =
   -6.5164
```

The above example shows that all global optimization solvers can give very close optimal results. It indicates that all of them are very good at single-variable global optimization in this case.

Also note that all optimization solvers except Surrogateopt showed the "value less than options.FunctionTolerance," which means that if we use "exitflag" as output after the x and y value, the exitflag equals 1. It indicates that the optimization has been converged to satisfy the stop criteria. Surrogateopt stopped because it exceeded the function evaluation limit set by 'options.MaxFunctionEvaluations'. If we use "exitflag" as output after the x and y value using Surrogateopt, the exitflag will equal 0. It indicates that the number of function evaluations exceeded options. MaxFunctionEvaluations is different from other global optimization solvers.

6.3 The Selection of Global Optimization Methods

6.3.1 Based on Constraints

Table 6.1 shows that different global optimization solvers may work for different constraint types. The particleswarm and simulannealbnd can be used for optimization with upper and lower bound constraints. The patternsearch, surrogateopt, and ga allow linear/nonlinear and upper and lower bound constraints. GlobalSearch and MultiStart can also be used for optimization with constraints, but they must be specified in the createOptimProblem. As far as a simple case like the above Example 6.1 is concerned, all global optimization solvers work and provide correct results, with only differences coming from different syntaxes.

However, some solvers (particleswarm and simulannealbnd) cannot be used directly for the cases with linear/nonlinear constraints in Chapter 5, such as Example 4.3, Example 4.4, Example 4.5, and Example 5.3. We use Example 5.3 here.

Example 6.2

(1)

$$min f(\overline{x}) = 2x_1^2 - 2x_1x_2 + 2x_2^2 - 6x_1 \tag{6.2}$$

$$\text{s.t. } g_1(\overline{x}) = 3x_1 + 4x_2 - 6 \le 0 \tag{6.3}$$

$$g_2(\overline{x}) = -x_1 + 4x_2 - 2 \le 0 \tag{6.4}$$

To use fmincon in MATLAB to solve the problem, x0=[0 0];

```
Solution:
≫ fun = @(x)2*x(1)^2-2*x(1)*x(2)+2*x(2)^2-6*x(1);
≫ x0 = [0,0];
≫ [x,fval] = fmincon(fun,x0,[3 4;-1 4],[6;2])
The results are as follows:
Local minimum found that satisfies the constraints.
Optimization completed because the objective function is non
  decreasing in feasible directions, to within the value of the
  optimality tolerance, and constraints are satisfied to within
  the value of the constraint tolerance.
```

```
<stopping criteria details>
x =
    1.4595    0.4054
fval =
   -5.3514
```

(2) To use ga in MATLAB to solve the problem, x0=[0 0];

```
Solution:
≫ fun = @(x)2*x(1)^2-2*x(1)*x(2)+2*x(2)^2-6*x(1);
≫ x0 = [0,0];
≫ [x,fval]  = ga(fun,2,[3 4;-1 4],[6;2],[],[],[],[])
The results are as follows:
Optimization terminated: average change in the fitness value
 less than options.FunctionTolerance.
x =
    1.4641    0.4021
fval =
   -5.3516
```

(3) To use patternsearch in MATLAB to solve the problem, x0=[0 0];

```
Solution:
≫ fun = @(x)2*x(1)^2-2*x(1)*x(2)+2*x(2)^2-6*x(1);
≫ x0 = [0,0];
≫ [x,fval]  = patternsearch(fun, x0,[3 4;-1 4],[6;2],[],[],...
 [],[])
The results are as follows:
Optimization terminated: mesh size less than
  options.MeshTolerance.
x =
    1.4596    0.4056
fval =
   -5.3517
```

(4) To use surrogateopt in MATLAB to solve the problem, x0=[0 0]. The surrogateopt needs lower and upper bound, i.e. lb=[-10;-10] and ub=[10; 10].

```
Solution:
≫ fun = @(x)2*x(1)^2-2*x(1)*x(2)+2*x(2)^2-6*x(1);
≫ ≫ [x,fval]  = surrogateopt(fun, [-10; -10],[10; 10],[],...
 [3 4;-1 4],[6;2])
The results are as follows:
 surrogateopt stopped because it exceeded the function
 evaluation limit set by 'options.MaxFunctionEvaluations'.
x =
    1.4592    0.4056
fval =
   -5.3514
```

The plot of surrogateopt results of Example 6.2 is shown in Figure 6.4.

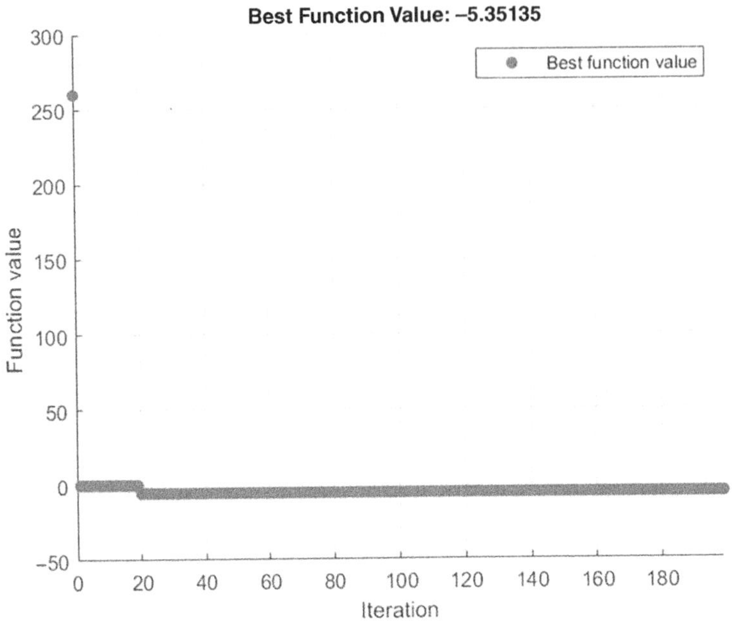

Figure 6.4 The plot of surrogateopt results of Example 6.2.

(5) To use MultiStart in MATLAB to solve the problem, x0=[0 0];

```
Solution:
≫ fun = @(x)2*x(1)^2-2*x(1)*x(2)+2*x(2)^2-6*x(1);
≫ x0 = [0,0];
≫ rng default % For reproducibility
≫ opts = optimoptions(@fmincon,'Algorithm','sqp');
≫ problem = createOptimProblem('fmincon','objective',fun,...
 'x0',x0, 'Aineq',[3 4;-1 4], 'bineq',[6;2]);
≫ ms = MultiStart;
≫ [x,y] = run(ms,problem,20)
```

```
The results are as follows:
MultiStart completed the runs from all start points.
All 20 local solver runs converged with a positive local solver
 exit flag.
x =
    1.4595    0.4054

y =
   -5.3514
```

(6) To use GlobalSearch in MATLAB to solve the problem, x0=[0 0];

```
Solution:
≫ fun = @(x)2*x(1)^2-2*x(1)*x(2)+2*x(2)^2-6*x(1);
≫ x0 = [0,0];
```

```
≫ rng default % For reproducibility
≫ opts = optimoptions(@fmincon,'Algorithm','sqp');
≫ problem = createOptimProblem('fmincon','objective',fun,...
'x0', x0, 'Aineq',[3 4;-1 4], 'bineq',[6;2]);
≫ ms = GlobalSearch;
≫ [x,y] = run(ms,problem)

The results are as follows:
GlobalSearch stopped because it analyzed all the trial points.
All two local solver runs converged with a positive local solver
 exit flag.
x =
    1.4595    0.4054
y =
   -5.3514
```

It can be seen that all five solvers (patternsearch, surrogateopt, ga, MultiStart, and GlobalSearch), which can allow linear/nonlinear constraints, provide the correct answers for Example 6.2. The answers are slightly different among them.

Learning Tips

- Due to incorporating randomization in all kinds of global optimization solvers, every run of the code may possible provide different answers or slightly different answers.
- You can use enhanced optimization in previous chapter to expand the existing MATLAB global optimization solvers' capability, not restricted by constraint requirements.

6.3.2 Based on the Number of Variables

It can be seen in Example 6.1 that all global optimization solvers can give very close optimal results for single variable optimization. To check multivariable optimization, let us optimize the S function in Example 5.2 to see how global optimization solvers perform.

Example 6.3

(1) To use ga to find the minimum point of the S function.

Solution:

```
≫ [x] = ga (@(x) ((2*(x(1)-2)-2*x(3))^2+(2*x(2)-x(3))^2+...
 (2*x(1)+x(2)-2)^2), 3)
The results are as follows:
Optimization terminated: average change in the fitness value
 less than options.FunctionTolerance.
x =
    1.3533   -0.9048   -0.6192
```

***Note**: You may get different answers for different runs.

(2) To use patternsearch to find the minimum point of the S function with the initial x0=[1 1 1].

```
Solution:
≫ [x] = patternsearch (@(x) ((2*(x(1)-2)-2*x(3))^2+(2*...
x(2)-x(3))^2+(2*x(1)+x(2)-2)^2), [1 1 1])
The results are as follows:
Optimization terminated: mesh size less than
 options.MeshTolerance.
x =
    1.2000    -0.4000    -0.8000
```

(3) To use surrogateopt to find the minimum point of the S function with the lb=[-10 -10 -10] and ub=[10 10 10].

```
Solution:
≫ [x] = surrogateopt (@(x) ((2*(x(1)-2)-2*x(3))^2+(2*...
x(2)-x(3))^2+(2*x(1)+x(2)-2)^2), [-10 -10 -10],[10 10 10])
The results are as follows:
Surrogateopt stopped because it exceeded the function evalua-
tion limit set by 'options.MaxFunctionEvaluations'.
x =
    1.1998    -0.4003    -0.8006
```

The plot of surrogateopt results of Example 6.3 is shown in Figure 6.5.

(4) To use simulannealbnd to find the minimum point of the S function with the initial x0=[1 1 1], the lb=[-10 -10 -10] and ub=[10 10 10].

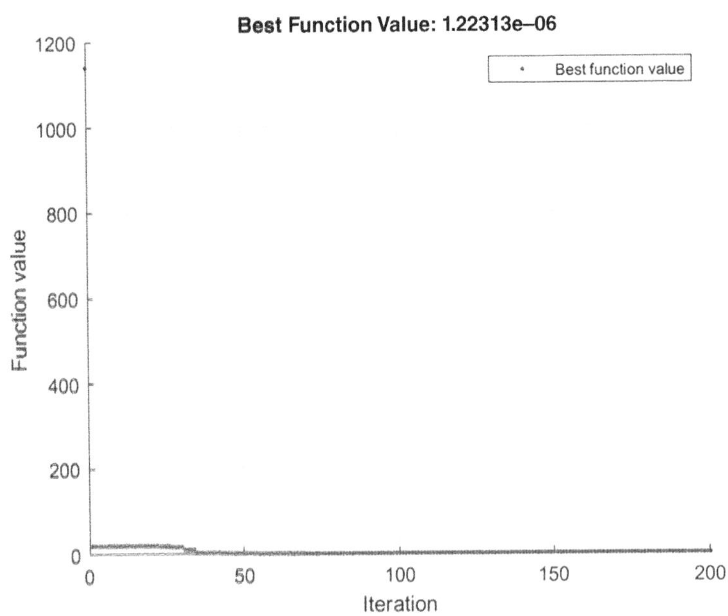

Figure 6.5 The plot of surrogateopt results of Example 6.3.

Solution:

```
≫ [x] = simulannealbnd (@(x) ((2*(x(1)-2)-2*x(3))^2+(2*...
x(2)-x(3))^2+(2*x(1)+x(2)-2)^2), [1 1 1], [-10 -10 -10],
 [10 10 10])
The results are as follows:
Optimization terminated: change in best function value less
 than options.FunctionTolerance.
x =
    1.2011   -0.4006   -0.7995
```

***Note**: You may get different answers for different runs.

(5) To use particleswarm to find the minimum point of the S function with the lb=[-10 -10 -10] and ub=[10 10 10].

Solution:

```
≫ [x] = particleswarm (@(x) ((2*(x(1)-2)-2*x(3))^2+(2*...
x(2)-x(3))^2+(2*x(1)+x(2)-2)^2), 3, [-10 -10 -10],[10 10 10])
The results are as follows:
Optimization ended: relative change in the objective value over
 the last OPTIONS.MaxStallIterations iterations is less than
 OPTIONS.FunctionTolerance.
x =
    1.2000   -0.4000   -0.8000
```

(6) To use MultiStart to find the minimum point of the S function with the initial x0=[1 1 1], the lb=[-10 -10 -10] and ub=[10 10 10].

Solution:

```
≫ fun = @(x) ((2*(x(1)-2)-2*x(3))^2+(2*x(2)-x(3))^2+(2*x(1)+...
 x(2)-2)^2);
≫ x0 =[1 1 1];
≫ rng default % For reproducibility
≫ opts = optimoptions(@fmincon,'Algorithm','sqp');
≫ problem = createOptimProblem('fmincon','objective',fun,...
 'x0',x0,'lb', [-10 -10 -10], 'ub', [10 10 10]);
≫ ms = MultiStart;
≫ [x] = run(ms,problem,20)

The results are as follows:
MultiStart completed the runs from all start points.
All 20 local solver runs converged with a positive local solver
 exit flag.
x =
 1.2000   -0.4000   -0.8000
```

(7) To use GlobalSearch to find the minimum point of the S function with the initial x0=[1 1 1], the lb=[-10 -10 -10] and ub=[10 10 10].

Solution:

```
≫ fun = @(x) ((2*(x(1)-2)-2*x(3))^2+(2*x(2)-x(3))^2+(2*...
x(1)+x(2)-2)^2);
≫ x0 =[1 1 1];
≫ rng default % For reproducibility
≫ opts = optimoptions(@fmincon,'Algorithm','sqp');
≫ problem = createOptimProblem('fmincon','objective',fun,...
 'x0',x0,'lb', [-10 -10 -10], 'ub', [10 10 10]);
≫ ms = GlobalSearch;
≫ [x] = run(ms,problem)
```

```
The results are as follows:
GlobalSearch stopped because it analyzed all the trial points.
All three local solver runs converged with a positive local
 solver exit flag.
x =
    1.2000   -0.4000   -0.8000
```

It can be seen from Example 6.3 that ga with three variables, surrogateopt and simulannealbnd could not provide accurate optimal results. In contrast, patternsearch, particleswarm, MultiStart, and GlobalSearch can achieve accurate optimal results. Of course, it just shows the results from this specific case, which will not cover all other kinds of cases. Because MultiStart and GlobalSearch are gradient-based solvers, they may not apply to optimization problems without gradients.

6.3.3 Based on Computational Time

As our readers may have noticed that different optimization solvers took different computational times to obtain their results, we will use an example to show their computational time. It can be a significant, time-consuming step to optimize large-scale systems. Let us optimize the *S* function in Example 5.3 to show different computational times and see how global optimization solvers perform.

Example 6.4

(1) To use ga to find the minimum point of the *S* function.

Solution:

```
≫ tic;
fun = @(x) ((4*x(1)-2*x(2)-6-3*x(5)+x(6))^2+(-2*x(1)+4*x(2)-...
 4*x(5)-4*x(6))^2+(2*x(5)*x(3))^2+(2*x(6)*x(4))^2+(3*x(1)+...
 4*x(2)-6+x(3)^2)^2+(-x(1)+4*x(2)-2+x(4)^2)^2);
x  = ga(fun,6)
time=toc
```

```
The results are as follows:
Optimization terminated: average change in the fitness value
 less than options.FunctionTolerance.
```

```
x =
     1.4609     0.3162     0.4687     1.4567    -0.2747    -0.1024
time =
     0.3208
```

***Note**: You may get different answers for different runs.

(2) To use patternsearch to find the minimum point of the *S* function with the initial x0= [0 0 1 1 1 1].

Solution:

```
>> tic;
fun = @(x) ((4*x(1)-2*x(2)-6-3*x(5)+x(6))^2+(-2*x(1)+4*x(2)-...
  4*x(5)-4*x(6))^2+(2*x(5)*x(3))^2+(2*x(6)*x(4))^2+(3*x(1)+...
  4*x(2)-6+x(3)^2)^2+(-x(1)+4*x(2)-2+x(4)^2)^2);
x0 = [0 0 1 1 1 1];
x  = patternsearch (fun,x0)
time=toc
```

```
The results are as follows:
Optimization terminated: mesh size less than
 options.MeshTolerance.
x =
     1.0000     0.7500          0          0    -0.8125     1.0625
time =
     0.2021
```

(3) To use surrogateopt to find the minimum point of the *S* function with the lb=[-10 -10 -10 -10 -10 -10] and ub=[10 10 10 10 10 10].

Solution:

```
>> tic;
fun = @(x) ((4*x(1)-2*x(2)-6-3*x(5)+x(6))^2+(-2*x(1)+4*x(2)-...
  4*x(5)-4*x(6))^2+(2*x(5)*x(3))^2+(2*x(6)*x(4))^2+(3*x(1)+...
  4*x(2)-6+x(3)^2)^2+(-x(1)+4*x(2)-2+x(4)^2)^2);
lb=[-10 -10 -10 -10 -10 -10];
ub=[10 10 10 10 10 10];
x  = surrogateopt (fun,lb,ub)
time=toc
```

```
The results are as follows:
Surrogateopt stopped because it exceeded the function evalua-
tion limit set by 'options.MaxFunctionEvaluations'.
x =
     1.3686     0.2003     1.1826     1.4513    -0.2098    -0.2017
time =
   46.0502
```

The plot of surrogateopt results of Example 6.4 is shown in Figure 6.6.

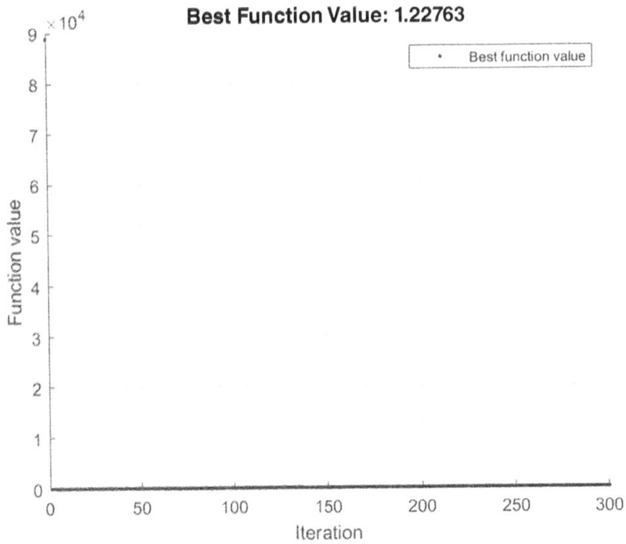

Figure 6.6 The plot of surrogateopt results of Example 6.4.

(4) To use simulannealbnd to find the minimum point of the *S* function with the initial x0=[0 0 1 1 1 1], the lb=[-10 -10 -10 -10 -10 -10] and ub=[10 10 10 10 10 10].

```
Solution:
>> tic;
fun = @(x) ((4*x(1)-2*x(2)-6-3*x(5)+x(6))^2+(-2*x(1)+4*x(2)-...
 4*x(5)-4*x(6))^2+(2*x(5)*x(3))^2+(2*x(6)*x(4))^2+(3*x(1)+...
 4*x(2)-6+x(3)^2)^2+(-x(1)+4*x(2)-2+x(4)^2)^2);
x0 = [0 0 1 1 1 1];
lb=[-10 -10 -10 -10 -10 -10];
ub=[10 10 10 10 10 10];
x  = simulannealbnd (fun,x0,lb,ub)
time=toc

The results are as follows:
Optimization terminated: change in best function value less
 than options.FunctionTolerance.
x =
    1.4473    0.4267    0.2133    1.2780   -0.3215    0.0204
time =
    0.8797
```

***Note**: You may get different answers for different runs.

(5) To use particleswarm to find the minimum point of the S function with the lb=[-10 -10 -10 -10 -10 -10] and ub=[10 10 10 10 10 10].

```
Solution:
>> tic;
```

```
fun = @(x) ((4*x(1)-2*x(2)-6-3*x(5)+x(6))^2+(-2*x(1)+4*x(2)-...
 4*x(5)-4*x(6))^2+(2*x(5)*x(3))^2+(2*x(6)*x(4))^2+(3*x(1)+...
 4*x(2)-6+x(3)^2)^2+(-x(1)+4*x(2)-2+x(4)^2)^2);
lb=[-10 -10 -10 -10 -10 -10];
ub=[10 10 10 10 10 10];
x = particleswarm (fun,6,lb,ub)
time=toc
```

```
The results are as follows:
Optimization ended: relative change in the objective value
 over the last OPTIONS.MaxStallIterations iterations is
 less than OPTIONS.FunctionTolerance.
x =
    1.4594    0.4054   -0.0001   -1.3556   -0.3244    0.0001
time =
    1.4503
```

(6) To use MultiStart (to run 20 starting points) to find the minimum point of the S function with the initial x0=[0 0 1 1 1 1], the lb=[-10 -10 -10 -10 -10 -10] and ub=[10 10 10 10 10 10].

Solution:

```
>> tic;
fun = @(x) ((4*x(1)-2*x(2)-6-3*x(5)+x(6))^2+(-2*x(1)+4*x(2)-...
 4*x(5)-4*x(6))^2+(2*x(5)*x(3))^2+(2*x(6)*x(4))^2+(3*x(1)+...
 4*x(2)-6+x(3)^2)^2+(-x(1)+4*x(2)-2+x(4)^2)^2);
x0 = [0 0 1 1 1 1];
lb=[-10 -10 -10 -10 -10 -10];
ub=[10 10 10 10 10 10];
rng default % For reproducibility
opts = optimoptions(@fmincon,'Algorithm','sqp');
problem = createOptimProblem('fmincon','objective',fun,...
 'x0',x0, 'lb', lb, 'ub', ub);
ms = MultiStart;
x = run(ms,problem,20)
time=toc
```

```
The results are as follows:
MultiStart completed the runs from all start points.
All 20 local solver runs converged with a positive local
 solver exit flag.
x =
    1.4595    0.4054   -0.0000   -1.3557   -0.3243    0.0000
time =
    2.2199
```

(7) To use GlobalSearch to find the minimum point of the S function with the initial x0=[0 0 1 1 1 1], the lb=[-10 -10 -10 -10 -10 -10] and ub=[10 10 10 10 10 10].

Solution:

```
>> tic;
fun = @(x) ((4*x(1)-2*x(2)-6-3*x(5)+x(6))^2+(-2*x(1)+4*x(2)-...
 4*x(5)-4*x(6))^2+(2*x(5)*x(3))^2+(2*x(6)*x(4))^2+(3*x(1)+...
 4*x(2)-6+x(3)^2)^2+(-x(1)+4*x(2)-2+x(4)^2)^2);
x0 = [0 0 1 1 1 1];
lb=[-10 -10 -10 -10 -10 -10];
ub=[10 10 10 10 10 10];
rng default % For reproducibility
opts = optimoptions(@fmincon,'Algorithm','sqp');
problem = createOptimProblem('fmincon','objective',fun,...
 'x0',x0, 'lb', lb, 'ub', ub);
ms = GlobalSearch;
x = run(ms,problem)
time=toc

The results are as follows:
GlobalSearch stopped because it analyzed all the trial
 points.
All four local solver runs converged with a positive local
 solver exit flag.
x =
    1.4595    0.4054    0.0000    1.3557   -0.3243   -0.0000
time =
    2.2298
```

It can be seen from Example 6.4 with six variables that ga, patternsearch, surrogateopt, and simulannealbnd could not provide accurate optimal results. However, particleswarm, MultiStart, and GlobalSearch can achieve accurate optimal results. Of course, it just shows the results from this specific case, which will not cover all other kinds of cases. It can also be seen that some solvers (Surrogateopt, particleswarm, MultiStart, and GlobalSearch) took more than one second to finish the optimization, while other solvers took less than one second. MultiStart and GlobalSearch are gradient-based solvers, so they may not apply to optimization problems without gradients. Therefore, particleswarm can be a good choice because it can provide accurate optimization results and take moderate time.

6.4 Application in Chemical Engineering

6.4.1 Optimization of a Three-Stage Compressor

Example 6.5

In a three-stage compressor, which is adiabatically reversible, one needs to optimize the discharge pressures from the first and second stages to minimize the total work needed. For a given diatomic gas and final pressure ($P_3 = 80$ bar), the optimal pressures are obtained as the solution to the following problem with a constraint $P_1 + P_2 \geq 50$:

$$minf(P_1, P_2) = P_1^{\frac{2}{7}} + \left(\frac{P_2}{P_1}\right)^{\frac{2}{7}} + \left(\frac{P_3}{P_2}\right)^{\frac{2}{7}} \tag{6.5}$$

$$\text{s.t. } g(P_1, P_2) = P_1 + P_2 \geq 50 \tag{6.6}$$

where P_1 and P_2 are the discharge pressures from the first and the second stages, respectively. Determine the optimal discharge pressures from the first and the second stages. The initial value $P_1 = 1$; $P_2 = 1$.

(1) To use ga to solve the problem

Solution:

```
% Example 6.5
clc; clear all; format compact;
P=80;
fun=@(x)x(1)^(2/7)+(x(2)/x(1))^(2/7)+(P/x(2))^(2/7);
A=[-1 -1];
B=[-50];
[x,fval]=ga(fun,2,A,B)
```

```
The results are as follows:
Optimization terminated: maximum number of generations exceeded.
x =
   10.8456   39.3536
fval =
    4.6459
```

(2) To use patternsearch to solve the problem

Solution:

```
% Example 6.5
clc; clear all; format compact;
P=80;
fun=@(x)x(1)^(2/7)+(x(2)/x(1))^(2/7)+(P/x(2))^(2/7);
x0=[1 1];
A=[-1 -1];
B=[-50];
[x,fval]= patternsearch (fun,x0,A,B)
```

```
The results are as follows:
Optimization terminated: mesh size less than
 options.MeshTolerance.
x =
    7.1005   42.8986
fval =
    4.6175
```

(3) To use MultiStart to solve the problem

Solution:

```
% Example 6.5
clc; clear all; format compact;
P=80;
fun=@(x)x(1)^(2/7)+(x(2)/x(1))^(2/7)+(P/x(2))^(2/7);
x0=[1 1];
A=[-1 -1];
B=[-50];
rng default % For reproducibility
opts = optimoptions(@fmincon,'Algorithm','sqp');
problem = createOptimProblem('fmincon','objective',fun,...
 'x0',x0, 'Aineq', A, 'bineq', B);
ms = MultiStart;
[x,fval] = run(ms,problem,20)
```

```
The results are as follows:
MultiStart completed some of the runs from the start points.
9 out of 20 local solver runs converged with a positive local
 solver exit flag.
x =
    7.1005   42.8995
fval =
    4.6175
```

(4) To use GlobalSearch to solve the problem

Solution:

```
% Example 6.5
clc; clear all; format compact;
P=80;
fun=@(x)x(1)^(2/7)+(x(2)/x(1))^(2/7)+(P/x(2))^(2/7);
x0=[1 1];
A=[-1 -1];
B=[-50];
rng default % For reproducibility
opts = optimoptions(@fmincon,'Algorithm','sqp');
problem = createOptimProblem('fmincon','objective',fun,...
 'x0',x0, 'Aineq', A, 'bineq', B);
ms = GlobalSearch;
[x,fval] = run(ms,problem)
```

```
The results are as follows:
GlobalSearch stopped because it analyzed all the trial
 points.
```

```
6 out of 22 local solver runs converged with a positive
 local solver exit flag.
x =
    7.1005    42.8995
fval =
    4.6175
```

(5) To construct the S function and perform optimization using particleswarm.

Solution:

Step 1: Use vector form, replace P_1 using x_1, and replace P_2 using x_2

$$f(\bar{x}) = x_1^{\frac{2}{7}} + \left(\frac{x_2}{x_1}\right)^{\frac{2}{7}} + \left(\frac{80}{x_2}\right)^{\frac{2}{7}} \text{ Subject to } g(\bar{x}) = 50 - x_1 - x_2 \leq 0 \tag{6.7}$$

Step 2: Convert inequality constraints to equality constraints

$$h_1(\bar{x}) = 50 - x_1 - x_2 + x_3^2 = 0 \tag{6.8}$$

Step 3: Create a Lagrange function

$$L(\bar{x}, \lambda_1) = f(\bar{x}) - \lambda_1 h(\bar{x}) = x_1^{\frac{2}{7}} + \left(\frac{x_2}{x_1}\right)^{\frac{2}{7}} + \left(\frac{80}{x_2}\right)^{\frac{2}{7}} - \lambda_1\left(50 - x_1 - x_2 + x_3^2\right) \tag{6.9}$$

Step 4: Construct S function

$$\begin{aligned} s &= (\nabla_{\bar{x}} L)^2 + (\nabla_{\bar{\lambda}} L)^2 = \sum_{i=1}^{n}\left(\frac{\partial L}{\partial x_i}\right)^2 + \sum_{j=1}^{l}(h_j(\bar{x}))^2 \\ &= \left(\frac{2}{7}(x_1)^{-\frac{5}{7}} - \frac{2}{7}x_2^{\frac{2}{7}}x_1^{-\frac{9}{7}} + \lambda_1\right)^2 + \left(\frac{2}{7}(x_2)^{-\frac{5}{7}}(x_1)^{-\frac{2}{7}} - \frac{2}{7}\cdot 80^{\frac{2}{7}}\cdot x_2^{-\frac{9}{7}} + \lambda_1\right)^2 \\ &\quad + (2\lambda_1 x_3)^2 + \left(50 - x_1 - x_2 + x_3^2\right)^2 \end{aligned} \tag{6.10}$$

To treat λ_1 as x_4 and use particleswarm to find the minimum point of the S function with lb=[0 0 -10 -10] and ub=[10 50 10 10]. Then, to calculate $f(\bar{x})$ using x_1 and x_2.

```
>> [x] = particleswarm (@(x) (2/7*x(1)^(-5/7)-2/7*x(2)^(2/7)*...
 x(1)^(-9/7)+x(4))^2+(2/7*x(2)^(-5/7)* x(1)^(-2/7) -2/7*...
80^(2/7)*x(2)^(-9/7)+x(4))^2+(2*x(4)*x(3))^2+(50-x(1)-...
x(2)+ x(3)^2)^2, 4, [0 0 -10 -10], [10 50 10 10])
>> f= x(1)^(2/7)+(x(2)/x(1))^(2/7)+(80/x(2))^(2/7)
```

```
The results are as follows:
Optimization ended: relative change in the objective value
 over the last OPTIONS.MaxStallIterations iterations is
 less than OPTIONS.FunctionTolerance.
x =
    7.9994    42.0031    -0.0499    -0.0051
f =
    4.6196
```

(6) To construct the S function and perform optimization using simulannealbnd.

Solution:

Based on question (5), use simulannealbnd to find the minimum point of the S function with the initial x0=[1 1 0 0], lb=[0 0 -10 -10] and ub=[10 50 10 10]. Then, to calculate $f(\overline{x})$ using x_1 and x_2.

```
>> [x] = simulannealbnd (@(x) (2/7*x(1)^(-5/7)-2/7*x(2)^(2/7)*...
x(1)^(-9/7)+x(4))^2+(2/7*x(2)^(-5/7)* x(1)^(-2/7) -2/7*...
80^(2/7)*x(2)^(-9/7)+x(4))^2+(2*x(4)*x(3))^2+(50-x(1)-...
x(2)+ x(3)^2)^2, [1 1 0 0] , [0 0 -10 -10], [10 50 10 10])
>> f= x(1)^(2/7)+(x(2)/x(1))^(2/7)+(80/x(2))^(2/7)
```

```
The results are as follows:
Optimization terminated: change in best function value less
than options.FunctionTolerance.
x =
    7.2696   42.7644   -0.1811   -0.0035
f =
    4.6176
```

(7) To construct the S function and perform optimization using surrogateopt.

Based on question (5), use surrogateopt to find the minimum point of the S function with lb=[0 0 -10 -10] and ub=[10 50 10 10]. Then, to calculate $f(\overline{x})$ using x_1 and x_2.

Solution:

```
>> [x] = surrogateopt (@(x) (2/7*x(1)^(-5/7)-2/7*x(2)^(2/7)*...
x(1)^(-9/7)+x(4))^2+(2/7*x(2)^(-5/7)* x(1)^(-2/7) -2/7*...
80^(2/7)*x(2)^(-9/7)+x(4))^2+(2*x(4)*x(3))^2+(50-x(1)-...
x(2)+ x(3) ^2)^2, [0 0 -10 -10], [10 50 10 10])
>> f= x(1)^(2/7)+(x(2)/x(1))^(2/7)+(80/x(2))^(2/7)
```

```
The results are as follows:
Objective or constraint function returned NaN.
Surrogateopt stopped because it exceeded the function evalua-
tion limit set by 'options.MaxFunctionEvaluations'.
x =
    6.0088   44.0115    0.1437    0.0008
f =
  4.6217
```

The plot of surrogateopt results of Example 6.5 is shown in Figure 6.7.

(8) To construct the S function and perform optimization using MultiStart.

Solution:

Based on question (5), use MultiStart to find the minimum point of the S function with the initial x0=[1 1 0 0], lb=[0 0 -10 -10] and ub=[10 50 10 10]. Then, to calculate $f(\overline{x})$ using x_1 and x_2.

```
>> % Example 6.5
```

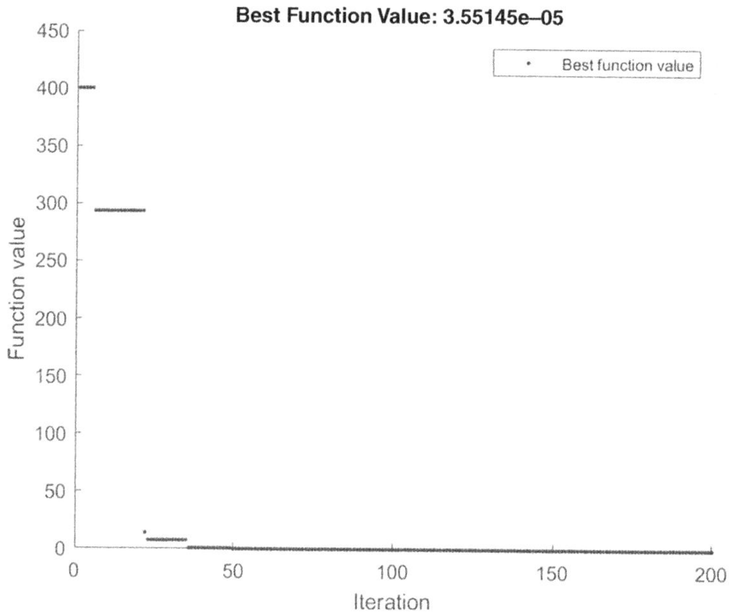

Figure 6.7 The plot of surrogateopt results of Example 6.5.

```
clc; clear all; format compact;
fun=@(x) (2/7*x(1)^(-5/7)-2/7*x(2)^(2/7)*x(1)^(-9/7)+x(4))^...
2+(2/7*x(2)^(-5/7)* x(1)^(-2/7) -2/7*80^(2/7)*x(2)^(-9/7)+...
x(4))^2+(2*x(4)*x(3))^2+(50-x(1)-x(2)+x(3)^2)^2;
x0= [1 1 0 0];
lb=[0 0 -10 -10];
ub=[10 50 10 10];
rng default % For reproducibility
opts = optimoptions(@fmincon,'Algorithm','sqp');
problem = createOptimProblem('fmincon','objective',fun,'x0',...
 x0, 'lb', lb, 'ub', ub);
ms = MultiStart;
x = run(ms,problem,20)
≫ f= x(1)^(2/7)+(x(2)/x(1))^(2/7)+(80/x(2))^(2/7)
```

```
The results are as follows:
MultiStart completed the runs from all start points.
All 20 local solver runs converged with a positive local
 solver exit flag.
x =
    7.1001    42.8999    -0.0014    -0.0032
f =
    4.6175
```

(9) To construct the S function and perform optimization using GlobalSearch.

Solution:

Based on question (5), use GlobalSearch to find the minimum point of the S function with the initial x0=[1 1 0 0], lb=[0 0 -10 -10] and ub=[10 50 10 10]. Then, to calculate $f(\overline{x})$ using x_1 and x_2.

```
≫ % Example 6.5
clc; clear all; format compact;
fun=@(x) (2/7*x(1)^(-5/7)-2/7*x(2)^(2/7)*x(1)^(-9/7)+x(4))^...
2+(2/7*x(2)^(-5/7)* x(1)^(-2/7) -2/7*80^(2/7)*x(2)^(-9/7)+...
x(4))^2+(2*x(4)*x(3))^2+(50-x(1)-x(2)+x(3)^2)^2;
x0= [1 1 0 0];
lb=[0 0 -10 -10];
ub=[10 50 10 10];
rng default % For reproducibility
opts = optimoptions(@fmincon,'Algorithm','sqp');
problem = createOptimProblem('fmincon','objective',fun,'x0',...
 x0, 'lb', lb, 'ub', ub);
ms = GlobalSearch;
x = run(ms,problem)
≫ f= x(1)^(2/7)+(x(2)/x(1))^(2/7)+(80/x(2))^(2/7)
```

```
The results are as follows:
GlobalSearch stopped because it analyzed all the trial
 points.
All 10 local solver runs converged with a positive local
 solver exit flag.
x =
    7.1018    42.8982    -0.0001    -0.0032
f =
    4.6175
```

It can be seen from the above results of Example 6.5 that the ga solver cannot provide accurate results. In contrast, patternsearch, MultiStart, and GlobalSearch can provide accurate results for constrained optimization. After constructing the S function, more global optimization solvers can be applied to the above problem. The particleswarm, simulannealbnd, MultiStart, and GlobalSearch can provide relatively accurate results for the *S* function optimization, while surrogateopt can only provide less accurate results.

6.4.2 Productivity Profit Optimization

Example 6.6

A chemical company sells three products: *A*, *B*, and *C*. Its revenue function can be represented as:

$$R(x) = 10 \cdot x_1 + 15 \cdot x_2 + 5.5 \cdot x_3 \ \$10^4/\text{month} \tag{6.11}$$

where x_1, x_2, and x_3 are the monthly production rates of *A*, *B*, and *C*. The production rates of *A*, *B*, and *C* must also satisfy the following restriction:

$$(x_1)^2 + 2 \cdot (x_2)^2 + 2 \cdot x_3 \leq 30 \tag{6.12}$$

An additional restriction must be considered due to the raw material availability:

$$x_1 + 3 \cdot x_2 + x_3 \leq 10 \tag{6.13}$$

For maximum revenue, determine the production rate per month for each chemical. What is the revenue at this optimum point?

(1) To use ga to solve the problem

Solution:

```
% Example 6.6
function [c,ceq]=noncon6_6(x)
c=x(1)^2+x(2)^2+2*x(3)-30;
ceq=[];
end
clc; clear all; format compact;
fun=@(x) -(10*x(1)+15*x(2)+5.5*x(3));
nonlcon=@noncon6_6;
A=[1 3 1];
b=[10];
Aeq=[];
beq=[];
lb=[0;0;0];
ub=[];
[x,fval]= ga(fun,3,A,b,Aeq,beq,lb,ub,nonlcon)
fmax=-fval
The results are as follows:
Optimization terminated: average change in the fitness value
 less than options.FunctionTolerance and constraint viola-
tion is less than options.ConstraintTolerance.
x =
    5.2191    1.5173    0.2300
fval =
  -76.2154
fmax =
   76.2154
```

(2) To use patternsearch to solve the problem

Solution:

```
% Example 6.6
function [c,ceq]=noncon6_6(x)
c=x(1)^2+x(2)^2+2*x(3)-30;
ceq=[];
end
clc; clear all; format compact;
fun=@(x) -(10*x(1)+15*x(2)+5.5*x(3));
nonlcon=@noncon6_6;
x0=[1 1 1];
```

```
A=[1 3 1];
b=[10];
Aeq=[];
beq=[];
lb=[0;0;0];
ub=[];
[x,fval]= patternsearch (fun,x0,A,b,Aeq,beq,lb,ub,nonlcon)
fmax=-fval

The results are as follows:
Optimization terminated: mesh size less than
 options.MeshTolerance and constraint violation is less than
 options.ConstraintTolerance.
x =
    5.2302    1.5490    0.1228
fval =
  -76.2124
fmax =
   76.2124
```

(3) To use MultiStart to solve the problem

Solution:

```
% Example 6.6
function [c,ceq]=noncon6_6(x)
c=x(1)^2+x(2)^2+2*x(3)-30;
ceq=[];
end
clc; clear all; format compact;
fun=@(x) -(10*x(1)+15*x(2)+5.5*x(3));
nonlcon=@noncon6_6;
x0=[1 1 1];
A=[1 3 1];
b=[10];
Aeq=[];
beq=[];
lb=[0;0;0];
ub=[];
rng default % For reproducibility
opts = optimoptions(@fmincon,'Algorithm','sqp');
problem = createOptimProblem('fmincon','objective',fun,'x0',...
 x0, 'Aineq', A, 'bineq', b, 'lb', lb, 'ub', ub, 'nonlcon',...
 nonlcon);
ms = MultiStart;
[x,fval] = run(ms,problem,20)
fmax=-fval
```

```
The results are as follows:
MultiStart completed the runs from all start points.
All 20 local solver runs converged with a positive local
 solver exit flag.
x =
    5.2419    1.5836    0.0075
fval =
  -76.2132
fmax =
   76.2132
```

(4) To use GlobalSearch to solve the problem

Solution:

```
% Example 6.6
function [c,ceq]=noncon6_6(x)
c=x(1)^2+x(2)^2+2*x(3)-30;
ceq=[];
end
clc; clear all; format compact;
fun=@(x) -(10*x(1)+15*x(2)+5.5*x(3));
nonlcon=@noncon6_6;
x0=[1 1 1];
A=[1 3 1];
b=[10];
Aeq=[];
beq=[];
lb=[0;0;0];
ub=[];
rng default % For reproducibility
opts = optimoptions(@fmincon,'Algorithm','sqp');
problem = createOptimProblem('fmincon','objective',fun,...
 'x0',x0, 'Aineq', A, 'bineq', b, 'lb', lb, 'ub', ub,...
 'nonlcon', nonlcon);
ms = GlobalSearch;
[x,fval] = run(ms,problem)
fmax=-fval
The results are as follows:
GlobalSearch stopped because it analyzed all the trial
 points.
All 67 local solver runs converged with a positive local
 solver exit flag.
x =
    5.2423    1.5846    0.0038
fval =
  -76.2132
fmax =
   76.2132
```

(5) To use the penalty function and simulannealbnd in MATLAB to solve the problem.

Solution:

Step 1: Construct a penalty function

$$\theta(\overline{x}, u) = f(\overline{x}) + u * \left(\sum_{1}^{m} ((max(0, g_i(\overline{x})))^2) \right) \tag{6.14}$$

Step 2: Give the initial value to u

$$u = u_0 = 1$$

Step 3: Optimize the penalty function

$$\begin{aligned} \min\, \theta(\overline{x}, u) = \min\, & \big(-(10 \cdot x_1 + 15 \cdot x_2 + 5.5 \cdot x_3) \\ & + u_0 * \left(\left(max\left(0, 30 - (x_1)^2 + 2 \cdot (x_2)^2 + 2 \cdot x_3\right)\right)^2 \right. \\ & \left. + \ (max(0, x_1 + 3 \cdot x_2 + x_3 - 10))^2 \right)\big) \end{aligned} \tag{6.15}$$

Step 4: Increase value u by multiplying a constant β such as $\beta = 10$ and return to Step 2 each iteration (i.e. $m+1 \rightarrow m$) and run optimization until convergence

$$u = u_0 * \beta^m \tag{6.16}$$

The code for the first iteration is as follows:

```
>> u0=1; beta=10; m=0; u=u0*beta^m;
>> fun = @(x)-(10*x(1)+15*x(2)+5.5*x(3))+u*((max(0,x(1)^2+...
 x(2)^2+2*x(3)-30))^2+(max(0,x(1)+3*x(2)+x(3)-10))^2);
>> x0=[1 1 1];
>> lb=[0;0;0];
>> ub=[];
>> [x,Val] = simulannealbnd (fun,x0,lb,ub)
>> ThetaVal=-Val
>> f=10*x(1)+15*x(2)+5.5*x(3)
The results for the first iteration are as follows:
Optimization terminated: change in best function value less
 than options.FunctionTolerance.
x =
    4.7616    1.6139    2.5232
Val =
  -81.0752
ThetaVal =
   81.0752
f =
   85.7020
```

The results for the first five iterations using a penalty function with simulannealbnd in MATLAB are shown in Table 6.2.

(6) To use the penalty function and particleswarm in MATLAB to solve the problem.

Table 6.2 Results of using a penalty function with simulannealbnd in MATLAB to solve Example 6.6.

u	x_{min}	f_{min}	θ_{min}
1	(4.7616,1.6139,2.5232)	85.7020	81.0752
10	(5.1650,1.5023,0.5479)	77.1979	76.7059
100	(5.1602,1.3719,0.7465)	76.2868	76.2353
1000	(5.2360,1.5676,0.0634)	76.2233	76.2180
10,000	(5.2244,1.5322,0.1792)	76.2125	76.2120

Solution:

The code for the first iteration is as follows:

```
>> u0=1; beta=10; m=0; u=u0*beta^m;
>> fun = @(x)-(10*x(1)+15*x(2)+5.5*x(3))+u*((max(0,x(1)^2+...
 x(2)^2+2*x(3)-30))^2+(max(0,x(1)+3*x(2)+x(3)-10))^2);
>> lb=[0;0;0];
>> ub=[10; 10; 10];
>> [x,Val] = particleswarm (fun,3,lb,ub)
>> ThetaVal=-Val
>> f=10*x(1)+15*x(2)+5.5*x(3)
The results for the first iteration are as follows:
Optimization ended: relative change in the objective value
 over the last OPTIONS.MaxStallIterations iterations is less
 than OPTIONS.FunctionTolerance.
x =
    4.7907    1.7311    2.1752
Val =
   -81.0862
ThetaVal =
    81.0862
f =
    85.8368
```

The results for the first five iterations using a penalty function with particleswarm in MATLAB are shown in Table 6.3.

(7) To use the penalty function and surrogateopt in MATLAB to solve the problem.

Solution:

The code for the first iteration is as follows:

```
>> u0=1; beta=10; m=0; u=u0*beta^m;
>> fun = @(x)-(10*x(1)+15*x(2)+5.5*x(3))+u*((max(0,x(1)^2+...
 x(2)^2+2*x(3)-30))^2+(max(0,x(1)+3*x(2)+x(3)-10))^2);
>> lb=[0;0;0];
>> ub=[10; 10; 10];
>> [x,Val] = surrogateopt (fun,lb,ub)
```

Table 6.3 Results of using a penalty function with particleswarm in MATLAB to solve Example 6.6.

u	x_{min}	f_{min}	θ_{min}
1	(4.7907,1.7311,2.1752)	85.8368	81.0862
10	(5.1976,1.5983,0.2285)	77.2079	76.7118
100	(5.1520,1.3511,0.8169)	76.2799	76.2296
1000	(5.2063,1.4824,0.3487)	76.2167	76.2118
10,000	(5.2056,1.4794,0.3565)	76.2075	76.2070

```
>> ThetaVal=-Val
>> f=10*x(1)+15*x(2)+5.5*x(3)
The results for the first iteration are as follows:
Surrogateopt stopped because it exceeded the function evalua-
tion limit set by 'options.MaxFunctionEvaluations'.
x =
     4.7242     1.5532     2.7881
Val =
   -81.0633
ThetaVal =
    81.0633
f =
    85.8747
```

The plot of surrogateopt results of Example 6.6 is shown in Figure 6.8.

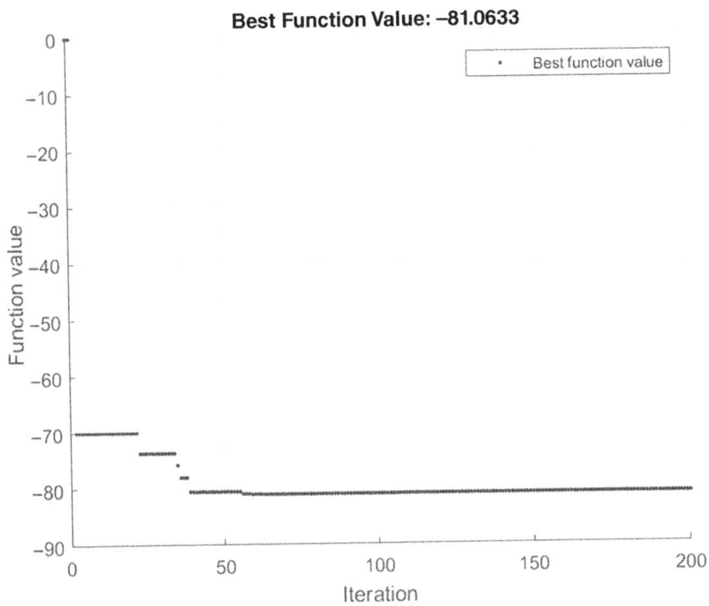

Figure 6.8 The plot of surrogateopt results of Example 6.6.

Table 6.4 Results of using a penalty function with surrogateopt in MATLAB to solve Example 6.6.

u	x_{min}	f_{min}	θ_{min}
1	(4.7242,1.5532,2.7881)	85.8747	81.0633
10	(4.7651,1.1968,1.9141)	76.1309	75.4038
100	(4.1176,0.1480,5.4440)	73.3383	73.3351
1000	(4.4145,1.1177,2.2367)	73.2120	73.1943
10,000	(4.2249,0.4307,4.4823)	73.3624	73.3624

The results for the first five iterations using a penalty function with surrogateopt in MATLAB are shown in Table 6.4.

It can be seen from the above example that ga, patternsearch, MultiStart, and GlobalSearch can provide accurate results for constrained optimization. After using the penalty function, more global optimization solvers can be applied, including simulannealbnd, particleswarm, and surrogateopt. When u equals 1000, simulannealbnd and particleswarm can provide accurate results, while the results from surrogateopt are a bit off from those of other methods.

6.4.3 Optimization for the Operation of a Continuous Stirred Tank Reactor (CSTR)

Example 6.7

In Figure 6.9, a CSTR reactor is used to convert liquid chemical A to product B with the following reaction:

$$2A \rightarrow B \tag{6.17}$$

The reaction rate is

$$-r_A = \frac{-dC_A}{dt} = 2.0 \cdot C_A^2 = 2.0 \cdot C_{A_0}^2 \cdot (1 - x_A)^2 \tag{6.18}$$

where x_A is the conversion rate, which is no more than 100%.

The price of chemical A is provided as

$$S_A = 4.0 \cdot C_{A_0}^{1.4} (\$/L) \tag{6.19}$$

Other costs, such as labor costs and depreciation of equipment,

$$S_{ld} = 0.4 \cdot V^{0.6} (\$) \tag{6.20}$$

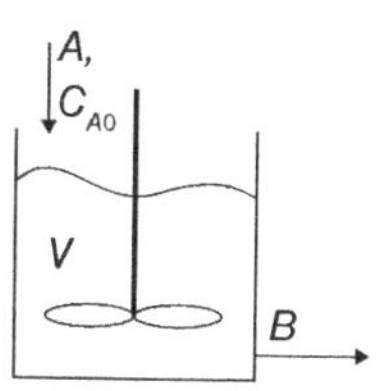

Figure 6.9 A CSTR reactor is used to convert liquid chemical A to product B.

where V is the load volume of liquid inside the reactor.

To perform the minimization of the costs for the operation of the CSTR reactor with the requirement of a production rate of B greater than 50 kg/h.

(1) To use ga with constraint to solve the problem

Solution:

Step 1: To formulate the optimization problem

Mass balance inside the CSTR reactor:

$$F_{A_0} \cdot C_{A_0} = F_{A_0} \cdot C_{A_0}(1-x_A) - r_A V = F_{A_0} \cdot C_{A_0}(1-x_A) + 2.0 \cdot C_{A_0}^2 \cdot (1-x_A)^2 \cdot V \tag{6.21}$$

where F_{A_0} is the feed rate of chemical A.

Then,

$$V = \frac{F_{A_0} \cdot C_{A_0} \cdot x_A}{2.0 \cdot C_{A_0}^2 \cdot (1-x_A)^2} = \frac{F_{A_0} \cdot x_A}{2.0 \cdot C_{A_0} \cdot (1-x_A)^2} \tag{6.22}$$

The production rate of B is:

$$P_B = \frac{1}{2} F_{A_0} \cdot C_{A_0} \cdot x_A \tag{6.23}$$

Therefore, the optimization problem is:

$$minf = 4.0 \cdot C_{A_0}^{1.4} \cdot F_{A_0} + 0.4 \cdot \left(\frac{F_{A_0} \cdot x_A}{2.0 \cdot C_{A_0} \cdot (1-x_A)^2} \right)^{0.6} \tag{6.24}$$

$$\text{s.t.} \ \frac{1}{2} F_{A_0} \cdot C_{A_0} \cdot x_A \geq 50 \tag{6.25}$$

To treat F_{A_0} as x_1, C_{A_0} as x_2, and x_A as x_3, then it becomes:

$$minf(\overline{x}) = 4.0 \cdot x_2^{1.4} \cdot x_1 + 0.4 \cdot \left(\frac{x_1 \cdot x_3}{2.0 \cdot x_2 \cdot (1-x_3)^2} \right)^{0.6} \tag{6.26}$$

$$\text{s.t.} \ \frac{1}{2} x_1 \cdot x_2 \cdot x_3 \geq 50 \tag{6.27}$$

$$x_3 \leq 1 \tag{6.28}$$

Step 2: To use ga to solve the problem

The code is as follows:

```
function [c,ceq]=nonlcon6_7(x)
c(1)=50-1/2* x(1)*x(2)*x(3);
c(2)=x(3)-1;
ceq=[];
end
clc; clear all; format compact;
fun=@(x)4.0*x(2)^1.4*x(1)+0.4*(((x(1)*x(3))/(2.0*x(2)*...
  (1-x(3))^2)))^0.6;
nonlcon=@nonlcon6_7;
A=[];
b=[];
```

```
Aeq=[];
beq=[];
lb=[0;0;0];
ub=[];
[x,fval]=ga(fun,3,A,b,Aeq,beq,lb,ub,nonlcon)
```

```
The results are as follows:
Optimization terminated: average change in the fitness value
 less than options.FunctionTolerance and constraint violation
 is less than options.ConstraintTolerance.
x =
   44.1636    2.4085    0.9401
fval =
  647.4783
```

(2) To use patternsearch with constraint to solve the problem

Solution:

```
clc; clear all; format compact;
fun=@(x)4.0*x(2)^1.4*x(1)+0.4*(((x(1)*x(3))/(2.0*x(2)*...
 (1-x(3))^2)))^0.6;
nonlcon=@nonlcon6_7;
A=[];
b=[];
Aeq=[];
beq=[];
lb=[0;0;0];
ub=[];
x0=[100;1;0.5];
[x,fval]= patternsearch (fun,x0,A,b,Aeq,beq,lb,ub,nonlcon)
```

```
The results are as follows:
Maximum number of function evaluations exceeded: increase
 options.MaxFunctionEvaluations.
x =
  568.6953
    0.2448
    0.7195
fval =
  421.3095
```

(3) To use MultiStart with constraint to solve the problem

Solution:

```
clc; clear all; format compact;
fun=@(x)4.0*x(2)^1.4*x(1)+0.4*(((x(1)*x(3))/(2.0*x(2)*...
 (1-x(3))^2)))^0.6;
nonlcon=@nonlcon6_7;
```

```
A=[];
b=[];
Aeq=[];
beq=[];
lb=[0;0;0];
ub=[];
x0=[100;1;0.5];
rng default % For reproducibility
opts = optimoptions(@fmincon,'Algorithm','sqp');
problem = createOptimProblem('fmincon','objective',fun,...
 'x0',x0, 'Aineq', A, 'bineq', b, 'lb', lb, 'ub', ub,...
 'nonlcon', nonlcon);
ms = MultiStart;
[x,fval] = run(ms,problem,20)
```

```
The results are as follows:
MultiStart completed some of the runs from the start points.
14 out of 20 local solver runs converged with a positive
 local solver exit flag.
x =
  587.6146
    0.2383
    0.7143
fval =
  420.6695
```

(4) To use GlobalSearch with constraint to solve the problem

Solution:

```
clc; clear all; format compact;
fun=@(x)4.0*x(2)^1.4*x(1)+0.4*(((x(1)*x(3))/(2.0*x(2)*...
 (1-x(3))^2)))^0.6;
nonlcon=@nonlcon6_7;
A=[];
b=[];
Aeq=[];
beq=[];
lb=[0;0;0];
ub=[];
x0=[100;1;0.5];
rng default % For reproducibility
opts = optimoptions(@fmincon,'Algorithm','sqp');
problem = createOptimProblem('fmincon','objective',fun,...
 'x0',x0, 'Aineq', A, 'bineq', b, 'lb', lb, 'ub', ub,...
 'nonlcon', nonlcon);
ms = GlobalSearch;
[x,fval] = run(ms,problem)
```

```
The results are as follows:
GlobalSearch stopped because it analyzed all the trial
 points.
All 40 local solver runs converged with a positive local
 solver exit flag.
x =
  587.6147
    0.2383
    0.7143
fval =
  420.6695
```

(5) To use simulannealbnd with penalty function method

Solution:

```
u0=1; beta=10; m=0; u=u0*beta^m;
fun = @(x)4.0*x(2)^1.4*x(1)+0.4*(((x(1)*x(3))/(2.0*x(2)*...
 (1-x(3))^2)))^0.6+u*((max(0,50-1/2*x(1)*x(2)*x(3)))^2+...
 (max(0, x(3)-1))^2);
x0 = [100;1;0.5];
lb=[0;0;0];
ub=[1000;10;1];
[x,ThetaVal] = simulannealbnd (fun,x0,lb,ub)
f=4.0*x(2)^1.4*x(1)+0.4*(((x(1)*x(3))/(2.0*x(2)*...
 (1-x(3))^2)))^0.6
```

```
The results are as follows:
Optimization terminated: change in best function value less
 than options.FunctionTolerance.
x =
  119.7621
    0.8488
    0.8729
ThetaVal =
  468.8723
f =
  437.1040
```

The results for the first five iterations using a penalty function with simulannealbnd in MATLAB are shown in Table 6.5.

(6) To use particleswarm with penalty function method

Solution:

```
u0=1; beta=10; m=0; u=u0*beta^m;
fun = @(x)4.0*x(2)^1.4*x(1)+0.4*(((x(1)*x(3))/(2.0*x(2)*...
 (1-x(3))^2)))^0.6+u*((max(0,50-1/2*x(1)*x(2)*x(3)))^2+...
 (max(0, x(3)-1))^2);
lb=[0;0;0];
ub=[1000;10;1];
```

Table 6.5 Results from using a penalty function with simulannealbnd in MATLAB to solve Example 6.7.

u	x_{min}	f_{min}	θ_{min}
1	(119.7621, 0.8488, 0.8729)	437.1040	468.8723
10	(125.4583, 0.8956, 0.8798)	490.3246	493.5878
100	(138.8593, 0.8251, 0.8718)	486.3247	486.6271
1000	(144.3875, 0.7952, 0.8709)	483.2433	483.2738
10,000	(242.4597, 0.5025, 0.8208)	445.2333	445.2355

```
[x,ThetaVal] = particleswarm (fun,3,lb,ub)
f=4.0*x(2)^1.4*x(1)+0.4*(((x(1)*x(3))/(2.0*x(2)*...
 (1-x(3))^2)))^0.6
```

```
The results are as follows:
Optimization ended: relative change in the objective value
 over the last OPTIONS.MaxStallIterations iterations is less
 than OPTIONS.FunctionTolerance.
x =
  532.0669    0.2430    0.7143
ThetaVal =
  406.2229
f =
  391.6595
```

The results for the first five iterations using a penalty function with particleswarm in MATLAB are shown in Table 6.6.

(7) To use surrogateopt with penalty function method

```
Solution:
u0=1; beta=10; m=0; u=u0*beta^m;
```

Table 6.6 Results from using a penalty function with particleswarm in MATLAB to solve Example 6.7.

u	x_{min}	f_{min}	θ_{min}
1	(532.0669,0.2430,0.7143)	391.6595	406.2229
10	(580.4903,0.2392,0.7147)	417.7970	419.2352
100	(542.6453,0.2540,0.7248)	420.4931	420.7295
1000	(581.5217,0.2409,0.7137)	420.6543	420.6730
10,000	(511.0510,0.2738,0.7149)	422.5868	422.5868

```
fun = @(x)4.0*x(2)^1.4*x(1)+0.4*(((x(1)*x(3))/(2.0*x(2)*...
  (1-x(3))^2)))^0.6+u*((max(0,50-1/2*x(1)*x(2)*x(3)))^2+...
  (max(0, x(3)-1))^2);
lb=[0;0;0];
ub=[1000;10;1];
[x,ThetaVal] = surrogateopt (fun,lb,ub)
f=4.0*x(2)^1.4*x(1)+0.4*(((x(1)*x(3))/(2.0*x(2)*...
  (1-x(3))^2)))^0.6
```

```
The results are as follows:
Objective or constraint function returned NaN.
Surrogateopt stopped because it exceeded the function evalua-
tion limit set by 'options.MaxFunctionEvaluations'.
x =
  583.9009    0.3064    0.4721
ThetaVal =
  539.8061
f =
  479.5169
```

The plot of surrogateopt results of Example 6.7 is shown in Figure 6.10.

The results for the first five iterations using a penalty function with surrogateopt in MATLAB are shown in Table 6.7.

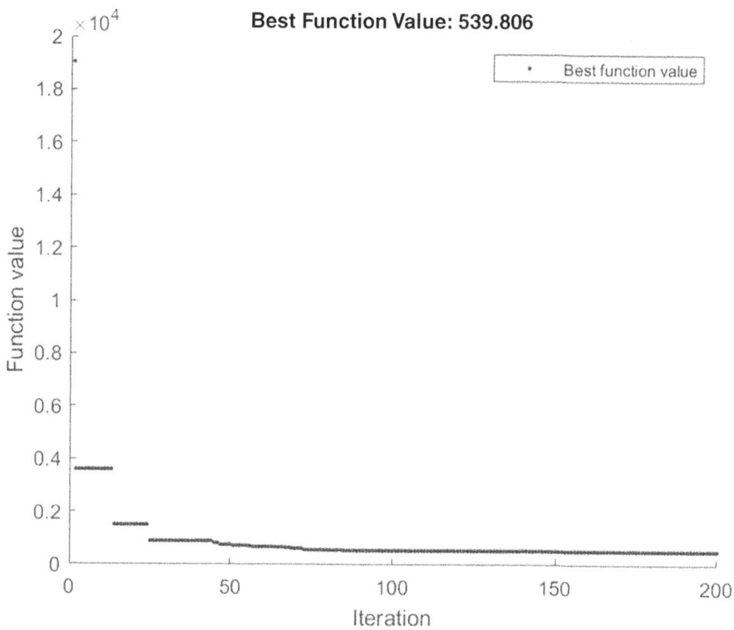

Figure 6.10 The plot of surrogateopt results of Example 6.7.

Table 6.7 Results from using a penalty function with surrogateopt in MATLAB to solve Example 6.7.

u	x_{min}	f_{min}	θ_{min}
1	(583.9009,0.3064,0.4721)	479.5169	539.8061
10	(589.7142,0.2163,0.8003)	460.3022	460.3022
100	(656.2500,0.3125,0.7813)	653.9715	653.9715
1000	(569.9242,0.2595,0.7228)	447.2971	447.2971
10,000	(656.2500,0.3125,0.7813)	653.9715	653.9715

It can be seen from the above example that MultiStart and GlobalSearch can provide accurate results for constrained optimization. In contrast, patternsearch can provide relatively good results but ga provide poor ones. After using the penalty function, more global optimization solvers can be applied, including simulannealbnd, particleswarm, and surrogateopt. The particleswarm can provide accurate results, while the results from surrogateopt and simulannealbnd provide poor results.

Learning Tips

- It will be very helpful to use more solvers (include solvers in previous chapter and global optimization solvers) to your optimization problem to crosscheck your results because any one MATLAB solver can have failures or give wrong results.
- The advantage of converting constrained optimization to unconstrained optimization will be able to use more solvers, even including other global optimization solvers.
- From all examples, MultiStart and GlobalSearch are best among all solvers, patternsearch and particleswarm are good and robust solvers, while ga, surrogateopt and simulannealbnd are not stable and often provide poor results.

6.5 Summary

In this chapter, we have presented practical global optimization methods in chemical engineering design and simulation. Any one of the solvers in The Global Optimization Toolbox from MATLAB has its limitations and may not give correct answers for the case of interest. Crosschecking your results by employing more solvers (including solvers in Chapter 4 and global optimization solvers) is very useful.

The enhanced optimization methods in the Chapter 5 can also be applied in global optimization, such as the Lagrange method, inequality constraints converted to equality constraints, the penalty function method, and the enhanced constraints method. These methods can help us use more solvers for our optimization problem to build our confidence.

MultiStart and GlobalSearch, based on fmincon, are the best among all solvers for global optimization. The patternsearch and particleswarm can also provide relatively good results. From some examples, ga, surrogateopt, and simulannealbnd solvers cannot provide accurate and stable results.

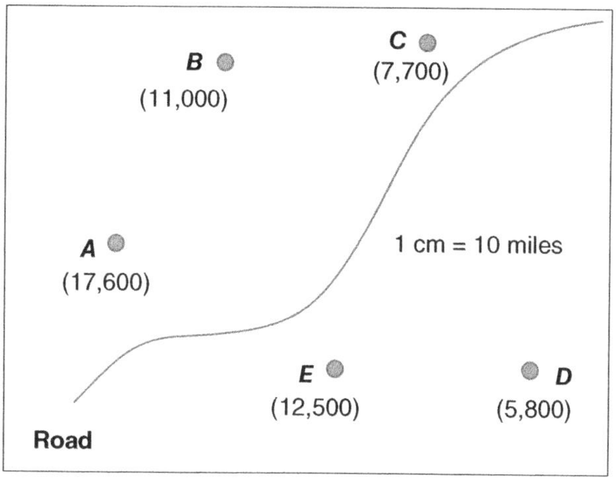

Figure 6.11 Build a gas station located on the road among five communities.

Problems

6.1 Optimize the location of a gas station along the road winding among five communities, as shown in Figure 6.11. You can choose optimization solvers from The Optimization Toolbox or The Global Optimization Toolbox from MATLAB and crosscheck your results. This problem incorporates features of weighted optimization and constraints under the following conditions:

(1) Minimize the sum of distances from the station to all communities.
(2) Consider the community populations given in parentheses as distance weights in the optimization.
(3) Consider the population distance weights, and constrain the distance between the station and community C to less than 20 miles. This additional constraint fosters a discussion of diversity, equity, and inclusion in engineering solutions that need to consider equitable access to petrol resources among communities with diverse needs and circumstances.

6.2

$$min f(\overline{x}) = x_1^3 + 2x_1^2 - 2x_1x_2 + 2x_2^2 - 6x_1 \tag{6.29}$$

$$\text{s.t. } g_1(\overline{x}) = 3x_1^2 + 4x_2 - 6 \leq 0 \tag{6.30}$$

$$g_2(\overline{x}) = -x_1^2 + 4x_2 - 2 \leq 0 \tag{6.31}$$

(1) To use ga in MATLAB to solve the problem, x0=[1 1].
(2) To use patternsearch in MATLAB to solve the problem, x0=[1 1].
(3) To use MultiStart in MATLAB to solve the problem, x0=[1 1].
(4) To use GlobalSearch in MATLAB to solve the problem, x0=[1 1].
(5) To construct the *S* function and use particleswarm in MATLAB to solve the problem.
(6) To construct the *S* function and use simulannealbnd in MATLAB to solve the problem.

(7) To construct the *S* function and use surrogateopt in MATLAB to solve the problem.
(8) To construct the *S* function and use MultiStart in MATLAB to solve the problem.
(9) To construct the *S* function and use GlobalSearch in MATLAB to solve the problem.

6.3

$$min f(\overline{x}) = (x_1 - 2)^2 - 6\ \ln\left((x_2 - 1)^2 + 1\right) \tag{6.32}$$

$$\text{s.t. } g_1(\overline{x}) = x_1^2 + x_2 - 3 \leq 0 \tag{6.33}$$

$$g_2(\overline{x}) = e^{-x_1} - x_2 \leq 0 \tag{6.34}$$

(1) To use ga in MATLAB to solve the problem, x0=[2 3].
(2) To use patternsearch in MATLAB to solve the problem, x0=[2 3].
(3) To use MultiStart in MATLAB to solve the problem, x0=[2 3].
(4) To use GlobalSearch in MATLAB to solve the problem, x0=[2 3].
(5) To use the penalty function and particleswarm in MATLAB to solve the problem.
(6) To use the penalty function and simulannealbnd in MATLAB to solve the problem.
(7) To use the penalty function and surrogateopt in MATLAB to solve the problem.
(8) To use the penalty function and MultiStart in MATLAB to solve the problem.
(9) To use the penalty function and GlobalSearch in MATLAB to solve the problem.

7

Optimal Experimental Design

Optimal experimental design in this chapter aims to help our readers learn how to properly design experiments to maximize the chance for success in experiments and further improve data analysis and optimization.

Experimental design aims to plan a series of experiments with systematic data collection considering validity, reliability, and replicability. It is essential for experimental design to choose the independent variables, reduce the measurement error, and analyze the sensitivity. A properly designed experiment can assist the data process in improving the quality of information, eliminating redundant data, and further accelerating the model-building process.

MATLAB has some functions for designing experiments, including Full Factorial Designs, Fractional Factorial Designs, Response Surface Designs, D-Optimal Designs, Latin Hypercube Designs, and Quasi-Random Designs that are introduced and discussed in this chapter. Orthogonal experimental design (OED) is introduced, which can be used to identify influencing factors, the levels of significance of each factor, and their combined effects. The OED method has wide applications in engineering disciplines.

We first introduce experimental design and error analysis, then use the OED method to improve data analysis and model building. As OED is an effective design, it is illustrated in more detail and applied to many cases.

7.1 Introduction

7.1.1 Experimental Design

To solve engineering problems, we need to determine the constitutive relations between the system's state variables and parameters. For example, the Navier–Stokes equations are derived from constitutive relations for fluid flow, and Maxwell's equations offer constitutive relations for electrodynamics.

However, if we cannot find enough constitutive relations, we need to develop more equations for the system. This is often the case when examining a problem for the first time, developing a new process, sometimes involving a new reaction pathway, developing or redesigning a new unit operation, or questioning the accuracy of an existing model or equation.

We may encounter problems, such as not knowing the form of the model or which variables should be included in the model.

Chemical Engineering Analysis and Optimization Using MATLAB®, First Edition.
Weiguo Xie, Sam Toan, and Richard Davis.

Companion website: www.wiley.com/go/xie/chemicalengineering

- If we know the kind of model equations (e.g. from theory) and which variables are important, then it is a model-fitting problem. In this case, we can conduct experiments to generate data to estimate the model's parameters.
- Sometimes, we may know the kind of equations we want to use but not which variables are essential. Then, we must carry out range-finding experiments. These experiments identify which variables have a significant effect on the desired output. They help guide the development of theory or further experimentation.
- Sometimes, we know very little about the model or even may not know the form of the model. Then, we need to perform the statistical design of experiments to observe the system phenomena and find out the trend of data.

Then, we need to use the experimental design to investigate the system. After experiments, we need to know how to analyze the data, including errors and significance of results, data regression methods, and goodness of fit criteria.

7.1.2 Experimental Error

Experimental errors cannot be avoided, and we must understand their sources and how to minimize them. There are two types of experimental error: random and systematic. The random error comes from fluctuating measurements, conditions, and observer and apparatus imprecision. The systematic error comes from miscalibration and observer bias.

7.1.2.1 Random Error

Figure 7.1 shows two kinds of random error, including temperature changes due to condition changes between morning and afternoon and the observer's imprecision due to the apparatus's design.

It can be reduced or eliminated by practicing the technique until becoming proficient or by reducing sources of fluctuation using better equipment. Estimating the average point value and error from individual readings is relatively easy by repeating the experiment.

Random errors are usually absolute; e.g. a thermocouple reads temperature to $\pm 0.1\,°C$, and with a stopwatch, you can time something to ± 1 s. This means that the error in small

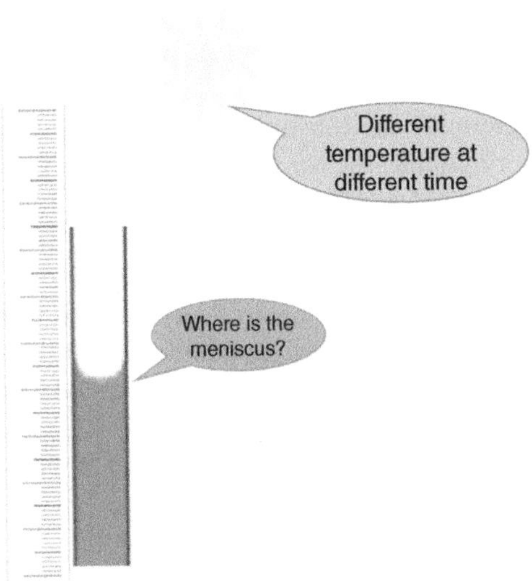

Figure 7.1 Two kinds of random error come from fluctuating conditions and the imprecision of the observer.

measurements is proportionally more significant than in larger measurements. Errors combine following the laws of variance: if value x_i has uncertainty $u(x_i)$, i.e. $x = x_i \pm u(x_i)$, then functions of x_i are made up of sums and differences.

Example 7.1

$$\text{To calculate } (121.35 \pm 0.24) + (2.1 \pm 0.02) + (36.0 \pm 0.4) \tag{7.1}$$

Solution:

$$(121.35 \pm 0.24) + (2.1 \pm 0.02) + (36.0 \pm 0.4) = (121.35 + 2.1 + 36.0) \pm [(0.24)^2 + (0.02)^2 + (0.4)^2]^{1/2} = (159.45 \pm 0.47) \tag{7.2}$$

If value x_i has uncertainty

$$u(x_i), \text{ i.e.} x = x_i \pm u(x_i) \tag{7.3}$$

then for functions containing products and quotients, Let

$$y = (x_1 \oplus x_2 \oplus x_3 \cdots \oplus x_i \cdots \oplus x_n) \tag{7.4}$$

where $\oplus$ can be multiplication $\times$ or division $\div$.

Then uncertainty

$$u(y) = y \cdot \left\{ \left[\left(\frac{u(x_1)}{x_1} \right)^2 + \left(\frac{u(x_2)}{x_2} \right)^2 + \cdots + \left(\frac{u(x_n)}{x_n} \right)^2 \right]^{1/2} \right\} \tag{7.5}$$

and

$$y \pm u(y) = y \cdot \left\{ 1 \pm \left[\sum_{i=1}^{n} \left(\frac{u(x_i)}{x_i} \right)^2 \right]^{1/2} \right\} \tag{7.6}$$

Example 7.2

$$\text{To calculate } \frac{2\pi \cdot (20.824 \pm 0.040)}{(223 \pm 15)} \tag{7.7}$$

Solution:

$$\frac{2\pi \cdot (20.824 \pm 0.040)}{(223 \pm 5)} = \frac{2\pi \cdot 20.824}{223} \left\{ 1 \pm \left[\left(\frac{0.040}{20.824} \right)^2 + \left(\frac{5}{223} \right)^2 \right]^{\frac{1}{2}} \right\} = (0.5867 \pm 0.0132) \tag{7.8}$$

If value x_i has uncertainty $u(x_i)$, i.e. $x = x_i \pm u(x_i)$ then for a general function $f(x_1, x_2, \ldots, x_n)$, the uncertainty is given by

$$u(f) = \left[\sum_{i=1}^{n} \left(\frac{\partial f}{\partial x_i} u(x_i) \right)^2 \right]^{1/2} \tag{7.9}$$

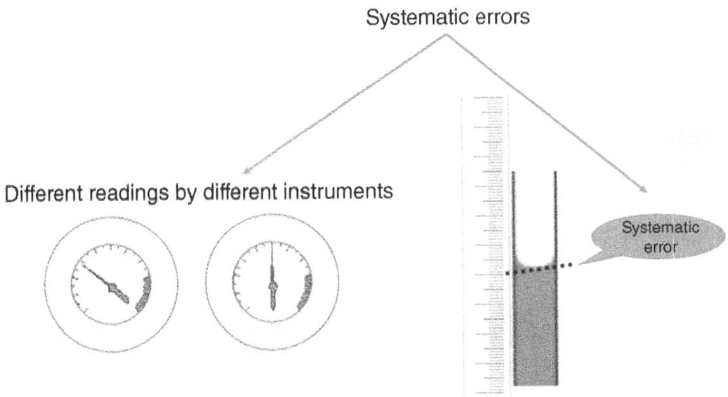

Figure 7.2 Two kinds of systematic error from instruments and observers.

Example 7.3

$$\text{To calculate } (3.45 \pm 0.16) \cdot \exp(-2.37 \pm 0.09) \tag{7.10}$$

Solution:

$$(2.31 \pm 0.26) \cdot \exp(-1.31 \pm 0.08) = 3.45 \cdot \exp(-2.37)$$
$$\pm \{[\exp(-2.37) \cdot 0.16]^2 + [3.45 \cdot \exp(-2.37) \cdot 0.09]^2\}^{\frac{1}{2}} = (0.3225 \pm 0.0327) \tag{7.11}$$

7.1.2.2 Systematic Error

Figure 7.2 shows two kinds of systematic error, including different instrument readings and systematic errors from an observer.

The systematic errors are tricky to estimate. To reduce or remove systematic error, we need to perform frequent re-calibration of measurement equipment randomization of experiments in advance to eliminate potential observer bias, blind testing (e.g. in pharmaceuticals clinical trials), repetition of results using a different apparatus, which can be expensive, but sometimes used for significant discoveries.

7.2 Design of Experiments (DOE)

7.2.1 Statistical Design of Experiments

We can use the statistical experimental design to investigate the system. For example, if we plan to study a reaction in a process reactor and we want to investigate the effect on yield of three variables: presence or absence of catalyst, pH, and reaction time. We should carry out a factorial experiment.

- We have eight possible combinations with three variables, each at two levels.
- Since $8 = 2^3$, this is called a 2^3 factorial experiment.
- If we had four variables at three levels, it would be a 3^4 factorial with 81 possible combinations.

Table 7.1 2^3 factorial experiment.

Experiment	Catalyst	pH	Reaction time
1	Absence	Low	Low
2	Presence	Low	Low
3	Absence	High	Low
4	Presence	High	Low
5	Absence	Low	High
6	Presence	Low	High
7	Absence	High	High
8	Presence	High	High

- We can analyze them using the Yates table method, as shown in Table 7.1.
- Analysis of these experiments allows us to test the independent significance of our three variables and the combined effects of any two or three variables.

7.2.2 MATLAB Functions for the Design of Experiments

Some MATLAB functions are available for the DOE, as shown in Table 7.2. The differences in the syntax of different MATLAB functions are clearly shown in the following table. Therefore, you need to pay attention to the syntax when using them. They give you ideas for arranging your experiments to collect your data. You also need to think about performing data analysis or plotting your data to observe any data trends.

Example 7.4

Create a full factorial design of three factors, with two levels for the first factor, three levels for the second factor, and two levels for the third factor.

Solution:

```
>> dFF = fullfact([2 3 2])
dFF =
     1     1     1
     2     1     1
     1     2     1
     2     2     1
     1     3     1
     2     3     1
     1     1     2
     2     1     2
     1     2     2
     2     2     2
     1     3     2
     2     3     2
```

Table 7.2 Syntax of MATLAB functions on design of experiments (https://www.mathworks.com/The MathWorks, Inc.).

MATLAB functions	Description
ff2n(n)	It gives factor settings for a two-level full factorial design with n factors.
fullfact(levels)	It gives factor settings for a full factorial design with n factors, where the vector levels of length n give the number of levels for each factor. It is an m-by-n matrix, where m is the number of treatments in the full-factorial design. Each row of it corresponds to a single treatment. Each column contains the settings for a single factor, with integer values from one to the number of levels.
fracfact(gen,Name=Value)	It creates a fractional factorial design with additional options specified by one or more name-value arguments.
fracfactgen(terms,k,R,basic)	A vector is specified for the indices of factors to be treated as basic. These factors receive full-factorial treatments in the design. The default includes factors that are part of the highest-order interaction in terms.
bbdesign(n,param,val)	It generates a Box–Behnken design for n factors.
ccdesign(n,'Name',value)	It is a central composite design and specifies one or more optional name/value pairs for the design.
candexch(C,nrows,Name= Value)	It generates a D-optimal design with additional options specified by one or more name-value arguments.
dcovary(nfactors,fixed,model)	It generates a D-optimal design with fixed covariates.
lhsdesign(n,p,Name,Value)	It returns a Latin hypercube sample matrix of size n-by-p. For each column of X, the n values are randomly distributed with one from each interval $(0,1/n)$, $(1/n,2/n)$, …, $(1 - 1/n,1)$, and randomly permuted. It also modifies the resulting design using one or more name-value pair arguments.
lhsnorm(mu,sigma,n,flag)	It provides a Latin hypercube sample from a normal distribution. It returns an n-by-p matrix, X, containing a Latin hypercube sample of size n from a p-dimensional multivariate normal distribution with mean vector, mu, and covariance matrix, sigma.

Example 7.5
To have a central composite design of three factors and visualize the design.

Solution:

```
>> dCC = ccdesign(3)
dCC =
   -1.0000   -1.0000   -1.0000
   -1.0000   -1.0000    1.0000
   -1.0000    1.0000   -1.0000
   -1.0000    1.0000    1.0000
    1.0000   -1.0000   -1.0000
    1.0000   -1.0000    1.0000
    1.0000    1.0000   -1.0000
    1.0000    1.0000    1.0000
   -1.6818         0         0
```

```
    1.6818         0         0
         0   -1.6818         0
         0    1.6818         0
         0         0   -1.6818
         0         0    1.6818
         0         0         0
         0         0         0
         0         0         0
         0         0         0
         0         0         0
         0         0         0
         0         0         0
         0         0         0
         0         0         0
         0         0         0
≫ dCC = ccdesign(3);
plot3(dCC(:,1),dCC(:,2),dCC(:,3),'ro','MarkerFaceColor','b')
axis square equal
```

Example 7.6
Create a D-optimal design with fixed covariates of three factors.

Solution:

```
≫ time = linspace(0,1,8)';
≫ [dCV1,X] = dcovary(3,time,'linear')
dCV1 =
   -1.0000   -1.0000    1.0000         0
   -1.0000    1.0000   -1.0000    0.1429
    1.0000    1.0000    1.0000    0.2857
    1.0000   -1.0000   -1.0000    0.4286
   -1.0000   -1.0000   -1.0000    0.5714
    1.0000    1.0000   -1.0000    0.7143
   -1.0000    1.0000    1.0000    0.8571
    1.0000   -1.0000    1.0000    1.0000

X =
    1.0000   -1.0000   -1.0000    1.0000         0
    1.0000   -1.0000    1.0000   -1.0000    0.1429
    1.0000    1.0000    1.0000    1.0000    0.2857
    1.0000    1.0000   -1.0000   -1.0000    0.4286
    1.0000   -1.0000   -1.0000   -1.0000    0.5714
    1.0000    1.0000    1.0000   -1.0000    0.7143
    1.0000   -1.0000    1.0000    1.0000    0.8571
    1.0000    1.0000   -1.0000    1.0000    1.0000
```

Note: It also returns the design matrix X associated with the design.

Example 7.7
Create a Latin Hypercube Design of three factors while minimizing the correlations results in a design with a much lower sum of squared correlations.

Solution:

```
>> rng default % For reproducibility
>> X = lhsdesign(10,3,'Criterion','correlation')
>> y = corr(X);
>> mini_correlations=(sum(y(:).^2) - 3)/2

X =
      0.6500      0.0500      0.4500
      0.4500      0.8500      0.0500
      0.0500      0.9500      0.8500
      0.9500      0.5500      0.9500
      0.5500      0.6500      0.6500
      0.1500      0.1500      0.7500
      0.3500      0.2500      0.5500
      0.2500      0.4500      0.2500
      0.7500      0.3500      0.3500
      0.8500      0.7500      0.1500

mini_correlations =
      0.0354
```

Where, for each column of X, the n values are randomly distributed with one from each interval $(0,1/n)$, $(1/n,2/n)$, … , $(1 - 1/n,1)$, and randomly permuted.

7.2.3 Orthogonal Experimental Design

OED is an efficient way to study the effect of several factors simultaneously using orthogonal arrays (OAs) and factor analysis (FA). Many design experiments use OED to determine which combinations of factor levels to use for each experiment and analyze the experimental results. The OED-based experiments can provide near-optimal quality characteristics for a specific objective. This results in a significant saving in the experimental effort.

The OED method must choose an OED table for experiments and data analysis. Some useful OED tables can be downloaded online (Bolboacă and Jäntschi 2007).

Table 7.3 Design of experiments for three factors with three levels on liquid iron example.

Level no.	Factor A	Factor B (kPa)	Factor C (m)
1	1:16	22.6	1.2
2	1:18	30.6	1.6
3	1:14	26.6	1.4

Example 7.8

Three factors can change the temperature of liquid iron, including coal ratio (Factor A), pressure (Factor B), and height of the load (Factor C) (Table 7.3).

If we use a full factorial design for three factors with three levels, we will need to perform $3^3 = 27$ runs. However, if we use an OED method such as in Figure 7.3, we only need to run nine specifically designed experiments instead. We will demonstrate how to use MATLAB code to perform data analysis, so readers who want to test our code, please create an Excel file and input data and words in a file titled OED_data_LiquidIron_9runs.xlsx, as shown in Figure 7.4. Mi (i from 1 to 3) means the mean value of the level *i* for the corresponding factor. The *DE* values are calculated from the difference between maximum *M* value and minimum *M*. The optimal scheme depends on the maximization (or minimization) problem to choose the level with max (or min) *M* value in that factor. The order of importance shows

Figure 7.3 The plot of the central composite design of Example 7.5.

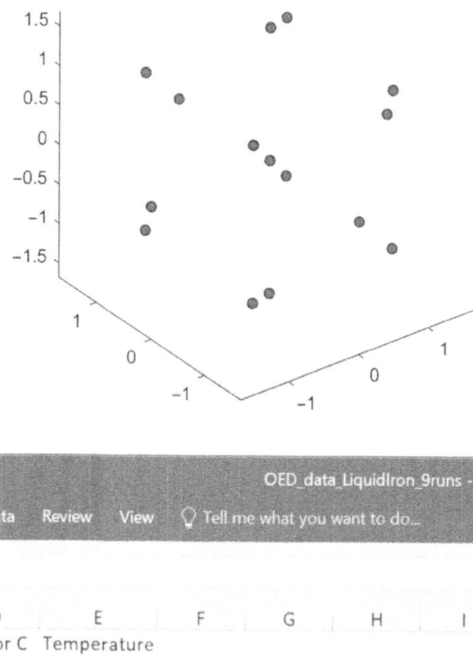

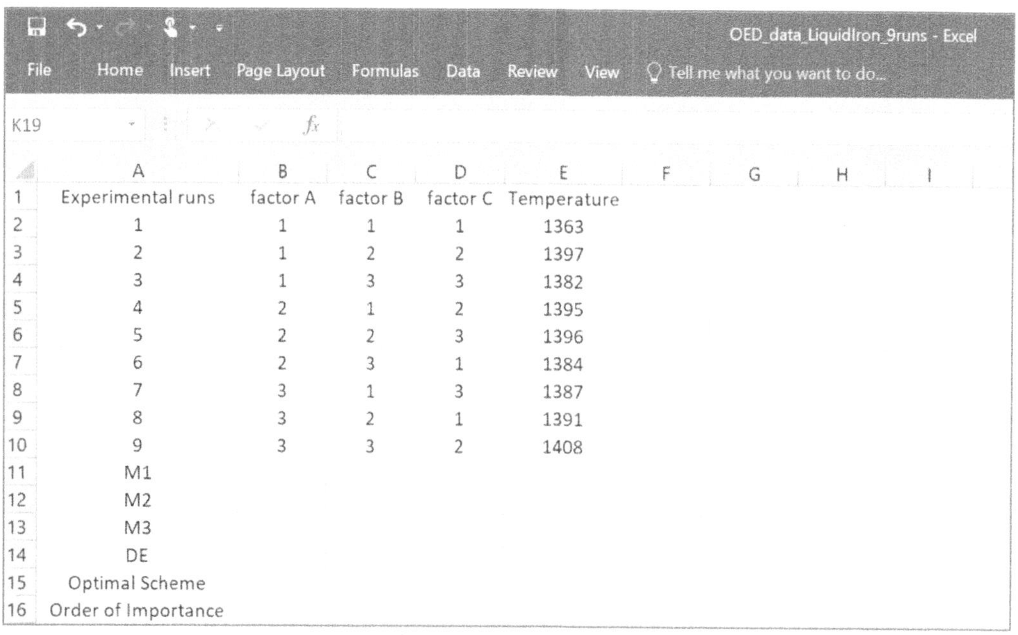

	A	B	C	D	E	F	G	H	I
1	Experimental runs	factor A	factor B	factor C	Temperature				
2	1	1	1	1	1363				
3	2	1	2	2	1397				
4	3	1	3	3	1382				
5	4	2	1	2	1395				
6	5	2	2	3	1396				
7	6	2	3	1	1384				
8	7	3	1	3	1387				
9	8	3	2	1	1391				
10	9	3	3	2	1408				
11	M1								
12	M2								
13	M3								
14	DE								
15	Optimal Scheme								
16	Order of Importance								

Figure 7.4 The screen copy of the OED_data_LiquidIron_9runs.xlsx file.

the comparison of *DE* values. The maximum *DE* implies the most important factor, and the second largest *DE* implies the second important factor, and so on. The order of importance can help us to first work on the most important factor to improve our design and pay less attention to the least important factor.

Solution:

```
MATLAB Code:
clc; clear all; close all
A=xlsread('OED_data_LiquidIron_9runs.xlsx','B2:E10');%read data
% from Excel file
K=zeros(3);%initial K values
%Calculate K values
for i=1:9
    for j=1:3
        for m=1:3
            if(A(i,j)==m)
                K(m,j)=K(m,j)+A(i,4);
            end
        end
    end
end
%Calculate M values
M=K/3;
xlswrite('OED_data_LiquidIron_9runs.xlsx', M, 1, 'B11:D13');
%Write M values
%Calculate DE values
DE=max(M)-min(M);
%Identify Optimal Scheme
[MT,OS]=max(M);
xlswrite('OED_data_LiquidIron_9runs.xlsx', DE, 1, 'B14:D14');
%Write DE values
xlswrite('OED_data_LiquidIron_9runs.xlsx', OS, 1, 'B15:D15');
% Optimal Scheme
%Rank the order of important factors
[B, Im]=sort(DE);
for m=1:3
    for i=1:3
    if(B(i)==DE(m))
        ImF(m)=3-i+1;
    end
    end
end
xlswrite('OED_data_LiquidIron_9runs.xlsx', ImF, 1, 'B16:D16');
% order of factors
```

After we executed the MATLAB code, the results are shown in Figure 7.5. The order of importance in this case shows that Factor C (height of the load) is the most important factor

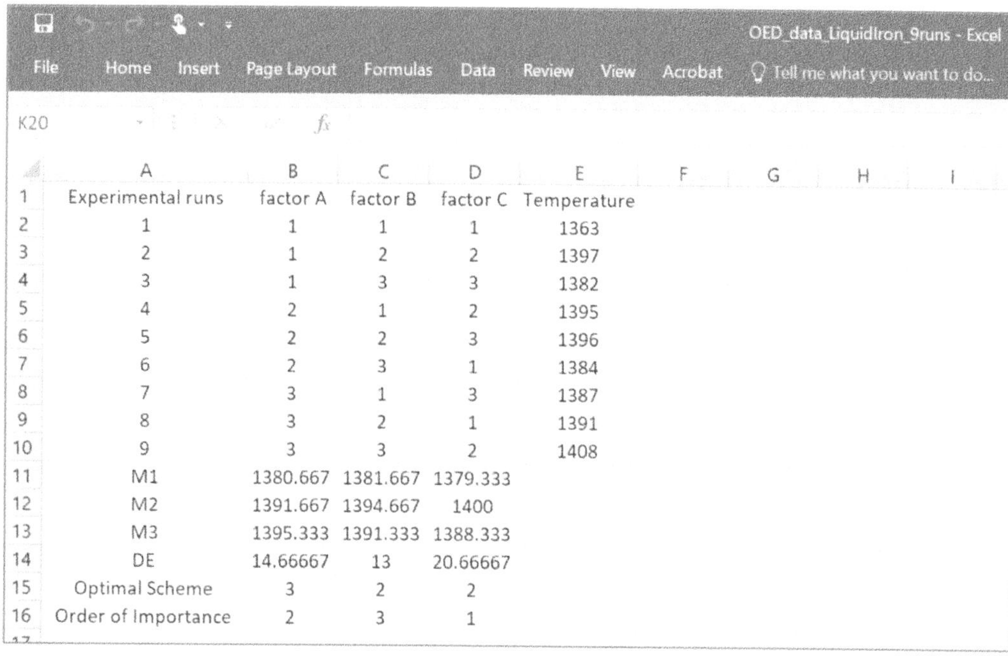

OED_data_LiquidIron_9runs - Excel

	A	B	C	D	E
1	Experimental runs	factor A	factor B	factor C	Temperature
2	1	1	1	1	1363
3	2	1	2	2	1397
4	3	1	3	3	1382
5	4	2	1	2	1395
6	5	2	2	3	1396
7	6	2	3	1	1384
8	7	3	1	3	1387
9	8	3	2	1	1391
10	9	3	3	2	1408
11	M1	1380.667	1381.667	1379.333	
12	M2	1391.667	1394.667	1400	
13	M3	1395.333	1391.333	1388.333	
14	DE	14.66667	13	20.66667	
15	Optimal Scheme	3	2	2	
16	Order of Importance	2	3	1	

Figure 7.5 The screen copy of the OED_data_LiquidIron_9runs.xlsx file after we executed the MATLAB code.

in improving the temperature of the liquid iron, Factor A (coal ratio) is the second important factor, and Factor B (pressure) is the least important factor. The comparison from *M* values leads to the optimal scheme: A3–B2–C2. The optimal scheme can be inside (or outside) the existing tests. In this case, it is outside the existing tests; therefore, the optimal scheme must be performed to check whether it is better than the existing maximum temperature case of the ninth run.

The trend on how each factor affects the temperature of the liquid iron can be seen from the plots of *M* values vs. the three factors in Figure 7.6. The MATLAB code is as follows:

```
MATLAB Code:
A1=[1/16;1/18;1/14];B1=[22.6;30.6;26.6];C1=[1.2;1.6;1.4];
tiledlayout(1,3);
% Tile 1
nexttile
plot(A1,M(:,1),'linestyle','none','marker','*')
xlim([0.05 0.08])
ylim([1375 1410])
xlabel('Coal ratio')
ylabel('Temperature')
% Tile 2
nexttile
plot(B1,M(:,2),'linestyle','none','marker','o')
xlim([20 35])
ylim([1375 1410])
xlabel('Pressure')
```

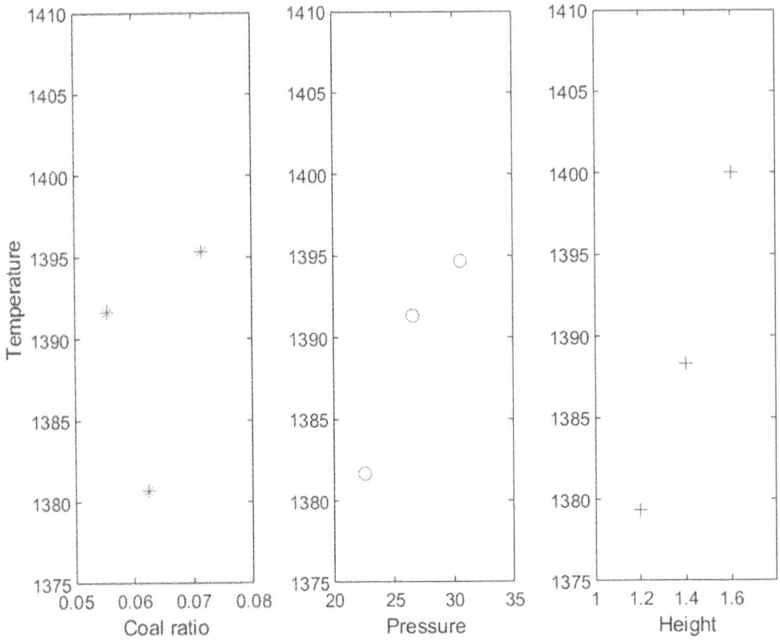

Figure 7.6 The plots of M values vs. three factors.

```
% Tile 3
nexttile
plot(C1,M(:,3),'linestyle','none','marker','+')
xlim([1.0 1.8])
ylim([1375 1410])
xlabel('Height')
```

It can be seen from Figure 7.6 that an increase in pressure or height leads to an increase in temperature. At the same time, the broader range of the temperature increase means that height plays a more critical role than pressure. The plot of the M values vs. coal ratio shows a "V" shape, which means there will be a minimum temperature when the coal ratio is between 1:18 and 1:14.

Learning Tips

- The choice of factors and levels needs some experience.
- The right OED table will help data processing.
- Experience of analysis and evaluation is coming from more practice.

7.3 Model Building with Experimental Data

Example 7.8 will demonstrate how to use the data to build mathematical models. The OED method was applied to conduct experimental runs, and data are shown in Figure 7.5. Three factors can change the temperature of liquid iron, including coal ratio (Factor A), pressure

(Factor B), and height of the load (Factor C), as shown in Figure 7.6. In this section, a method for model building will be based on the relationships among *M* values vs. three factors.

7.3.1 The Relationships Among *M* Values vs. Three Factors

The model-fitting skills learned from Chapter 3 will be performed using *M* values and their corresponding levels of the three factors from the data in Figure 7.5 and Table 7.3.

***Note**: It does not require the levels of each factor to have ascending order for data analysis in the Section 7.2.3. However, to use the MATLAB model fitting toolbox, the levels of each factor need to have an ascending order, e.g. the levels and *M* values for factor A need to be transferred as follows:

Factor A			Factor A
Level 1	1:16	M1	1380.667
Level 2	1:18	M2	1391.667
Level 3	1:14	M3	1395.333

→

Factor A			Factor A
Level 2	1:18	M2	1391.667
Level 1	1:16	M1	1380.667
Level 3	1:14	M3	1395.333

(1) *M* values vs. coal ratio (Factor A)

The results are as follows using a polynomial relationship, and a screenshot of the model fitting is shown in Figure 7.7:

```
Linear model Poly2:
     f(x)  = p1*x^2 + p2*x + p3
Coefficients:
       p1  =  2.033e+05
       p2  = -2.558e+04
       p3  =     2185
Goodness of fit:
  SSE: 1.241e-24
  R-square: 1
  Adjusted R-square: NaN
  RMSE: NaN
```

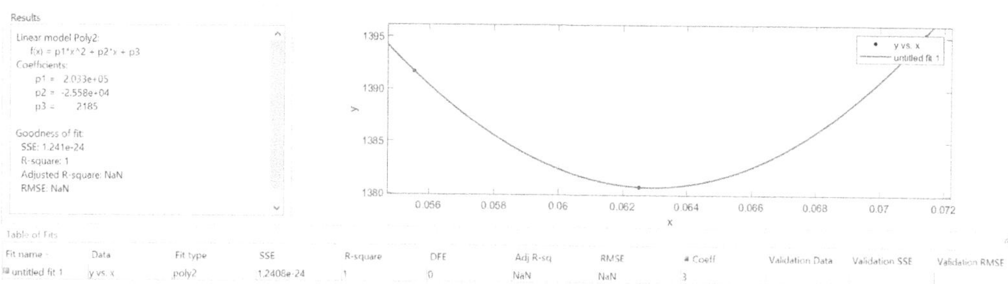

Figure 7.7 The model fitting of *M* values vs. coal ratio (Factor A).

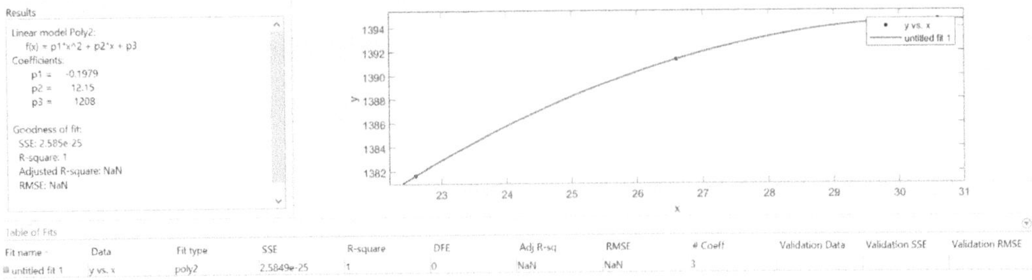

Figure 7.8 The model fitting of *M* values vs. pressure (Factor B).

(2) *M* values vs. pressure (Factor B)

The results are as follows using a polynomial relationship, and a screenshot of the model fitting is shown in Figure 7.8:

```
Linear model Poly2:
     f(x)  = p1*x^2 + p2*x + p3
Coefficients:
       p1  = -0.1979
       p2  =  12.15
       p3  =  1208
Goodness of fit:
  SSE: 2.585e-25
  R-square: 1
  Adjusted R-square: NaN
  RMSE: NaN
```

(3) *M* values vs. height of the load (Factor C)

The results are as follows using a polynomial relationship, and a screenshot of the model fitting is shown in Figure 7.9:

```
Linear model Poly2:
     f(x)  = p1*x^2 + p2*x + p3
Coefficients:
       p1  =  33.33
       p2  = -41.67
       p3  =  1381
```

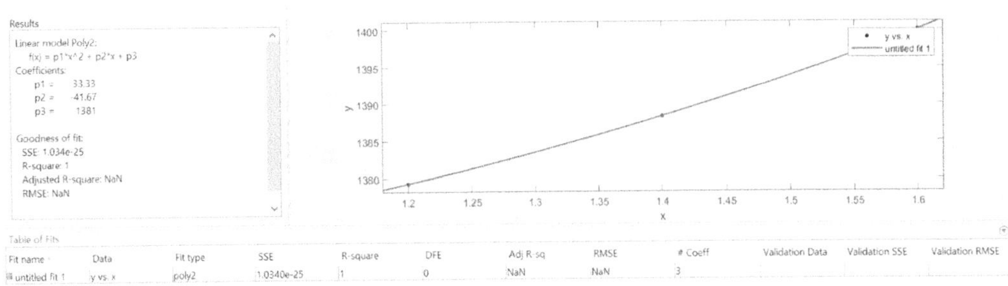

Figure 7.9 The model fitting of *M* values vs. height of the load (Factor C).

```
Goodness of fit:
  SSE: 1.034e-25
  R-square: 1
  Adjusted R-square: NaN
  RMSE: NaN
```

7.3.2 The Power Relationships and Multiplication used for Model Building

The proposed model of the liquid iron temperature will have the following expression:

$$T = \left(\frac{f(A)}{DE_A}\right)^{\alpha} \cdot \left(\frac{g(B)}{DE_B}\right)^{\beta} \cdot \left(\frac{h(C)}{DE_C}\right)^{\gamma} \tag{7.12}$$

where $f(A)$ is the relationship between M values vs. coal ratio (Factor A) as shown in the Section 7.2.3; $g(B)$ is the above relationship between M values vs. pressure (Factor B) as shown in the Section 7.2.3; $h(C)$ is the above relationship between M values vs. the height of the load (Factor C) as shown in the Section 7.2.3. The DE values correspond to their factors in Figure 7.4, i.e. $DE_A = 14.66667$; $DE_B = 13$; $DE_C = 20.66667$.

There are only three unknown parameters α, β, and γ. Model-fitting exercises from the experimental data can determine them.

7.3.3 The Model Fitting to Determine the Unknown Parameters

Figure 7.5 shows the experimental data have all experimental conditions and the corresponding temperature. The method introduced in Chapter 3 on model fitting via optimization algorithms will be used to perform model fitting to determine the unknown parameters. To revisit the Eq. (3.38) (same as 7.13): to find $\{a_i\}$ such that the least squares difference between the observed value and model predicted value:

$$\min_i \left[\sum (y_{obs} - y_{pred})^2\right] \tag{7.13}$$

where y_{obs} is the observed temperature value; y_{pred} is the temperature function $T = \left(\frac{f(A)}{DE_A}\right)^{\alpha} \cdot \left(\frac{g(B)}{DE_B}\right)^{\beta} \cdot \left(\frac{h(C)}{DE_C}\right)^{\gamma}$.

(1) The main MATLAB code using fmincon solver is as follows:

```
MATLAB Code:
global T A1 B1 C1 A
A=xlsread('OED_data_LiquidIron_9runs.xlsx','B2:E10');
a=A(:,4);
T=a;
A1=[1/16;1/18;1/14];B1=[22.6;30.6;26.6];C1=[1.2;1.6;1.4];
x0=[0;0;0];
lb=[-1;-1;-1];
ub=[1;1;1];
%Use fmincon solver
[x,fval,exitflag]  = fmincon(@myfun_OED1,x0,[],[],[],[],lb,ub)
for i=1:length(a)
```

```
    ai=A(i,1);
    bi=A(i,2);
    ci=A(i,3);
a_cal(i)=((2.033e+05*A1(ai)^2 + -2.558e+04*...
A1(ai)+2185)/14.66667)^x(1)*((-0.1979*B1(bi)^2 + 12.15*...
B1(bi)+1208)/13)^x(2)*((33.33*C1(ci)^2 + -41.67*...
C1(ci)+1381)/20.66667)^x(3);
end
figure;plot(a_cal,a,'o',a,a,'-');
xlabel('Tcal');ylabel('T');
```

The function code is as follows:

```
function ft1 = myfun_OED1(x)
global T A1 B1 C1 A
sum1=0;
for i=1:length(T)
    ai=A(i,1);
    bi=A(i,2);
    ci=A(i,3);
    sum1= sum1+(((2.033e+05*A1(ai)^2 + -2.558e+04*...
A1(ai)+2185)/14.66667)^x(1)*((-0.1979*B1(bi)^2 + 12.15*...
B1(bi)+1208)/13)^x(2)*((33.33*C1(ci)^2 + -41.67*...
C1(ci)+1381)/20.66667)^x(3)- T(i))^2;
end
ft1=sum1;
```

The results using fmincon are as follows:
Local minimum possible. Constraints satisfied.
fmincon stopped because the size of the current step is less than the value of the step size tolerance and constraints are satisfied to within the value of the constraint tolerance.

```
<stopping criteria details>
x =
     0.5154
     0.3650
     0.7572
fval =
   231.3334
exitflag =
     2
```

The plot of the result using fmincon solver is shown in Figure 7.10.

(2) The main MATLAB code using ga solver is as follows:

The ga solver will be tried first to cross-check the optimization results using other optimization solvers. In the above main MATLAB code, the line of fmincon will

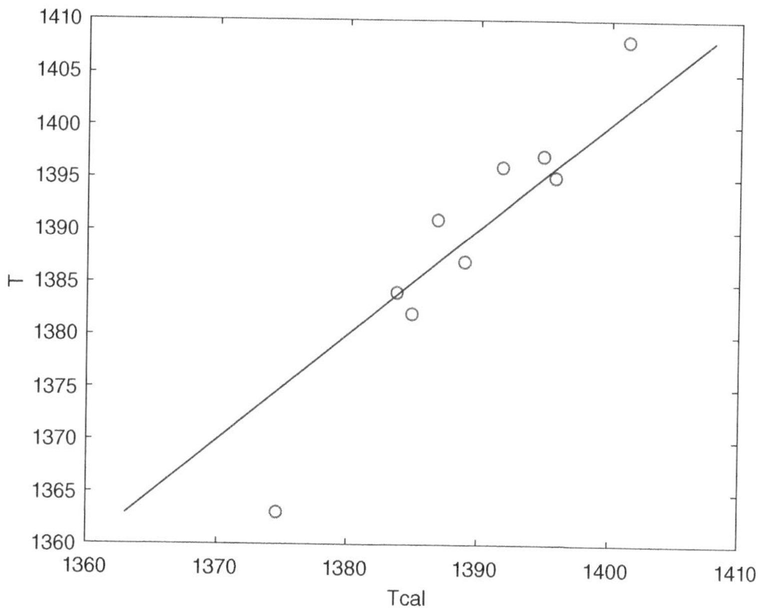

Figure 7.10 The plot of the result using fmincon solver.

be replaced by the following command:

```
[x,fval,exitflag]  = ga(@myfun_OED1,3,[],[],[],[],lb,ub)
```

The results using ga are as follows:

Optimization terminated: maximum number of generations exceeded.

```
x =
    0.1697    0.8388    0.6051
fval =
  349.8471
exitflag =
     0
```

The plot of the result using ga solver is shown in Figure 7.11.

(3) The main MATLAB code using simulannealbnd solver is as follows:

The simulannealbnd solver will be tried to cross-check the optimization results with other optimization solvers. In the above main MATLAB code, the line of fmincon will be replaced by the following command:

```
[x,fval,exitflag]  = simulannealbnd(@myfun_OED1,x0,lb,ub)
```

The results using simulannealbnd are as follows:

Maximum number of function evaluations exceeded: increase options.MaxFunction-Evaluations.

```
x =
    0.0059
    0.6835
    0.9546
```

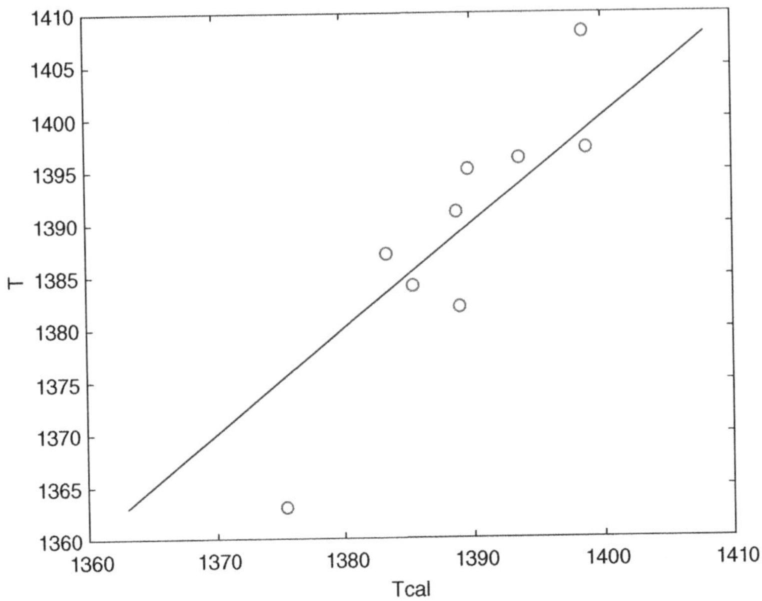

Figure 7.11 The plot of the result using ga solver.

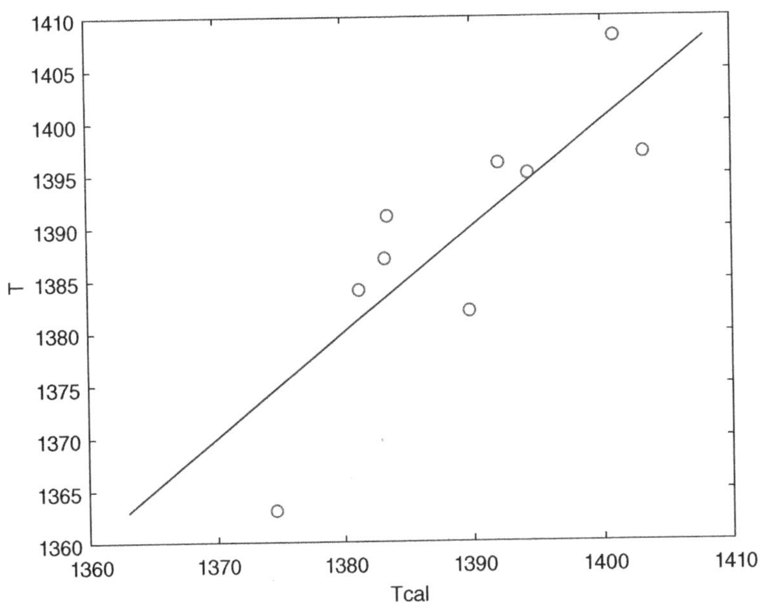

Figure 7.12 The plot of the result using simulannealbnd solver.

```
fval =
  376.0403
exitflag =
     0
```

The plot of the result using simulannealbnd solver is shown in Figure 7.12.

(4) The main MATLAB code using GlobalSearch is as follows:

To cross-check the optimization results using other optimization solvers, the GlobalSearch will be tried. In the above main MATLAB code, the line of fmincon will be replaced by the following command:

```
rng default % For reproducibility
gs = GlobalSearch;
problem = createOptimProblem('fmincon','x0',x0,...
    'objective',@myfun_OED1,'lb',lb,'ub',ub);
[x,fval,exitflag] = run(gs,problem)
```

The results using GlobalSearch are as follows:
GlobalSearch stopped because it analyzed all the trial points.
All 12 local solver runs converged with a positive local solver exit flag.

```
x =
     0.5154
     0.3651
     0.7571
fval =
  231.3333
exitflag =
     1
```

The plot of the result using GlobalSearch solver is shown in Figure 7.13.

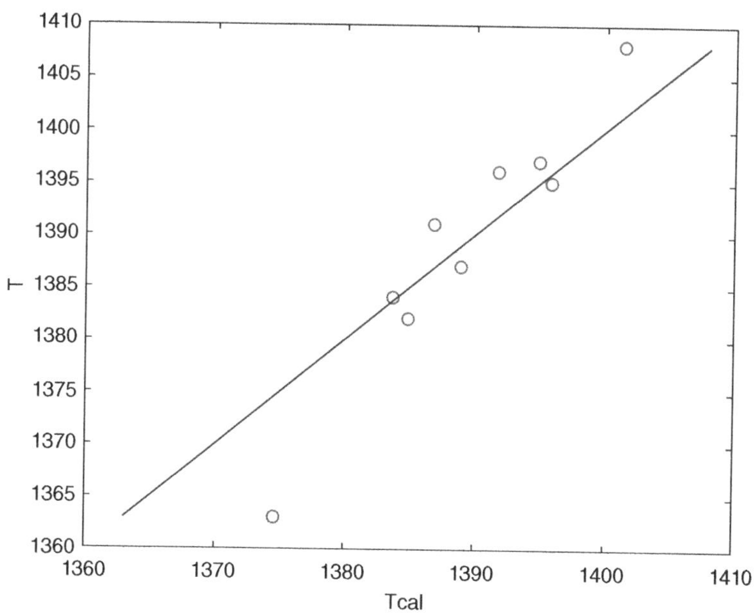

Figure 7.13 The plot of the result using GlobalSearch solver.

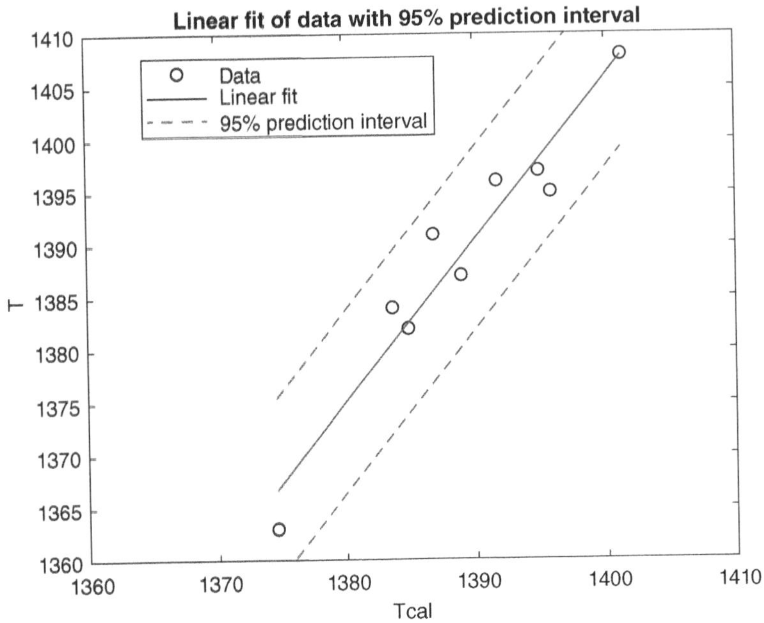

Figure 7.14 The model is based on the OED design within a 95% prediction interval.

Other optimization solvers can be applied in this model-building case; some will be listed in the exercise of the latter section of the problems in this chapter.

Comparing the results of the above four optimization solvers, fmincon and GlobalSearch can provide better-optimized results than ga and simulannealbnd. Please note that the ga and simulannealbnd results can differ for different runs. The model based on the OED design using fmincon optimization gives good results within a 95% prediction interval as shown in Figure 7.14.

7.4 Application in Chemical Engineering

7.4.1 Experimental Design for Nanoparticle-Reinforced Silica Gels

Example 7.9 In the nanoparticle-incorporated sol–gel process, many factors could affect the quality of the final gel product. The OED has been applied to determine the best experimental conditions (Xie et al. 2020). The four factors and their three levels considered in this study are shown in Table 7.4. The range of each factor level was based on the results of preliminary experiments. The experiments were conducted using an L9 (3^4) orthogonal array (OA), and the response values (Vickers hardness) obtained are given in Table 7.5. The order of the experiments was randomized to avoid noise sources. The ultrasonic waves are introduced to make the uniform mixing for the hydrolysis process.

It can be seen from Table 7.5 that the factors that influence hardness are listed in decreasing order as follows: C > A > D > B, according to the *DE* values. Therefore, the maximum

Table 7.4 Factors and levels of design for nanoparticle-reinforced silica gels.

Experimental parameters	Designed factor	Level 1	Level 2	Level 3
Ultrasonic time (min)	A	5	10	15
Amount of nanoparticle (g)	B	0.05	0.10	0.15
Gelation temperature (°C)	C	40	50	60
Gelation time (d)	D	1	2	3

Table 7.5 L9 (3^4) orthogonal array with factors and response.

Experimental runs	A	B	C	D	Vickers hardness
1	1	1	1	1	16.0
2	1	2	2	2	51.0
3	1	3	3	3	52.3
4	2	1	2	3	56.8
5	2	2	3	1	47.9
6	2	3	1	2	39.0
7	3	1	3	2	76.2
8	3	2	1	3	40.2
9	3	3	2	1	59.7
M1	39.8	49.7	31.7	41.2	
M2	47.9	46.4	55.8	55.4	
M3	58.7	50.3	58.8	49.8	
DE	18.9	4.0	27.1	14.2	
Optimal scheme	3	3	3	2	
Order of importance	2	4	1	3	

hardness of the nanoparticle-reinforced gel will be obtained when the gelation temperature, ultrasonic time, gelation time, and the amount of nanoparticles are C3, A3, D2, and B3 (60 °C, 15 min, 2 d, and 0.15 g), respectively. According to the *DE* values, the gelation temperature was the most important determinant of hardness. The ultrasonic time (min) for mixing is the second most important factor. The gelation time (d) is the third most important factor. The amount of nanoparticle (g) is the least important factor. The trend on how each factor affects the hardness of the nanoparticle-reinforced gel can be seen from the plots of *M* values vs. four factors in Figure 7.15. This OED L9 (3^4) applied in this case achieved a time saving of 88.9% effort with 9 trials, compared to the total combinations of 81 trials from 4 factors at 3 levels. After the OED design, further experiments can be conducted to improve the hardness of nanoparticle-reinforced silica gels.

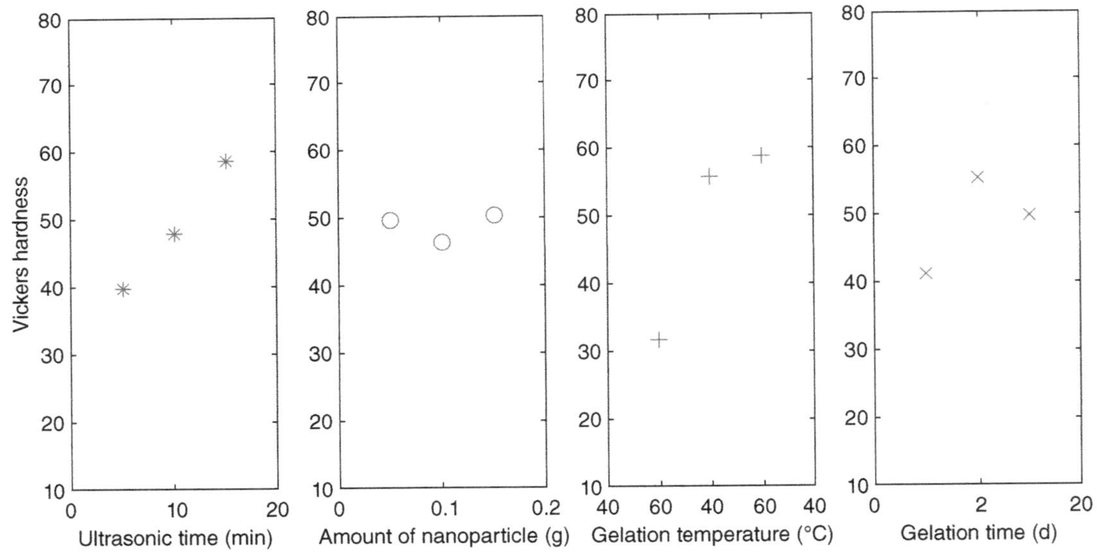

Figure 7.15 The plots of M values vs. four factors.

7.4.2 Experimental Design for Turbulence Model Development of a Flotation Cell

Example 7.10 Turbulence in multiphase technical flows is key to the dispersion and mixing of the multiphase constituents and various multi-physics processes. Turbulence is an important factor affecting flotation performance since it affects air dispersion, particle suspension and bubble-particle collection interactions, and entrainment of fine particles. Figure 7.16 shows a schematic of a stirred flotation cell with the froth phase on the top and the pulp phase on the bottom. The systematic OED design method saves time and effort in optimization. For example, the OED L16 (4^5) has been applied in turbulence model development of a stirred flotation cell achieved a time saving of 98.4% effort with 16 trials, shown in Table 7.6, compared to the total combinations of 1024 trials from 5 factors at 4 levels (Meng et al. 2016). After initial analysis from experience, there are four factors will influence the turbulence in the flotation cell, including Factor A (Impeller speed, rpm), Factor B (Superficial gas velocity, cm/s), Factor C (Water level, cm), and Factor D (Surfactant concentration, wt%). The OED L16 (4^5) can still be applied for those four factors and make the Factor 5 constant (i.e. the last column of Table 7.6 can be deleted). Four levels for each factor are shown in Table 7.7. The experimental responses and data analysis of the turbulence volume (L) in a stirred flotation cell are shown in Table 7.8.

It can be seen from Table 7.8 that the factors that influence the turbulence volume are listed in decreasing order as follows: A > D > C > B, according to the *DE* values. Therefore, the maximum turbulence volume in the stirred flotation cell will be obtained when impeller speed, surfactant concentration, water level, and superficial gas velocity are A4, D2, C3, and B2 (270 rpm, 2%, 100 cm, and 0.8 cm/s), respectively. According to the *DE* values, the impeller speed was the most important determinant of the turbulence volume. The surfactant concentration is the second most important factor. The water level is the third most important factor. The superficial gas velocity is the least important factor. The trend on how each factor affects the turbulence volume can be seen from the plots of *M* values vs. four factors in Figure 7.17. It can be seen that impeller speed has a much more significant influence than other factors.

Figure 7.16 A schematic of a stirred flotation cell.

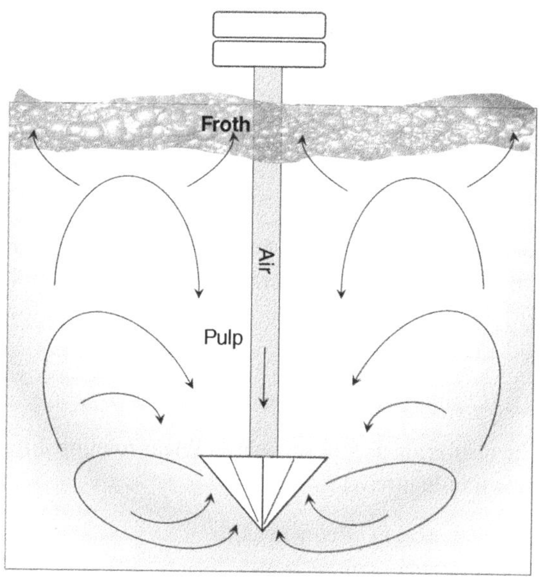

Table 7.6 The OED L16 (4^5) table.

Experiment	Factor 1	Factor 2	Factor 3	Factor 5	Factor 5
1	1	1	1	1	1
2	1	2	2	2	2
3	1	3	3	3	3
4	1	4	4	4	4
5	2	1	2	3	4
6	2	2	1	4	3
7	2	3	4	1	2
8	2	4	3	2	1
9	3	1	3	4	2
10	3	2	4	3	1
11	3	3	1	2	4
12	3	4	2	1	3
13	4	1	4	2	3
14	4	2	3	1	4
15	4	3	2	4	1
16	4	4	1	3	2

Table 7.7 Factors and levels of design for turbulence model development of a flotation cell.

	Factor A (impeller speed, rpm)	Factor B (superficial gas velocity, cm/s)	Factor C (water level, cm)	Factor D (surfactant concentration, wt%)
Level 1	150	0.6	70	0
Level 2	190	0.8	85	2
Level 3	230	1.0	100	4
Level 4	270	1.2	106	6

7.4.2.1 The Relationships Among *M* Values vs. Four Factors

The model-fitting skills learned from Chapter 3 will be used to build a mathematical model with *M* values and their corresponding levels of the four factors from the data in Table 7.7 and Table 7.8.

(1) *M* values vs. impeller speed (Factor A)

The results are as follows using a linear relationship, and a screenshot of the model fitting is shown in Figure 7.18:

```
Linear model Poly1:
     f(x) = p1*x + p2
```

Table 7.8 The experimental response and data analysis of the turbulence volume (L) in the stirred flotation cell.

Experimental runs	Factor A	Factor B	Factor C	Factor D	Volume
1	1	1	1	1	35.1
2	1	2	2	2	139.4
3	1	3	3	3	92.5
4	1	4	4	4	80.3
5	2	1	2	3	288.8
6	2	2	1	4	359.1
7	2	3	4	1	317
8	2	4	3	2	373
9	3	1	3	4	627.3
10	3	2	4	3	607.1
11	3	3	1	2	629.8
12	3	4	2	1	664.8
13	4	1	4	2	1001.6
14	4	2	3	1	1032.7
15	4	3	2	4	900.5
16	4	4	1	3	801.4
M1	86.825	488.2	456.35	512.4	
M2	334.475	534.575	498.375	535.95	
M3	632.25	484.95	531.375	447.45	
M4	934.05	479.875	501.5	491.8	
DE	847.225	54.7	75.025	88.5	
Optimal scheme	4	2	3	2	
Order of importance	1	4	3	2	

```
Coefficients (with 95% confidence bounds):
       p1 =   7.099   (6.113, 8.084)
       p2 = -993.8   (-1205, -782.2)
Goodness of fit:
  SSE: 839.3
  R-square: 0.9979
  Adjusted R-square: 0.9969
  RMSE: 20.49
```

(2) M values vs. superficial gas velocity (Factor B)

The results are as follows using a polynomial relationship, and a screenshot of the model fitting is shown in Figure 7.19:

```
Linear model Poly3:
     f(x) = p1*x^3 + p2*x^2 + p3*x + p4
```

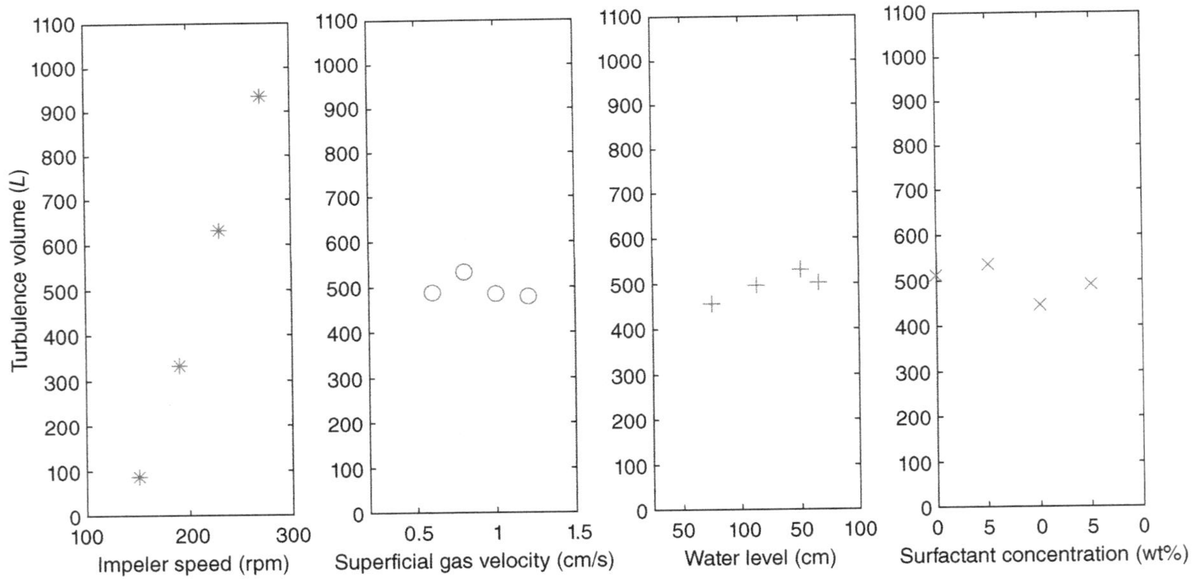

Figure 7.17 The plots of M values vs. four factors.

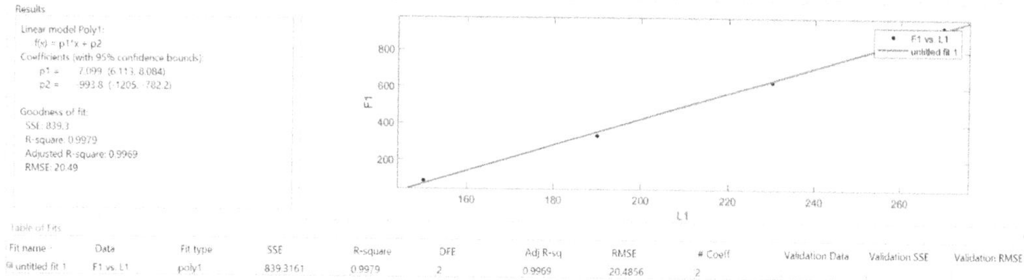

Figure 7.18 The model fitting of *M* values vs. impeller speed (Factor A).

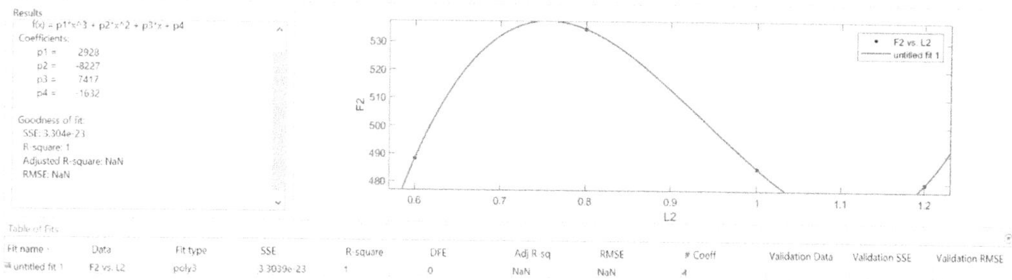

Figure 7.19 The model fitting of *M* values vs. superficial gas velocity (Factor B).

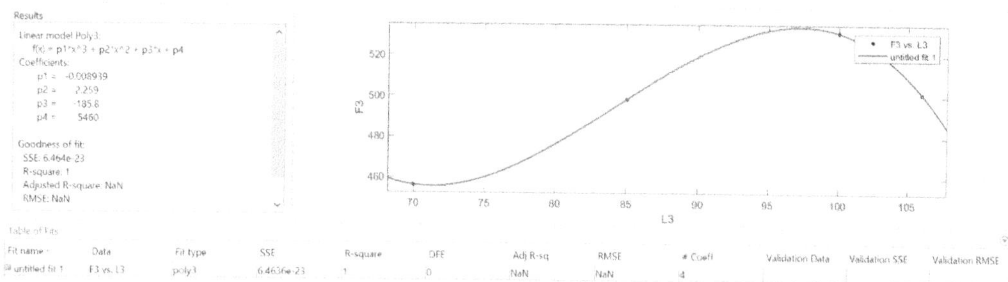

Figure 7.20 The model fitting of *M* values vs. water level (Factor C).

```
Coefficients:
       p1 =  2928
       p2 = -8227
       p3 =  7417
       p4 = -1632
Goodness of fit:
  SSE: 3.304e-23
  R-square: 1
  Adjusted R-square: NaN
  RMSE: NaN
```

(3) *M* values vs. water level (Factor C)

The results are as follows using a polynomial relationship, and a screenshot of the model fitting is shown in Figure 7.20:

```
Linear model Poly3:
     f(x) = p1*x^3 + p2*x^2 + p3*x + p4
Coefficients:
       p1 = -0.008939
       p2 =  2.259
       p3 = -185.8
       p4 =  5460
Goodness of fit:
  SSE: 6.464e-23
  R-square: 1
  Adjusted R-square: NaN
  RMSE: NaN
```

(4) M values vs. surfactant concentration (Factor D)

The results are as follows using a polynomial relationship, and a screenshot of the model fitting is shown in Figure 7.21:

```
Linear model Poly3:
     f(x) = p1*x^3 + p2*x^2 + p3*x + p4
Coefficients:
       p1 =  5.102
       p2 = -44.62
       p3 =  80.6
       p4 =  512.4
Goodness of fit:
  SSE: 1.292e-25
  R-square: 1
  Adjusted R-square: NaN
  RMSE: NaN
```

7.4.2.2 The Power Relationships and Multiplication Used for Model Building

The proposed model of the turbulence model development of a flotation cell will have the following expression:

$$V = \left(\frac{f(A)}{DE_A}\right)^{\alpha} \cdot \left(\frac{g(B)}{DE_B}\right)^{\beta} \cdot \left(\frac{h(C)}{DE_C}\right)^{\gamma} \cdot \left(\frac{s(D)}{DE_D}\right)^{\delta} \tag{7.14}$$

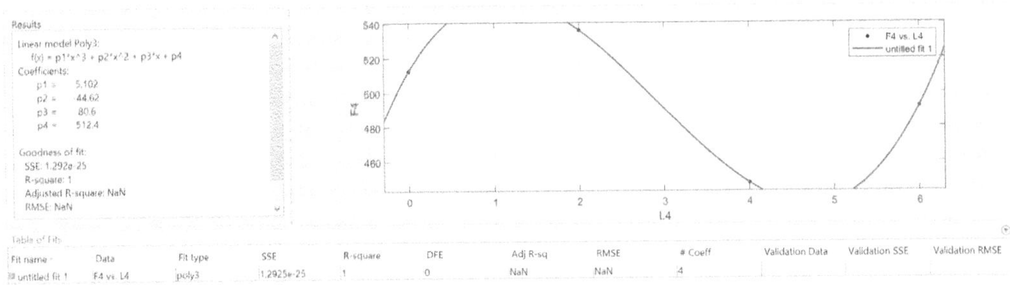

Figure 7.21 The model fitting of M values vs. surfactant concentration (Factor D).

where $f(A)$ is the relationship between M values vs. impeller speed (Factor A) as shown in previous section; $g(B)$ is the above relationship between M values vs. superficial gas velocity (Factor B) as shown in previous section; $h(C)$ is the above relationship between M values vs. water level (Factor C) as shown in previous section; $s(D)$ is the above relationship between M values vs. surfactant concentration (Factor D) as shown in previous section. The DE values correspond to their factors in Table 7.8, i.e. $DE_A = 847.225$; $DE_B = 54.7$; $DE_C = 75.025$; $DE_D = 88.5$.

There are only four unknown parameters α, β, γ, and δ. Model-fitting exercises from the experimental data can determine them.

7.4.2.3 The Model Fitting to Determine the Unknown Parameters

As shown in Table 7.8, the experimental data have all experimental conditions and the corresponding temperature. The method introduced in Chapter 3 on model fitting via optimization algorithms will be used to perform model fitting to determine the unknown parameters. To use Eqs. (7.13) and (7.14): to find $\{a_i\}$ such that the least squares difference between the observed value and model predicted value: y_{obs}, the observed temperature value; y_{pred}, the turbulence volume function

The main MATLAB code using GlobalSearch optimization is as follows:

```
MATLAB Code:
global V A1 B1 C1 D1 A C DE
A=xlsread('OED_data_turbulence.xlsx','B2:F17');
C=xlsread('OED_data_turbulence.xlsx','I2:L5');
DE=xlsread('OED_data_turbulence.xlsx','B22:E22');
V=A(:,5);
A1=C(:,1);B1=C(:,2);C1=C(:,3); D1=C(:,4);
x0=[0;0;0;0];
lb=[-1;-1;-1;-1];
ub=[10;10;10;10];
%Use GlobalSearch optimization
rng default % For reproducibility
gs = GlobalSearch;
problem = createOptimProblem('fmincon','x0',x0,...
    'objective',@myfun_OED1_turbulence,'lb',lb,'ub',ub);
[x,fval,exitflag] = run(gs,problem)
% Plot results within 95% prediction interval
for i=1:length(V)
    ai=A(i,1);
    bi=A(i,2);
    ci=A(i,3);
    di=A(i,4);
       p1 =7.099;
       p2 =-993.8;
       za=((p1*A1(ai) + p2)/DE(1))^x(1);
       p1 = 2928;
       p2 = -8227;
       p3 = 7417;
```

```
        p4 = -1632;
        zb=((p1*B1(bi)^3 + p2*B1(bi)^2 + p3*B1(bi) + p4)/...
DE(2))^x(2);
        p1 =-0.008939;
        p2 =2.259;
        p3 =-185.8;
        p4 =5460;
        zc=((p1*C1(ci)^3 + p2*C1(ci)^2 + p3*C1(ci) + p4)/...
DE(3))^x(3);
        p1 =5.102;
        p2 =-44.62;
        p3 =80.6;
        p4 =512.4;
        zd=((p1*D1(di)^3 + p2*D1(di)^2 + p3*D1(di) + p4)/...
DE(4))^x(4);
V_cal(i)=za*zb*zc*zd;
end
figure
[p,S] = polyfit(V_cal,V,1);
[y_fit,deltam] = polyval(p,V_cal,S);
delta=max(deltam);
plot(V_cal,V,'bo')
hold on
plot(V_cal,y_fit,'r-')
plot(V_cal,y_fit+2*delta,'m-',V_cal,y_fit-2*delta,'m-')
title('Linear fit of data with 95% prediction interval')
legend('Data','Linear fit','95% prediction interval')
xlabel('Vcal');ylabel('V');
xlim([0 1100])
ylim([0 1100])
```

The function code is as follows:

```
function ft1 = myfun_OED1_turbulence(x)
global V A1 B1 C1 D1 A C DE
sum1=0;
for i=1:length(V)
    ai=A(i,1);
    bi=A(i,2);
    ci=A(i,3);
    di=A(i,4);
       p1 =7.099;
       p2 =-993.8;
       za=((p1*A1(ai) + p2)/DE(1))^x(1);
       p1 = 2928;
       p2 = -8227;
       p3 = 7417;
       p4 = -1632;
```

```
        zb=((p1*B1(bi)^3 + p2*B1(bi)^2 + p3*B1(bi) + p4)/...
DE(2))^x(2);
        p1 =-0.008939;
        p2 =2.259;
        p3 =-185.8;
        p4 =5460;
        zc=((p1*C1(ci)^3 + p2*C1(ci)^2 + p3*C1(ci) + p4)/...
DE(3))^x(3);
        p1 =5.102;
        p2 =-44.62;
        p3 =80.6;
        p4 =512.4;
        zd=((p1*D1(di)^3 + p2*D1(di)^2 + p3*D1(di) + p4)/...
DE(4))^x(4);
    sum1= sum1+(za*zb*zc*zd - V(i))^2;
end
ft1=sum1;
```

The results using GlobalSearch are as follows:
GlobalSearch stopped because it analyzed all the trial points.
All 15 local solver runs converged with a positive local solver exit flag.

```
x =
    1.0158
    1.6073
    0.4456
    1.3578
fval =
   2.1134e+04
exitflag =
     1
```

Many other optimization solvers can be applied in this model-building case. Some will be listed in the exercise of the latter section of the problems in this chapter. The model built based on the OED design using GlobalSearch optimization gives good results within a 95% prediction interval, as shown in Figure 7.22.

Learning Tips

- The OED method can be applied to efficiently identify the significance of each factor.
- The power relationships and multiplication can be used for model building with the OED design.
- The model fitting with optimization can be used to determine the unknown parameters.

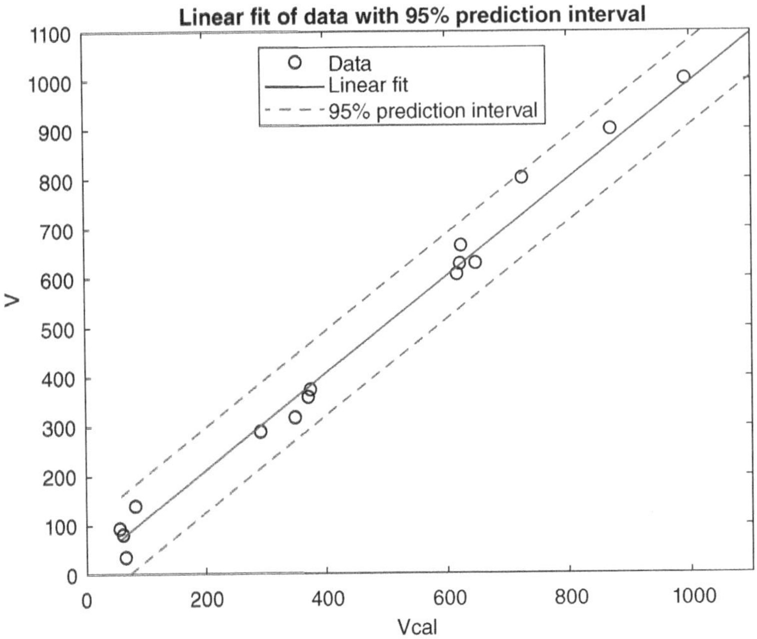

Figure 7.22 The model was built based on the OED design within a 95% prediction interval.

7.5 Summary

This chapter presents optimal experimental design and its application in chemical engineering design and simulation. Properly designed experiments will lead to successful experiments and improve data analysis and optimization.

There are some existing MATLAB functions for the DOE, including Full Factorial Designs, Fractional Factorial Designs, Response Surface Designs, D-Optimal Designs, Latin Hypercube Designs, and Quasi-Random Designs, in this chapter. OED has been introduced in more detail with some chemical engineering applications. In some cases, the OED method has been applied for model-building. The MATLAB Optimization toolbox in Chapter 4 and Global Optimization toolbox in Chapter 6 can be used to determine the unknown parameters for model-fitting with the OED.

Problems

7.1 A gold-bearing pyrite mine uses a processing circuit, as Figure 7.18 shows. Because the equipment has been fixed, the only way to optimize the circuit will be to optimize the operational conditions (for the seven pieces of equipment with the red circle highlighted in Figure 7.23). The target is to maximize the flotation concentrate tonnage per hour. The OED table L15 (3^7) will be applied in this task. Parameters can be changed as follows: Jaw crusher CSS (mm); %solid setting for Water feeder1; SAG mill ball load; Cone crusher CSS (mm); Water feeder2 new water addition (tph); Ball mill fraction critical speed; Water feeder3 new water addition (tph). Table 7.9 shows the factors and

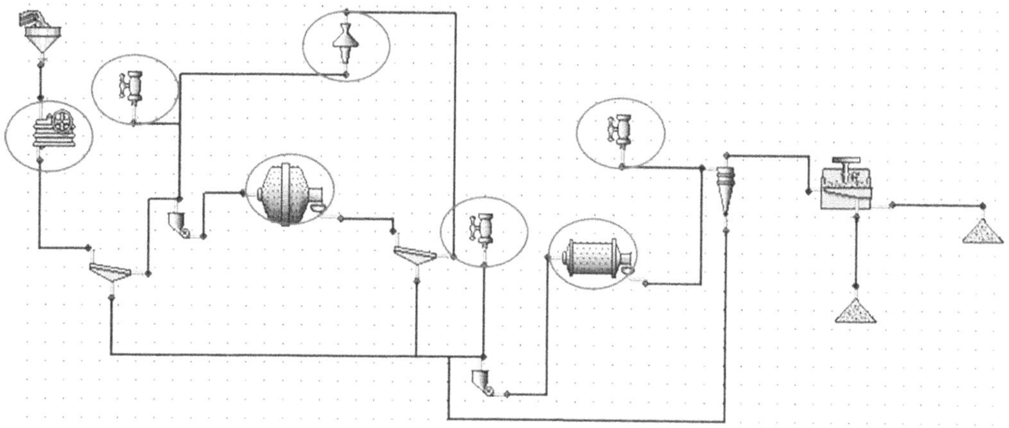

Figure 7.23 A mineral processing circuit (the seven pieces of equipment with the red circle highlighted).

Table 7.9 The factors and levels of design of the mineral processing circuit.

	Factor 1 Jaw CSS	Factor 2 Water1 % solids	Factor 3 SAG ball load	Factor 4 Cone CSS	Factor 5 Water 2 volume	Factor 6 Ball mill speed	Factor 7 Water 3 volume
Level 1	100	70	6	10	100	0.7	100
Level 2	105	73.5	6.3	10.5	105	0.735	105
Level 3	95	66.5	5.7	9.5	95	0.665	95

levels of design of the mineral processing circuit. Table 7.10 shows the experimental response and data analysis of the mineral processing circuit.

(1) Use OED data analysis to fill in Table 7.10.
(2) Plot M values vs. seven factors.

7.2 Use both Tables 7.7 and 7.8 of the OED design for turbulence model development of a flotation cell, and modify the existing MATLAB code of GlobalSearch optimization in the previous section and compare results:

(1) Use fmincon to determine the unknown parameters in the newly built model.
(2) Use ga to determine the unknown parameters in the newly built model.
(3) Use simulannealbnd to determine the unknown parameters in the newly built model.
(4) Use MultiStart to determine the unknown parameters in the newly built model.

7.3 Use both Tables 7.4 and 7.5 of the experimental data from the OED design for nanoparticle-reinforced silica gels and apply the power relationships and multiplication used for model building.

(1) MATLAB model fitting toolbox to find M values vs. four factors.
(2) Use fmincon to determine the unknown parameters in the newly built model.
(3) Use ga to determine the unknown parameters in the newly built model.
(4) Use GlobalSearch to determine the unknown parameters in the newly built model.

Table 7.10 The experimental response and data analysis of the mineral processing circuit using the OED table L15 (3^7) (Bolboacă and Jäntschi 2007/MDPI/CC BY 4.0).

Tests	Factor 1	Factor 2	Factor 3	Factor 4	Factor 5	Factor 6	Factor 7	Recovery
1	1	1	1	1	1	1	2	85.53
2	2	1	1	1	3	3	1	84.49
3	3	1	1	3	1	2	3	86.59
4	1	1	3	3	1	3	2	84.30
5	2	1	3	1	3	1	2	85.70
6	3	2	1	3	3	1	2	86.85
7	1	2	3	3	3	1	3	86.91
8	2	2	2	3	2	3	1	84.31
9	3	2	3	1	2	3	3	85.34
10	1	3	1	2	3	3	3	82.92
11	2	3	2	1	1	2	3	85.42
12	3	2	3	2	2	2	1	87.72
13	1	3	2	2	2	2	1	85.25
14	2	3	2	2	1	1	1	84.26
15	3	3	2	2	2	2	2	85.18
M1								
M2								
M3								
DE								
Optimal scheme								
Order of importance								

8

Data Statistics

Data statistics presented in this chapter aim to develop the readers' capability to analyze and model data. With the surging availability of massive data in chemical engineering, it is crucial to have skills in data statistics, such as dimensional analysis, graphical analysis, and statistical analysis (Xie and Davis 2022). The goal of this chapter is to help you build your confidence for optimized data-driven decision-making.

MATLAB has a built-in Statistics and Machine Learning Toolbox to describe, analyze, and model data. Machine learning is a big topic that requires intensive knowledge, which is out of the scope of this book and can be a separate book. Therefore, we focus on data statistics in this chapter. Some useful MATLAB tools are introduced, such as regressions, principal component analysis (PCA), dimensional reduction, and statistical visualization.

In chemical engineering, dimensional analysis is fundamental because measured data have numerical values and associated units. Therefore, it is helpful for our readers to understand the importance and meaning of dimensionless groups. The Buckingham Pi Method is introduced to correlate experimental data in dimensionless groupings of parameters and variables to generate valuable models. The Buckingham Pi Method has wide applications in engineering disciplines.

This chapter introduces the Data Statistics Toolbox in MATLAB and how it can be applied in chemical engineering. We also focus on the Buckingham PI Method for dimensional analysis to improve data analysis and model building, which is illustrated in more detail and applied to many cases.

8.1 Introduction

8.1.1 Data Types

With the surging availability of massive data, chemical engineering graduates face design, control, or operations optimization problems that involve handling data sets. Solving these problems requires proficiency in computer programming for data processing, analysis, visualization, modeling, and optimization. MATLAB generally has the following data types: numeric arrays, characters and strings, tables, structures, and cell arrays.

In MATLAB, there are functions to identify the data type of a variable, as shown in Table 8.1.

Chemical Engineering Analysis and Optimization Using MATLAB®, First Edition.
Weiguo Xie, Sam Toan, and Richard Davis.

Companion website: www.wiley.com/go/xie/chemicalengineering

Table 8.1 Functions to identify the data type of a variable in MATLAB (https://www.mathworks.com/The MathWorks, Inc.).

Functions	Description
Numeric data types	
isfloat	Determine if the input is a floating-point array.
isinteger	Determine whether the input is an integer array.
islogical	Determine if the input is a logical array.
isnumeric	Determine whether the input is a numeric array.
isreal	Determine whether the array uses complex storage.
issparse	Determine whether the input is sparse.
Characters and strings	
isstring	Determine if the input is a string array.
ischar	Determine if the input is a character array.
iscellstr	Determine if the input is a cell array of character vectors.
Dates and times	
isdatetime	Determine if the input is a date/time array.
isduration	Determine if the input is a duration array.
iscalendarduration	Determine if the input is a calendar duration array.
Cells, structures, and tables	
iscell	Determine if the input is a cell array.
isstruct	Determine if the input is a structure array.
istable	Determine if the input is a table.
istimetable	Determine if the input is a timetable.
istabular	Determine if the input is a table or timetable (since R2021b).
Other types	
isa	Determine if the input has a specified data type.
iscategorical	Determine whether the input is a categorical array.
isenum	Determine if the variable is enumeration.
isgraphics	True for valid graphics object handles.
isjava	Determine if the input is a Java object.
isobject	Determine if the input is a MATLAB object.
Information about variables	
whos	List variables in the workspace, with sizes and types.
class	Class of object.
underlyingType	Type of underlying data determining array behavior (since R2020b).
isUnderlyingType	Determine whether input has specified underlying data type (since R2020b).
validateattributes	Check the validity of the array.

Example 8.1
To type the following functions in the Command Window to see the answers:

Solution:

```
≫ pi
ans =
 3.1416

≫ isinteger(pi)
ans =
  logical
   0

≫ isreal(pi)
ans =
  logical
   1

≫ isa(pi,'double')
ans =
  logical
   1

≫ magic(5)
ans =
    17    24     1     8    15
    23     5     7    14    16
     4     6    13    20    22
    10    12    19    21     3
    11    18    25     2     9
```

*Note: Although all elements in the matrix look like integers, they are treated as 'real' 'double' data by default in MATLAB.

```
≫ isinteger(magic(5))
ans =
  logical
   0

≫ isreal(magic(5))
ans =
  logical
   1

≫ A= int32(magic(5))
```

```
A =
  5×5 int32 matrix
   17   24    1    8   15
   23    5    7   14   16
    4    6   13   20   22
   10   12   19   21    3
   11   18   25    2    9

≫ isa(A,'integer')
ans =
  logical
   1
```

*Note: We specified the A=magic(5) using int32 (i.e. 32-bit integers), and then A will be an integer data type.

```
≫ whos A
  Name      Size            Bytes  Class    Attributes
  A         5x5               100  int32

≫ b=magic(5);
≫ whos b
  Name      Size            Bytes  Class     Attributes
  b         5x5               200  double

≫ istable(magic(5))
ans =
  logical
   0

≫ iscell(magic(5))
ans =
  logical
   0

≫ isa(magic(5),'double')
ans =
  logical
   1
```

8.1.2 Physical Units

Figure 8.1 illustrates that a shoe does not equal a cupcake, but 1 = 1 by value; 2 m does not equal two minutes, but 2 = 2 by value. Therefore, you must distinguish between the numerical values and their associated units. Most computer languages (MATLAB, Python, Java, C, VB, R) can only work with data without units because they are general languages that are

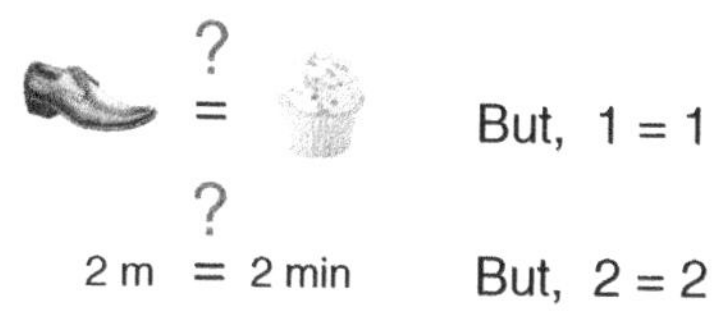

Figure 8.1 Data with or without units.

Table 8.2 SI base quantities.

Base quantity	Unit name	Symbol	Dimension symbol
Length	Meter	M	L
Mass	Kilogram	Kg	M
Time	Second	S	T
Electric current	Ampere	A	I
Thermodynamic temperature	Kelvin	K	θ
Amount of substance	Mole	Mol	N
Luminous intensity	Candela	Cd	J

not specific to any engineering discipline. All contents in previous chapters require you to know all the correct physical units before you use MATLAB functions and commands.

However, in real-world problems, especially chemical engineering design, physical units of any measurement data must be considered, which also requires chemical engineering knowledge.

The SI base units are the standard units of measurement defined by the International System of Units (SI) for the seven base quantities, as shown in Table 8.2. They form a set of mutually independent dimensions as required by dimensional analysis commonly employed in science and technology.

In the real world, the dimension of a quantity Q can be written as

$$\dim Q = L^a M^b T^c I^d \theta^e N^f J^g \tag{8.1}$$

where a, b, c, d, e, f, and g are the dimensional exponents.

For example, the dimension of the physical quantity speed v is

$$\dim v = \frac{\text{displacement}}{\text{time}} = \frac{L}{T} = LT^{-1} \tag{8.2}$$

The dimension of the physical quantity acceleration a is

$$\dim a = \frac{\text{speed}}{\text{time}} = \frac{LT^{-1}}{T} = LT^{-2} \tag{8.3}$$

The dimension of the physical quantity force F is

$$\dim F = \text{mass} \times \text{acceleration} = M \times LT^{-2} = LMT^{-2} \tag{8.4}$$

The dimension of the physical quantity pressure P is

$$\dim P = \frac{\text{force}}{\text{area}} = \frac{LMT^{-2}}{L^2} = L^{-1}MT^{-2} \tag{8.5}$$

The dimension of the physical quantity energy E is

$$\dim E = \text{force} \times \text{displacement} = LMT^{-2} \times L = L^2MT^{-2} \tag{8.6}$$

8.1.3 Dimensional Analysis

Dimensional analysis is a technique for restructuring the original dimensional variables of a problem into a set of dimensionless products using the constraints imposed upon them by their dimensions. It is based on the simple requirement for dimensional homogeneity in any relationship between the variables.

Example 8.2
Determine whether the following equation form with constants (a and b) is correct:

$$\Delta p = a \cdot \rho \cdot v^2 + b \cdot \Delta h \tag{8.7}$$

where p is the pressure drop; r is the fluid density; v is the fluid velocity; Δh is the height change.
***Hint**: Constants must be dimensionless.

Solution:
In the above expression, a is dimensionless and truly a fitting parameter. However, b has dimensions [ML^{-2} T^{-2}] and could, therefore, be decomposed into other known variables, e.g. $b = c \cdot \rho \cdot g$ where g is the acceleration due to gravity, and c is a dimensionless constant.

We can further modify the pressure drop equation using physics in the term for liquid height, i.e.

$$c = 1 \text{ and } \Delta p = a \cdot \rho \cdot v^2 + \rho \cdot g \cdot \Delta h \tag{8.8}$$

8.1.4 Some Dimensionless Groups

Some helpful dimensionless groups developed historically include:

(1) Biot number:

$$Bi = \frac{hL}{k} \tag{8.9}$$

where k is the thermal conductivity of the body [W/(m·K)]; h is a convective heat transfer coefficient [W/(m^2·K)]; L is a characteristic length [m] of the geometry considered. The Biot number is the ratio of the thermal resistance for conduction inside a body to the resistance for convection at the body's surface. It indicates whether the temperature inside a body varies significantly in space when the body is heated or cooled over time by a heat flux at its surface.

(2) Mass transfer Biot number:

$$Bi_m = \frac{k_c L}{D} \tag{8.10}$$

where k_c is the convective mass transfer coefficient [mol/(s·m^2)/(mol/m^3)], or [m/s]; D is the mass diffusivity [m^2/s]; L is a characteristic length [m] of the geometry considered.

(3) Fourier number:

$$Fo = \frac{\alpha t}{L^2} \tag{8.11}$$

where α is the thermal diffusivity [m^2/s]; t is the time [s]; L is a characteristic length [m] through which conduction occurs. The Fourier number is the ratio of time, t, to a characteristic time scale for heat diffusion, t_d.

(4) Mass transfer Fourier number:

$$Fo_m = \frac{Dt}{L^2} \tag{8.12}$$

where D is the mass diffusivity [m^2/s]; t is the time [s]; L is a characteristic length [m] through which conduction occurs. The mass transfer Fourier number is the ratio of time, t, to a characteristic time scale for mass diffusion, t_d.

(5) Lewis number:

$$Le = \frac{\alpha}{D} \tag{8.13}$$

where α is the thermal diffusivity [m^2/s]; D is the mass diffusivity [m^2/s]. The Lewis number is the ratio of thermal diffusivity to mass diffusivity.

(6) Nusselt number:

$$Nu = \frac{h}{k/L} \tag{8.14}$$

where k is the thermal conductivity of the body [W/(m·K)]; h is a convective heat transfer coefficient [W/(m^2·K)]; L is a characteristic length [m]. The Nusselt number is the ratio of conductive thermal resistance to the convective thermal resistance of the fluid.

(7) Péclet number for heat transfer:

$$Pe = \frac{Lu}{\alpha} \tag{8.15}$$

where L is a characteristic length [m]; u is the local flow velocity [m/s]; α is the thermal diffusivity [m^2/s]. The Péclet number for heat transfer is the ratio of the rate of advection of heat by the flow to the rate of diffusion of the same heat driven by an appropriate gradient.

(8) Péclet number for mass transfer:

$$Pe_m = \frac{Lu}{D} \tag{8.16}$$

where L is a characteristic length [m]; u is the local flow velocity [m/s]; D is the mass diffusivity [m^2/s]. The Péclet number for mass transfer is the ratio of the rate of advection of a mass by the flow to the rate of diffusion of the same mass driven by an appropriate gradient.

(9) Prandtl number:

$$Pr = \frac{\nu}{\alpha} \tag{8.17}$$

where α is the thermal diffusivity [m^2/s]; ν is momentum diffusivity (kinematic viscosity) [m^2/s]. The Prandtl number is the ratio of momentum diffusivity to thermal diffusivity.

(10) Reynolds number:

$$Re = \frac{uL}{\nu} = \frac{\rho uL}{\mu} \tag{8.18}$$

where ρ is the density of the fluid [kg/m^3]; ν is the momentum diffusivity (kinematic viscosity) [m^2/s]; u is the flow speed [m/s]; μ is the dynamic viscosity of the fluid [Pa·s] or [N·s/m^2]or [kg/(m·s)]. The Reynolds number is the ratio of inertial and viscous forces. It can be used to predict the transition from laminar to turbulent flow.

(11) Particle Reynolds number:

$$Re' = \frac{xU\rho}{\mu} \tag{8.19}$$

where ρ is the density of the fluid [kg/m^3]; x is particle diameter [m]; U is the velocity of the particle [m/s]; μ is the dynamic viscosity of the fluid [Pa·s] or [N·s/m^2] or [kg/(m·s)]. The particle Reynolds number is the ratio of inertial and viscous forces. It can be used to predict the transition from laminar to turbulent flow. If particle Reynolds numbers are greater than 0.2, the degree of turbulence becomes more significant.

(12) Modified Reynolds number:

$$Re_1 = \frac{\rho U_o}{(1-\varepsilon)S_v\mu} \tag{8.20}$$

where ρ is the density of the fluid [kg/m^3]; superficial (or empty tube) velocity $U_o = \frac{Q}{A}$, where Q is the volumetric flow rate (m$^{3/}$s), and the bed cross-sectional area is A (m^2); ε is the bed porosity or void fraction; S_v is the specific surface area per unit volume of the particle [1/m]; μ is the dynamic viscosity of the fluid [Pa·s] or [N·s/m^2]or [kg/(m·s)]. The modified Reynolds number can be used to determine the flow regime of the fluid within the porous medium if $Re_1 > 2$, indicating significant turbulence.

(13) Schmidt number:

$$Sc = \frac{v}{D} \tag{8.21}$$

where v is the kinematic viscosity [m^2/s]; D is the mass diffusivity [m^2/s]. The Schmidt number is the ratio of the shear component for diffusivity (viscosity divided by density) to the diffusivity for mass transfer.

(14) Sherwood number:

$$Sh = \frac{k_c L}{D} \tag{8.22}$$

where k_c is the mass transfer coefficient [m/s]; D is the mass diffusivity [m^2/s]; L is a characteristic length [m]. The Sherwood number is the ratio of convective mass transfer to diffusive mass transfer in a system.

(15) Stanton number for heat transfer:

$$St = \frac{h}{\rho u c_p} = \frac{Nu}{RePr} \tag{8.23}$$

where h is a convective heat transfer coefficient [W/(m^2·K)]; ρ is the density of the fluid [kg/m^3]; u is the local flow velocity [m/s]; c_p is specific heat of the fluid [m^2/(K·s^2)]. The Stanton number is the ratio of heat transferred into a fluid to the thermal capacity of the fluid.

(16) Weber number:

$$We = \frac{\rho u^2 L}{\sigma} \tag{8.24}$$

where ρ is the density of the fluid [kg/m^3]; u is the local flow velocity [m/s]; σ is the surface tension [N/m] or [kg/s^2]; L is a characteristic length [m]. The Weber number is the inertial and surface tension forces ratio.

8.1.5 Buckingham Pi Method

The Buckingham Pi Method (Buckingham 1914) can correlate experimental data in dimensionless groupings of parameters and variables to generate valuable models. It is a method with theoretical considerations to deduce which parameters should be grouped.

Buckingham Pi (π) theorem states that if the original unknown relationship is represented by $f(x_1, x_2, \dots, x_n) = 0$, where the x_i are the variables, it can be transformed into a new function $\phi(\pi_1, \pi_2, \dots, \pi_{n-m}) = 0$ of $n - m$ independent dimensionless products π_j of the original x_i variables. m is the number of fundamental dimensions out of which the dimensions of the original physical variables are ultimately composed.

Example 8.3

The head loss in a horizontal pipe in turbulent flow is related to the pressure drop Δp and is a measure of the resistance to flow in the pipe. It depends on the diameter of pipe D, viscosity, μ, density ρ, length of pipe l, the velocity of the flow v, and surface roughness η. To use the Buckingham Pi (π) method to deduce Pis with D, μ, and ρ as three bases: where all these parameters (Δp, D, μ, ρ, l, v, η) have the dimensions of ($ML^{-1}T^{-2}$), (L), ($ML^{-1}T^{-1}$), (ML^{-3}), (L), (LT^{-1}), and ([], without unit), respectively.

Solution:

There are seven variables: Δp, D, μ, ρ, l, v, η.
There are three base units: L, T, M.
Therefore, it has four Pis according to the Buckingham Pi (π) theorem. The three bases from the seven variables, e.g. D, μ, and ρ.

$$\pi_1 = D^a \cdot \mu^b \cdot \rho^c \cdot \Delta p = M^0 \cdot T^0 \cdot L^0$$

Then,

$$\pi_1 = L^a \cdot (ML^{-1}T^{-1})^b \cdot (ML^{-3})^c \cdot ML^{-1}T^{-2} = M^0 \cdot T^0 \cdot L^0$$

So, $a = 2, b = -2, c = 1$,

$$\text{i.e. } \pi_1 = D^2\mu^{-2}\rho\Delta p$$

$$\pi_2 = D^a \cdot \mu^b \cdot \rho^c \cdot l = M^0 \cdot T^0 \cdot L^0$$

Then,

$$\pi_2 = L^a \cdot (ML^{-1}T^{-1})^b \cdot (ML^{-3})^c \cdot L = M^0 \cdot T^0 \cdot L^0$$

So, $a = -1, b = 0, c = 0$,

$$\text{i.e. } \pi_2 = D^{-1}l$$

$$\pi_3 = D^a \cdot \mu^b \cdot \rho^c \cdot v = M^0 \cdot T^0 \cdot L^0$$

Then,

$$\pi_3 = L^a \cdot (ML^{-1}T^{-1})^b \cdot (ML^{-3})^c \cdot LT^{-1} = M^0 \cdot T^0 \cdot L^0$$

So, $a = 1, b = -1, c = 1$,

$$\text{i.e. } \pi_3 = D\mu^{-1}\rho v$$

$$\pi_4 = D^a \cdot \mu^b \cdot \rho^c \cdot \eta = M^0 \cdot T^0 \cdot L^0$$

Then,

$$\pi_4 = L^a \cdot (ML^{-1}T^{-1})^b \cdot (ML^{-3})^c \cdot [] = M^0 \cdot T^0 \cdot L^0$$

So, $a = 0$, $b = 0$, $c = 0$,

$$\text{i.e. } \pi_4 = \eta$$

It can be seen that the π_3 is the Reynolds number, an essential dimensionless group for the flow. The π_4 is the surface roughness η, which is itself because it is no unit.

8.1.6 Model Analysis

After building the model, we need to perform model analysis to check the model's validity or at least how confident we are with the model. There are three kinds of model analysis approaches that check the fit between the model and data: analysis of limiting conditions, graphical analysis, and statistical analysis.

(1) Analysis of limiting conditions
This simple approach can be quite useful. Many conditions can be checked, such as whether the model satisfies all the boundary conditions and limits, including initial state ($t = 0$), steady state, equilibrium state, asymptotic time limits, space, data value, and conditions at unique boundaries. Asymptotic limits can be both zero and infinity. We can also try zero and infinite values of parameters, such as density, diffusivity, or rate constant, to ensure that the model gives physically reasonable answers. It is also worthwhile to perform sensitivity analysis to check the model sensitivity to all the above conditions.
This method can screen the equations used for the system and determine whether a model is suitable for extrapolation beyond the range where it was fitted to data. If the limiting behavior is incorrect, then the model should not be extrapolated.
***Note**: We cannot conclude that the model is correct by analyzing limiting conditions. However, we are satisfied that it has the correct behavior for limiting values of essential variables.

(2) Graphical analysis
It is also simple and sometimes insightful, but it is not quantitative. It is much easier to spot a discrepancy between data and a model from a graph than just looking at the data. The simplest graphical method is the straight-line plot. However, modifications or rearrangements are required to obtain a linear model. There are two commonly used plots: parity plot and residuals plot.
A parity plot compares the model against measured experimental results under the same conditions (e.g. same variables). The result plotted could be anything of interest, e.g. reactor conversion, reactor yield of desired product or by-product, mole fraction at equilibrium, mole flow rate in distillate, or pressure drop.
The parity plot in Figure 8.2 compares the previous Example 7.8 temperatures. The y-axis is the measured temperature, and the x-axis is the calculated temperature from the model in Chapter 7: $T = \left(\frac{f(A)}{DE_A}\right)^\alpha \cdot \left(\frac{g(B)}{DE_B}\right)^\beta \cdot \left(\frac{h(C)}{DE_C}\right)^\gamma$. It is pretty easy to spot a discrepancy between data and the model from the parity plot.
A residual plot shows the difference between the experimental and predicted result by the model vs. a critical system parameter. The system parameter may be the same thing as a result or maybe something different. For example, we might plot for a reactor problem (X = conversion, Y = yield of desired product) using Xexpt – Xmodel vs. Xexpt,

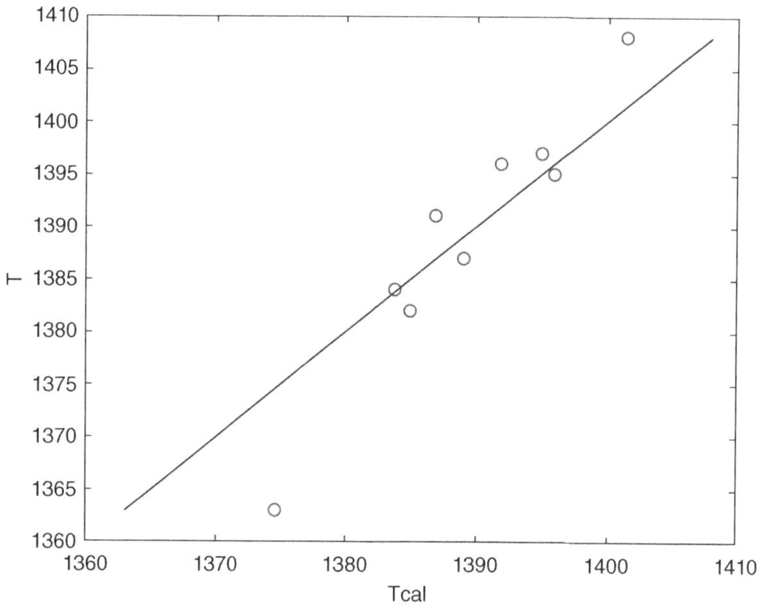

Figure 8.2 The parity plot of previous Example 7.8 on the temperature of liquid iron.

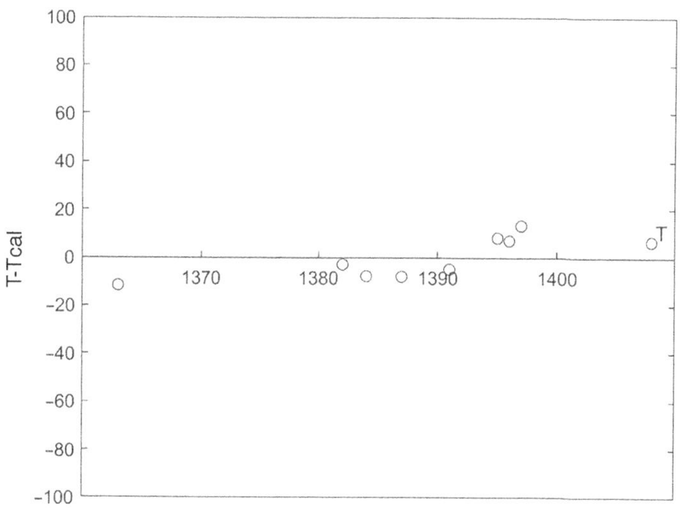

Figure 8.3 The parity plot of previous Example 7.8 on the temperature of liquid iron.

or Xexpt – Xmodel vs. temperature, or Xexpt – Xmodel vs. initial concentration of reagent, or Xexpt – Xmodel vs. the amount of catalyst, or Yexpt – Ymodel vs. Xexpt.

The residuals plot using the data of Figure 8.2, T–Tcal vs. T, is shown in Figure 8.3.

The residuals plot can be used to identify whether there is a systematic variation between the model and data as a function of the variable plotted.

Residual and parity plots should be used in concert to verify a model over the range of parameters to which it is to be applied.

(3) Statistical analysis
It is a quantitative approach that is especially useful for large datasets but can often give misleading results. Various statistical methods can be used to establish the goodness of model fit. Statistical methods are essential because they give a quantitative rather than qualitative assessment of goodness of fit. However, they do not show if there is a significant deviation between the model and the data, mainly if a significant deviation is at one end of an extensive data set with a good overall regression. Therefore, a wise choice is to use statistical methods to differentiate among models with similar parity and then draw parity and residual plots to eliminate models that show unacceptable systematic deviations between the model and data in the range of interest.
In comparing two alternative models, the most important is the chi-squared (χ^2) test (whether there is a statistically significant difference between them), which can be used to examine the hypothesis that one model is a better fit than another.

8.2 Data Statistics Toolbox in MATLAB

8.2.1 Statistics and Machine Learning Toolbox

Statistics and Machine Learning Toolbox in MATLAB is one of the leading products by MathWorks Inc. (https://www.mathworks.com/products/statistics.html), which provides functions and apps to describe, analyze, and model data. The User's Guide for this toolbox is about 8000 pages, which is too large to cover them all in this book. We introduce some commonly used functions and leave others to readers interested in digging deeper to access the above weblink.

MATLAB provides descriptive statistics, visualizations, and clustering for exploratory data analysis, fits probability distributions to data, generates random numbers for Monte Carlo simulations, and performs hypothesis tests. Regression and classification algorithms let you draw inferences from data and build predictive models interactively, using the Classification and Regression Learner apps or programmatically, using AutoML. For multidimensional data analysis and feature extraction, the toolbox provides PCA, regularization, dimensionality reduction, and feature selection methods that let you identify variables with the best predictive power.

The toolbox also provides supervised, semi-supervised, and unsupervised machine-learning algorithms, including support vector machines (SVMs), boosted decision trees, shallow neural nets, k-means, and other clustering methods. Many toolbox algorithms can be used on data sets that are too big to be stored in memory.

8.2.2 Statistics Visualization

There are three types of plots for statistical visualization in MATLAB: scatter plots using grouped data, box plots, and distribution plots.

8.2.2.1 Scatter Plots

MATLAB functions plot and scatter can produce scatter plots. MATLAB also has the function plotmatrix to produce a matrix of such plots showing the relationship between several pairs of variables. The Statistics and Machine Learning Toolbox functions gscatter and gplotmatrix

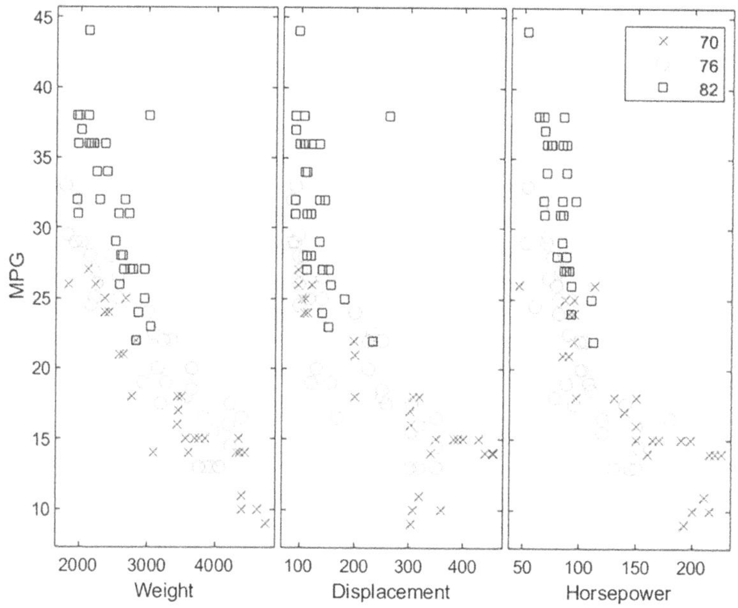

Figure 8.4 The weight, displacement, and horsepower vs. MPG for the MATLAB database "carsmall".

produce grouped versions of these plots. These functions help determine whether the values of two variables or the relationship between those variables is the same in each group. These functions use different plotting symbols to indicate group membership. The function gname can label points on the plots with a text label or an observation number.

The following example shows how to create scatter plots using grouped sample data.

Example 8.4

To examine the weight, displacement, and horsepower vs. MPG (miles per gallon) of cars from three different model years. The data are downloaded from the MATLAB database "carsmall." It can be seen from Figure 8.4 that newly manufactured cars have higher MPGs compared to older cars. It is a good way to make sense by visually comparing data points.

Solution:

```
load carsmall
xvars = [Weight Displacement Horsepower];
yvars = [MPG];
xnames = {'Weight', 'Displacement', 'Horsepower'};
ynames = {'MPG'};
gplotmatrix(xvars,yvars,Model_Year,'rgb','xos',[],'on',[],xnames,ynames)
```

8.2.2.2 Box Plots

A box plot can be applied to visualize summary statistics, especially useful for side-by-side comparisons of two distributions. In the plot, the tops and bottoms of each "box" are the 25th and 75th percentiles of the samples, respectively. The distances between the tops and bottoms are the interquartile ranges. The line in the middle of each box is the sample median. If the median is not centered in the box, it shows sample skewness. The whiskers are lines extending above and below each box. Whiskers are drawn from the ends of the interquartile ranges to the furthest observations within the whisker length (the adjacent values). Observations beyond the whisker length are marked as outliers. By default, an outlier is a value more than 1.5 times the interquartile range away from the top or bottom of the box, but this value can be adjusted with additional input arguments. Outliers are displayed with a red + sign. Notches display the variability of the median between samples. The width of a notch is computed so that box plots whose notches do not overlap (as above) have different medians at the 5% significance level. The significance level is based on a normal distribution assumption, but comparisons of medians are reasonably robust for other distributions. Comparing box-plot medians is like a visual hypothesis test, analogous to the t test used for means.

The following example shows how to create box plots using the sample data. The medians of the "First" and "Second" are not centered in the box, which means sample skewness.

Example 8.5

Solution:

```
rng('default')  % For reproducibility
x1 = rand(10,1);
x2 = rand(20,1);
x3 = rand(30,1);
x4 = rand(40,1);
x = [x1; x2; x3; x4];
g1 = repmat({'First'},10,1);
g2 = repmat({'Second'},20,1);
g3 = repmat({'Third'},30,1);
g4 = repmat({'Fourth'},40,1);
g = [g1; g2; g3; g4];
boxplot(x,g)
```

8.2.2.3 Distribution Plots

A few distribution plot options in the MATLAB Statistics and Machine Learning Toolbox include normal probability plots, quantile–quantile plots, and cumulative distribution plots. They can be used to visually assess the distribution of sample data by comparing the empirical distribution of the data with the theoretical values expected from a specified distribution. The normal probability plots use normplot to assess whether sample data comes from a normal distribution. The probability plots use probplot to create plots for distributions other than normal or to explore the distribution of censored data. The quantile–quantile plots use qqplot to assess whether two sample data sets come from the same distribution family. This plot is robust with respect to differences in location and scale. The cumulative distribution plots use cdfplot or ecdf to display the sample data's empirical cumulative distribution function (cdf) for visual comparison to the theoretical cdf of a specified distribution.

The following example shows how to create a normal distribution plot using randomly generated sample data. In Figure 8.5, the plus signs plot the empirical probability vs. the data

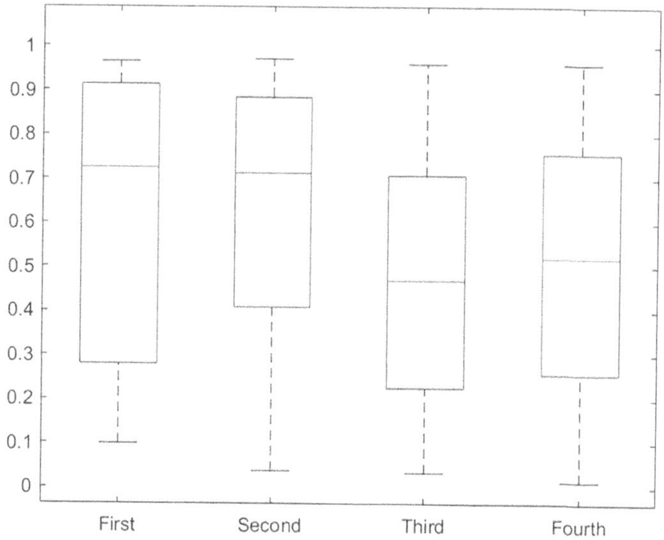

Figure 8.5 The box plot of the four sets of randomly generated data.

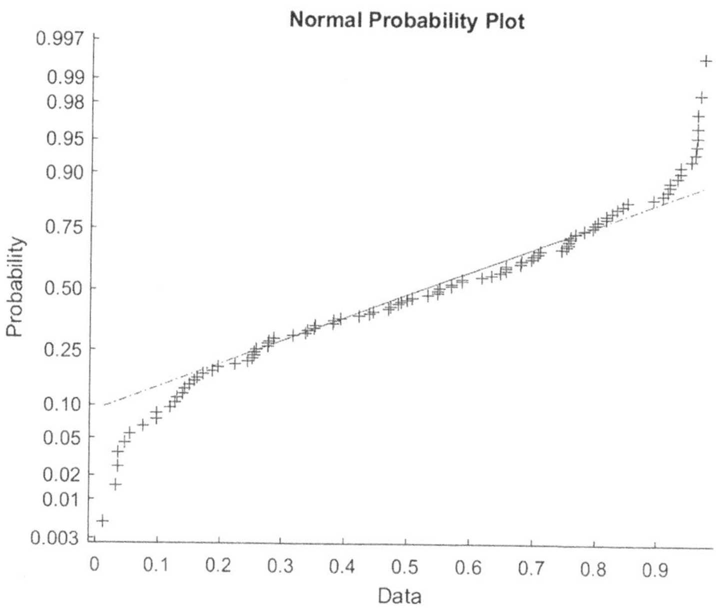

Figure 8.6 The normal probability plot using randomly generated sample data.

value for each point in the data. A solid line connects the 25th and 75th percentiles in the data, and a dashed line extends it to the ends of the data. The y-axis values are probabilities from zero to one, but the scale is not linear. The distance between tick marks on the y-axis matches the distance between the quantiles of a normal distribution. The quantiles are close together near the median (50th percentile) and stretch out symmetrically as you move away from the median. It can be seen from Figure 8.6 that the underlying distribution of the randomly generated sample data is not normal.

Example 8.6

Solution:

```
rng('default')  % For reproducibility
x = rand(100,1);
normplot(x)
```

8.2.3 Analysis of Variance

Analysis of variance (ANOVA) is a statistical method to analyze the differences among means. ANOVA is based on the law of total variance, where the observed variance in a particular variable is partitioned into components attributable to different sources of variation. In its simplest form, ANOVA provides a statistical test of whether two or more population means are equal and generalizes the t test beyond two means. It involves assigning sample variance to different sources and deciding whether the variation arises within or among different population groups. Samples are described in terms of variation around group means and variation of group means around an overall mean.

8.2.3.1 One-Way ANOVA

Example 8.7 shows how to perform one-way ANOVA to determine whether the randomly generated data have a common mean.

Example 8.7

Solution:

```
rng('default')  % For reproducibility
x = rand(10,3);
[p,tbl,stats] = anova1(x);
figure
multcompare(stats)
```

The p value in the table of about 0.8817 from Figure 8.7a indicates that the randomly generated data in this example are very similar. It is also confirmed by both the overlapping of the notched box plot in Figure 8.7b and the multiple comparison test in Figure 8.7c.

8.2.3.2 ANOVA with Multiple Responses

There are cases with multiple response variables, and often, we are interested in determining whether the entire set of means is different from one group to the next. MATLAB has a multivariate version of the analysis of variance (MANOVA) that can address the problem. Example 8.8 shows how to perform MANOVA using the carsmall data set to check whether the characteristics of the cars have changed over time from 1970, 1976, and 1982.

ANOVA Table

Source	SS	df	MS	F	Prob>F
Columns	0.02657	2	0.01329	0.13	0.8817
Error	2.83687	27	0.10507		
Total	2.86344	29			

(a)

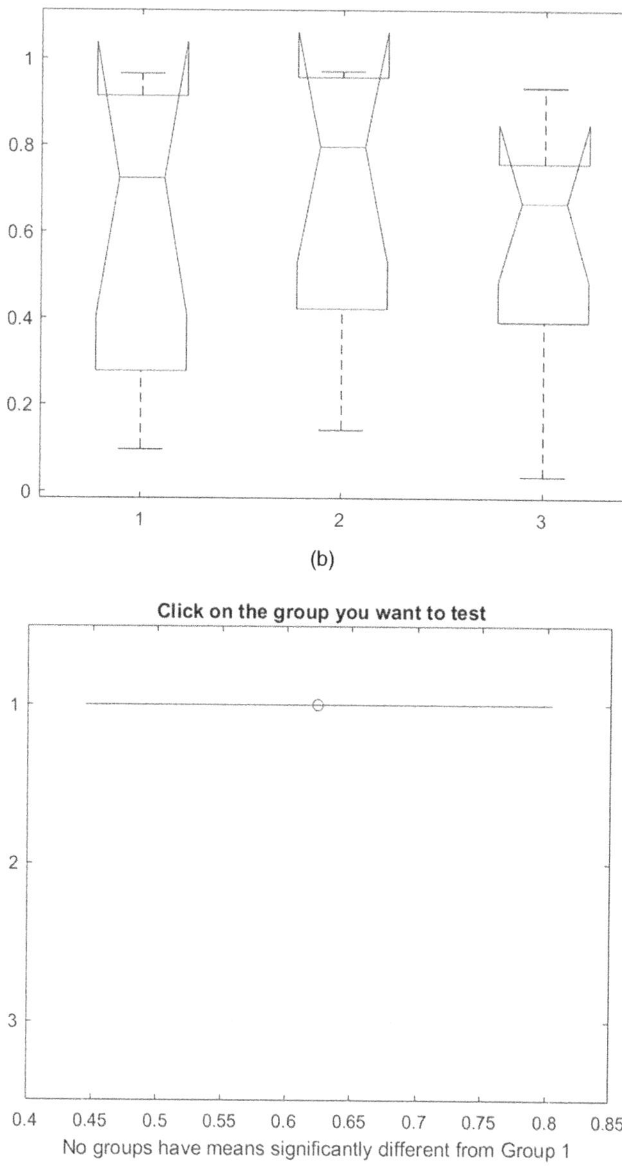

(c)

Figure 8.7 The results from anova1: (a) the standard ANOVA table, (b) the box plots, and (c) the multiple comparison test.

Example 8.8

Solution:

```
x = [MPG Horsepower Displacement Weight];
[d,p,stats] = manoval(x,Model_Year);
figure
c2=stats.canon(:,2);
c1=stats.canon(:,1);
gscatter(c2,c1,Model_Year,[],'oxs')
md=max(stats.mdist)
n=find(stats.mdist == md)
Model=Model(n,:)
gmd=stats.gmdist

Results:
md =
   31.5273
n =
    20
Model =
 'buick estate wagon (sw)
gmd =
          0    3.8277   11.1106
     3.8277         0    6.1374
    11.1106    6.1374         0
```

The grouped scatter plot of the first two canonical variables is shown in Figure 8.8.

In Example 8.8, the first canonical variable, c1, separates the 1982 cars (which have high values of c1) from the older cars. The second canonical variable, c2, reveals some separation between the 1970 and 1976 cars. The largest distance from the group's mean, i.e.

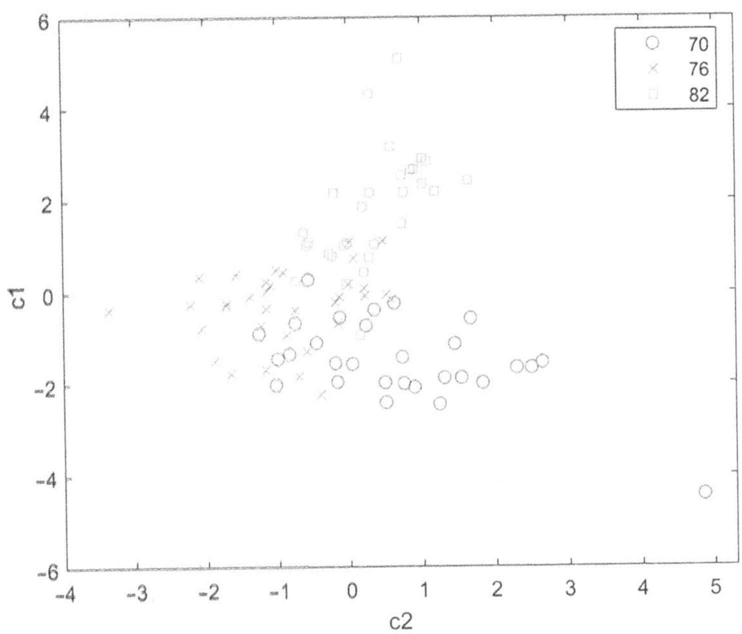

Figure 8.8 The grouped scatter plot of the first two canonical variables.

max(stats.mdist), is 31.5273, corresponding to the car model of "buick estate wagon (sw)." The gmdist field measures the distances between each pair of group means. The multivariate distance between the extreme years 1970 and 1982 (11.1106) is larger than the difference between more closely spaced years (3.8277 and 6.1374). It indicates that the points progress as the year changes from 1970 through 1976 to 1982.

8.2.4 Regression

Regression is a statistical process to fit models to response data. It needs to have a dependent variable (often called the "outcome" or "response" variable, or a "label" in machine learning parlance) and one or more independent variables (often called "predictors," "covariates," "explanatory variables," or "features"). The most common form of regression analysis is linear regression. The regression process depends on the model. If a model is parametric, regression estimates the parameters from the data. If a model is linear in the parameters, estimation is based on linear algebra methods that minimize the residual vector's norm. If a model is nonlinear in the parameters, estimation is based on search methods from optimization that minimize the norm of a residual vector.

MATLAB has functions for regression problems, as shown in Table 8.3.

Example 8.9

To use fitlm for the MPG (mills per gallon) of the carsmall case:

Solution:

```
load carsmall
%Store the variables in a table.
tbl = table(Weight,Horsepower,MPG,'VariableNames',{'Weight',...
'Horsepower','MPG'});
%Display the first five rows of the table.
tbl(1:5,:)
lm = fitlm(tbl,'MPG~Weight+Horsepower')
```

```
Results:
ans =
  5x3 table
    Weight     Horsepower     MPG
    ______     __________     ___
     3504          130         18
     3693          165         15
     3436          150         18
     3433          150         16
     3449          140         17
lm =
Linear regression model:
    MPG ~ 1 + Weight + Horsepower
Estimated coefficients:
                      Estimate          SE          tStat         pValue
                     __________     __________     _______     __________
    (Intercept)          47.769         1.7417      27.427      1.751e-45
    Weight           -0.0065651      0.0010507     -6.2484     1.3519e-08
    Horsepower        -0.042018       0.018671     -2.2504        0.02686
Number of observations: 93, Error degrees of freedom: 90
Root mean squared error: 4.07
R-squared: 0.752, adjusted R-squared: 0.747
F-statistic vs. constant model: 136, p value = 5.57e-28
```

Table 8.3 MATLAB functions for a regression problem (https://www.mathworks.com/The MathWorks, Inc.).

MATLAB functions	Description	Syntax
fitlm	Fit model coefficients for continuous or categorical predictors, continuous response, linear model	• mdl = fitlm(tbl) • mdl = fitlm(X,y) • mdl = fitlm(___,modelspec) • mdl = fitlm(___,Name,Value)
fitglm	Fit generalized linear model coefficients for continuous or categorical predictors, response possibly with restrictions such as nonnegative or integer-valued, generalized linear model	• mdl = fitglm(tbl) • mdl = fitglm(X,y) • mdl = fitglm(___,modelspec) • mdl = fitglm(___,Name,Value)
fitnlm	Fit nonlinear model coefficients for continuous predictors with a continuous nonlinear response, parameterized nonlinear model	• mdl = fitnlm(tbl,modelfun,beta0) • mdl = fitnlm(X,y,modelfun,beta0) • mdl = fitnlm(___,modelfun,beta0,Name,Value)
plsregress	Fit model and coefficients using partial least-squares regression for correlated continuous predictors, continuous response, linear model	• [XL,YL] = plsregress(X,Y,ncomp) • [XL,YL,XS,YS,BETA,PCTVAR,MSE,stats] = plsregress(X,Y,ncomp) • [XL,YL,XS,YS,BETA,PCTVAR,MSE,stats] = plsregress(___,Name,Value)
ridge	Ridge regression for continuous predictors, continuous response, linear model	• B = ridge(y,X,k) • B = ridge(y,X,k,scaled)
lasso	Lasso or elastic net regularization for continuous predictors, continuous response, linear model	• B = lasso(X,y) • B = lasso(X,y,Name,Value) • [B,FitInfo] = lasso(___)
stepwiselm	Fit model and coefficients for continuous or categorical predictors, continuous response, linear model of unknown complexity	• mdl = stepwiselm(tbl) • mdl = stepwiselm(X,y) • mdl = stepwiselm(___,modelspec) • mdl = stepwiselm(___,Name,Value)
mvregress	Fit multivariate regression model coefficients for continuous predictors, multivariable response, linear model	• beta = mvregress(X,Y) • beta = mvregress(X,Y,Name,Value) • [beta,Sigma] = mvregress(___) • [beta,Sigma,E,CovB,logL] = mvregress(___)

Table 8.3 (Continued)

MATLAB functions	Description	Syntax
nlmefit	Fit mixed-effects model coefficients for continuous predictors, continuous response, mixed-effects model	• beta = nlmefit(X,y,group,V,fun,beta0) • [beta,PSI] = nlmefit(X,y,group,V,fun,beta0) • [beta,PSI,stats] = nlmefit(X,y,group,V,fun,beta0) • [beta,PSI,stats,B] = nlmefit(X,y,group,V,fun,beta0) • [beta,PSI,stats,B] = nlmefit(X,y,group,V,fun,beta0,'Name',value)
nlmefitsa	Fit mixed-effects model coefficients with stochastic EM algorithm for continuous predictors, continuous response, and mixed-effects model	• [BETA,PSI,STATS,B] = nlmefitsa(X,Y,GROUP,V,MODELFUN,BETA0) • [BETA,PSI,STATS,B] = nlmefitsa(X,Y,GROUP,V, MODELFUN,BETA0,'Name',Value)
fitrtree	Fit binary decision tree for regression for continuous or categorical predictors, continuous response, unknown model	• tree = fitrtree(Tbl,ResponseVarName) • tree = fitrtree(Tbl,formula) • tree = fitrtree(Tbl,Y) • tree = fitrtree(X,Y) • tree = fitrtree(___,Name,Value)
fitrensemble	Fit ensemble of learners for regression for continuous or categorical predictors, continuous response, unknown model	• Mdl = fitrensemble(Tbl,ResponseVarName) • Mdl = fitrensemble(Tbl,formula) • Mdl = fitrensemble(Tbl,Y) • Mdl = fitrensemble(X,Y) • Mdl = fitrensemble(___,Name,Value)

It can be seen from the results that both the R-squared and adjusted R-squared are not close to one, which means it is not a good linear fit. In addition, the *p* value is very small, which means it is not a constant model.

We can draw the parity plot and the residuals plot with the data from the above example using the following code:

```
plot(lm.Fitted,MPG,'*')
xlabel('MPGc')
ylabel('MPG')
figure
plot(MPG,MPG-lm.Fitted,'*')
xlabel('MPG')
ylabel('MPG-MPGc')
```

Figure 8.9a shows the parity plot of the above example to calculate MPG in terms of linear regression of weight and horsepower. The *y*-axis is the measured MPG, and the *x*-axis is the calculated MPG from linear regression. The scatter data points also show that the data did not have a linear fit. Figure 8.9b shows the residual plot of the above example. The *y*-axis is the difference between the measured MPG and the linear regression of the MPG; the *x*-axis is the measured MPG. The scatter data points also show that the data are far from zero.

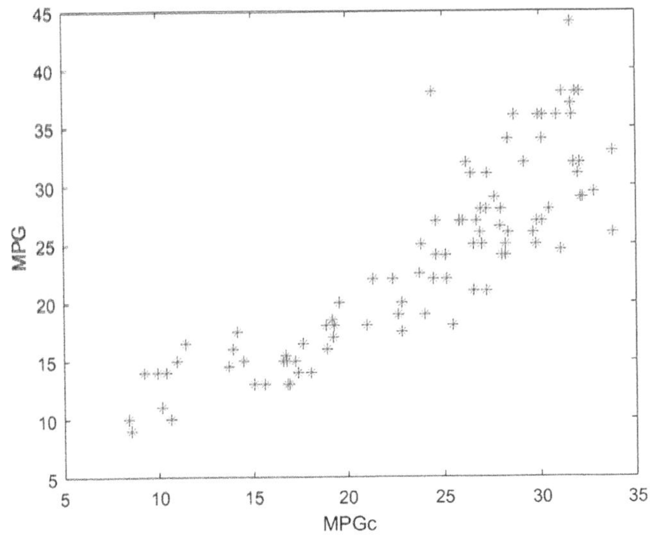

(a)

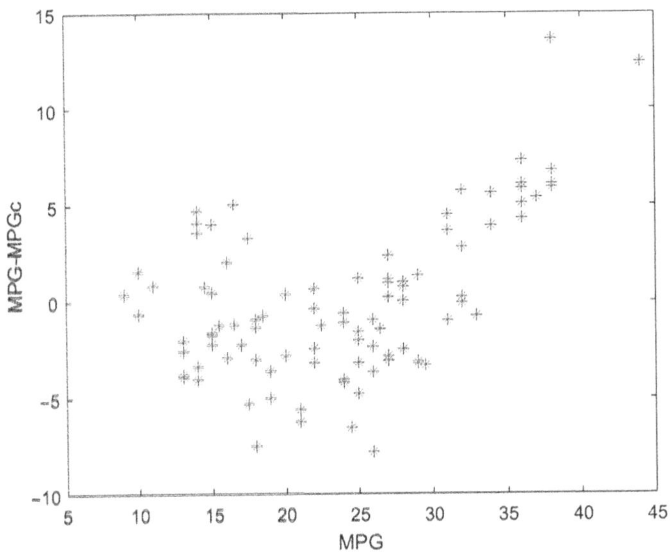

(b)

Figure 8.9 The plots of the cars' MPG: (a) the parity plot and (b) the residuals plot.

Example 8.10
To use plsregress for the displacement of the carsmall case:

Solution:

```
load carsmall
X=[Weight,Acceleration,Cylinders,Model_Year];
```

```
Y=[Displacement];
[XL,yl,XS,YS,beta,PCTVAR] = plsregress(X,Y,4);
plot(1:4,cumsum(100*PCTVAR(2,:)),'-bo');
xlabel('Number of PLS components');
ylabel('Percent variance explained in y');
figure
yfit = [ones(size(X,1),1) X]*beta;
TSS = sum((Y-mean(Y)).^2);
RSS = sum((Y-yfit).^2);
R_squared = 1 - RSS/TSS
plot(yfit,Y,'*')
xlabel('Fitted displacement');
ylabel('Measured displacement');
figure
plot(Y,Y-yfit,'*')
xlabel('Displacement')
ylabel('Displacement - the fitted displacement')

Results:
XL =
   1.0e+03 *
    8.0292   -0.0002    0.0000   -0.0000
   -0.0163   -0.0231   -0.0175    0.0029
    0.0158    0.0051    0.0017    0.0064
   -0.0242   -0.0350    0.0236   -0.0010
yl =
  989.2315  318.1517  168.4300  213.6498
beta =
  -36.1441
    0.0402
   -5.9762
   36.2976
    0.1122
PCTVAR =
    1.0000    0.0000    0.0000    0.0000
    0.7912    0.0818    0.0229    0.0369
R_squared =
    0.9329
```

It can be seen from the result that the R_squared is 0.9329, which means it is a good linear fit. It is a critical step for choosing the number of components in a PLS model. In Figure 8.10a, the plot gives a rough indication, showing nearly 79% of the variance in y explained by the first component (weight), with three additional components reaching 93% contributions. Figure 8.10b shows the parity plot of the above example to calculate the displacement in terms of partial least squares regression of weight, acceleration, cylinders, and Model_Year.

The y-axis is the measured displacement, and the x-axis is the calculated displacement from PLS regression. The data points also show that the data have a good linear fit. Figure 8.10c shows the residual plot of the above example. The y-axis is the difference between the measured displacement and the PLS regression of the displacement; the x-axis is the measured displacement. Most data points are in the range of ± 50.

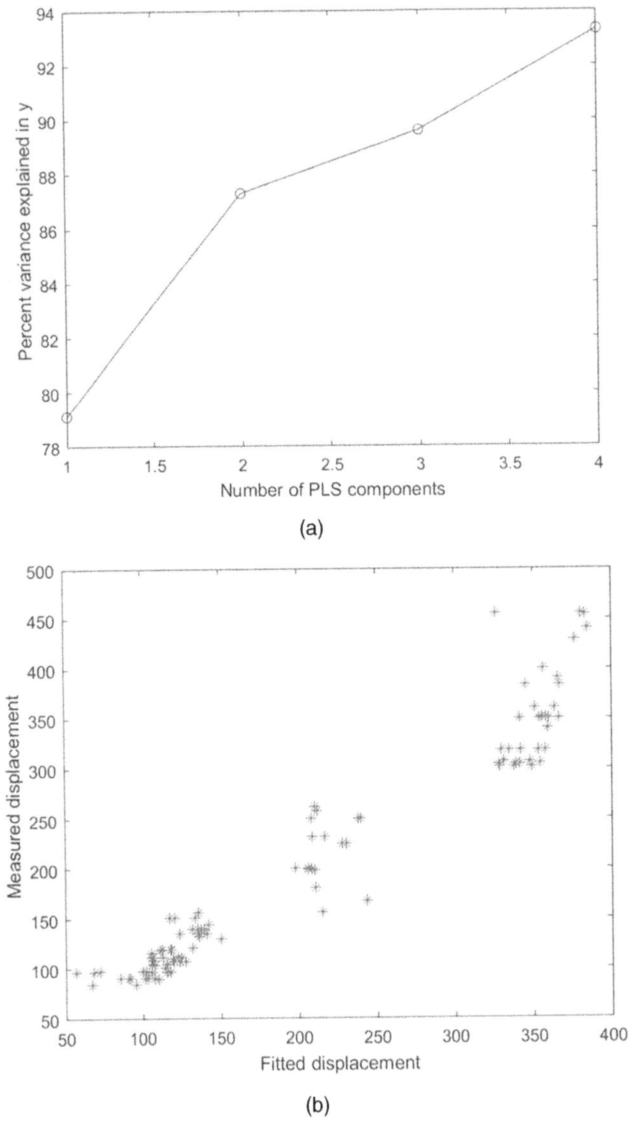

Figure 8.10 The plots of the displacement of the cars: (a) the percentage variance as a function of the number of components; (b) the parity plot; (c) the residuals plot.

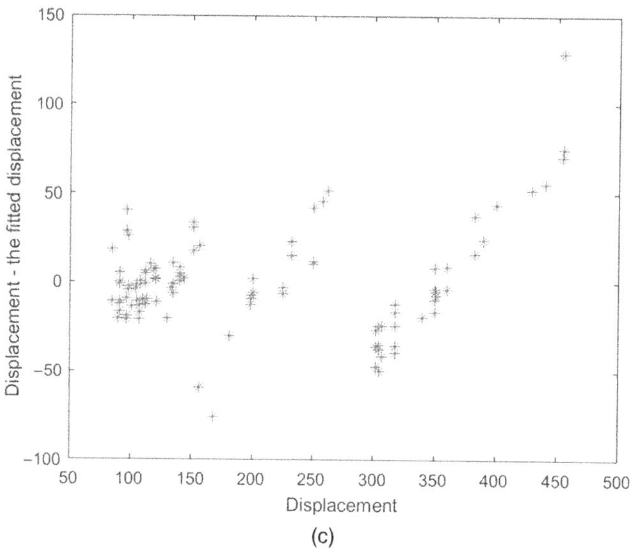

Figure 8.10 (*Continued*)

Example 8.11
To use fitnlm for the displacement of the carsmall case:

Solution:

```
load carsmall
tbl = table(Weight,Acceleration,Displacement);
modelfun = @(b,x)b(1) + b(2)*x(:,1).^b(3) + b(4)*x(:,2).^b(5);
beta0 = [-50 500 -1 500 -1];
mdl = fitnlm(tbl,modelfun,beta0)
```

```
Results:
mdl =
Nonlinear regression model:
    Displacement ~ b1 + b2*Weight^b3 + b4*Acceleration^b5
Estimated coefficients:
              Estimate          SE           tStat         pValue
              ________     _________       _______     __________

    b1          4179.4        106.82        39.124     8.4826e-61
    b2         -7151.4         140.4       -50.936      2.722e-71
    b3       -0.068738     0.0059145       -11.622     5.1569e-20
    b4          2794.4        2546.3        1.0974         0.2752
    b5         -1.0724       0.56634       -1.8936        0.06129
Number of observations: 100, Error degrees of freedom: 96
Root mean squared error: 37.2
R-squared: 0.893,  adjusted R-squared 0.889
F-statistic vs. constant model: 266, p value = 2.12e-46
```

It can be seen from the results that both the R-squared and adjusted R-squared are close to 0.9, which means it is a good nonlinear fit. In addition, the p value is very small, which means it is not a constant model.

We can draw the parity plot and the residuals plot with the data from the above example using the following code:

```
plot(mdl.Fitted, Displacement,'*')
xlabel('Displacement fitted')
ylabel('Displacement')
figure
plot(Displacement, Displacement -mdl.Fitted,'*')
xlabel('Displacement')
ylabel('Displacement - displacement fitted')
```

Figure 8.11a shows the parity plot of the above example to calculate displacement in terms of nonlinear regression of weight and acceleration. The y-axis is the measured displacement, and the x-axis is the calculated displacement from nonlinear regression. The data points also show that the data have a good linear fit. Figure 8.11b shows the residual plot of the above example. The y-axis is the difference between the measured displacement and the nonlinear regression; the x-axis is the measured displacement. Most data points are in the range of ± 100.

8.2.5 Principal Component Analysis

The MATLAB function plot displays a graph of the relationship between two variables, and the plot3 and surf commands display different three-dimensional views. However, when there are more than three variables, it is more difficult to visualize their relationships. PCA is a linear dimensionality reduction technique for multivariable systems to generate a new set of variables called principal components. Each principal component is a linear combination of the original variables. All the principal components are orthogonal, so there is no redundant information – the principal components form an orthogonal basis for the data space.

The complete set of principal components is as large as the original set of variables. In general, the number of principal components can be chosen based on the sum of the variances of the first few principal components to exceed 80% of the total variance of the original data. Then, the plots of the reduced model from these few new variables against the original complete data can be used to develop a deeper understanding of the driving forces that generated the original data.

The MATLAB function pca can be used to find the principal components. To use pca, you must have the measured data you want to analyze.

Example 8.12

To use PCA for the normally distributed random numbers:

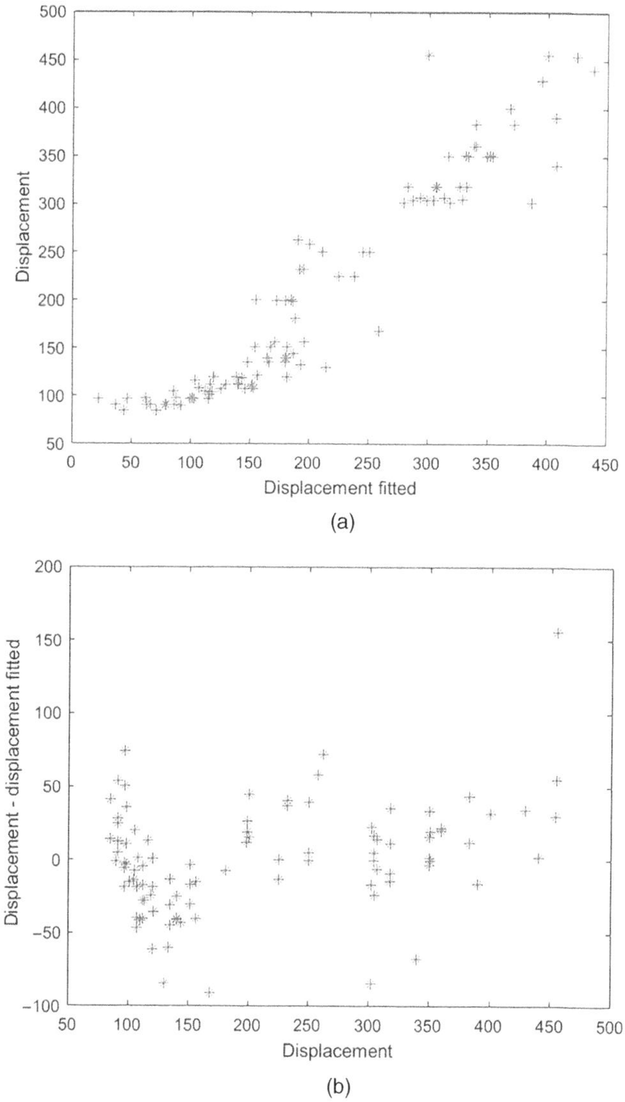

Figure 8.11 The plots of the displacement of the cars: (a) the parity plot; (b) the residuals plot.

Solution:

```
rng('default')  % For reproducibility
data = randn(50, 6);
% Perform PCA
[coeff,score,latent,tsquared,explained,mu] = pca(data);
% Plot the explained variance
plot(cumsum(explained));
xlabel('Number of principal components');
ylabel('Cumulative explained variance (%)');
% Choose a number of principal components based on the plot around 80%
num_components = 4;
% Reconstruct data using the selected number of principal components
reconstructed_data = score(:, 1:num_components) * coeff(:, 1:num_components)';
```

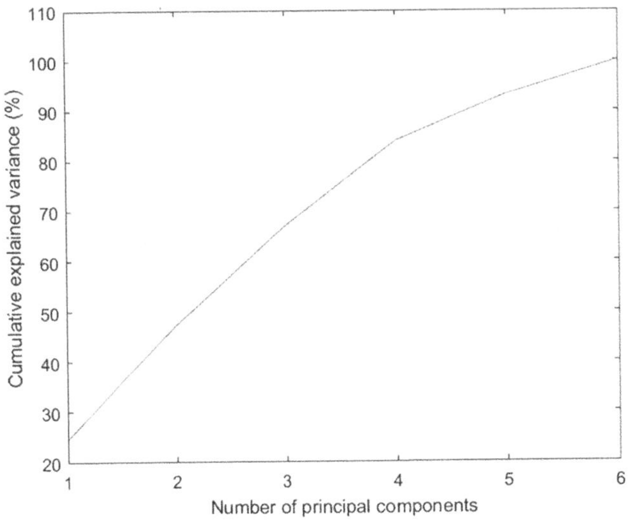

Figure 8.12 The plots of the percentage variance as a function of the number of components.

```
% Calculate the reconstruction error
reconstruction_error = norm(data - reconstructed_data)
```

```
Results:
reconstruction_error =
    5.6590
```

It can be seen from Figure 8.12 that the plot gives a rough indication, showing nearly 80% of the variance explained by the first four components. With the four components reconstruction, the Euclidean norm of the error is 5.6590.

8.3 Application in Chemical Engineering

8.3.1 Application of Buckingham Pi (π) Method for a Flotation Rate Model

Example 8.13

To deduce a flotation rate model using Buckingham Pi (π) method:

$$k = f(v, \varepsilon, d_b, \rho, S_b) \tag{8.25}$$

where k is the flotation rate, v is the kinematic viscosity, ε is the energy dissipation rate, d_b is the bubble diameter, ρ is the density, and S_b is the bubble surface area flux. These parameters have the dimensions of (T^{-1}), (L^2T^{-1}), (L^2T^{-3}), (L), (ML^{-3}), and (T^{-1}), respectively.

Solution:

There are six variables: $k, v, \varepsilon, d_b, \rho, S_b$

There are three base units: L, T, M

Therefore, it has three Pis according to the Buckingham Pi (π) theorem. We can choose three bases from the six variables, e.g. bubble diameter (d_b), density (ρ), and bubble surface area flux (S_b).

$$\pi_1 = d_b^a \cdot \rho^b \cdot S_b^c \cdot k = M^0 \cdot T^0 \cdot L^0$$

Then,

$$\pi_1 = L^a \cdot (ML^{-3})^b \cdot (T^{-1})^c \cdot T^{-1} = M^0 \cdot T^0 \cdot L^0$$

So, $a = 0, b = 0, c = -1$,

$$\text{i.e. } \pi_1 = \frac{k}{S_b}$$

$$\pi_2 = d_b^a \cdot \rho^b \cdot S_b^c \cdot \upsilon = M^0 \cdot T^0 \cdot L^0$$

Then,

$$\pi_2 = L^a \cdot (ML^{-3})^b \cdot (T^{-1})^c \cdot L^2T^{-1} = M^0 \cdot T^0 \cdot L^0$$

So, $a = -2, b = 0, c = -1$,

$$\text{i.e. } \pi_2 = \frac{\upsilon}{d_b^2 S_b}$$

$$\pi_3 = d_b^a \cdot \rho^b \cdot S_b^c \cdot \varepsilon = M^0 \cdot T^0 \cdot L^0$$

Then,

$$\pi_3 = L^a \cdot (ML^{-3})^b \cdot (T^{-1})^c \cdot L^2T^{-3} = M^0 \cdot T^0 \cdot L^0$$

So, $a = -2, b = 0, c = -3$,

$$\text{i.e. } \pi_3 = \frac{\varepsilon}{d_b^2 S_b^3}$$

We can get the relationship among those three Pis:

$$\pi_1 = g(\pi_2, \pi_3)$$

$$\text{i.e. } \frac{k}{S_b} = g\left(\frac{\upsilon}{d_b^2 S_b}, \frac{\varepsilon}{d_b^2 S_b^3}\right),$$

$$\text{Therefore } k = g\left(\frac{\upsilon}{d_b^2 S_b}, \frac{\varepsilon}{d_b^2 S_b^3}\right) \cdot S_b$$

8.3.2 Statistics Visualization for Particle Size Distributions

The following example shows how to create scatter plots using grouped sample data for particle size distributions.

Example 8.14

The JK mini drop weight tester (mini-DWT) was used to measure the breakage characteristics of small-size samples (Yu et al. 2018). The specific energy Ecs levels are 0.10, 0.50, 1.00, 1.50, 2.00, and 2.50 kWh/t. The original mean sample sizes are 0.92, 1.84, and 3.66 mm. The data of Mini-DWT are in Table 8.4. The MATLAB code for scatter plots using grouped sample data for particle size distributions is as follows:

Solution:

(1) Scatter plots for particles' percentage retained

```
% load particle size data
A=[0.00   0.00   0.00   0.00   0.00   0.00   0.00   0.00   0.00   0.00   0.00   0.00   39.96  5.68   0.00    0.00   0.00   0.00
0.00      0.00   0.00   0.00   0.00   0.00   0.00   0.00   0.00   0.00   0.00   0.00   36.26  29.93  7.27    5.56   0.89   0.00
0.00      0.00   0.00   0.00   0.00   0.00   68.60  27.38  16.19  9.26   7.01   1.00   11.42  27.15  25.67  14.30   9.28   4.93
0.00      0.00   0.00   0.00   0.00   0.00   21.08  30.24  26.12  18.77  14.76  13.45  3.28   12.06  24.03  23.29  17.00  14.80
79.84     44.35  31.86  26.61  20.10  6.54   3.03   16.18  17.68  19.75  14.21  15.07  2.43   6.61   12.19  14.30  19.13  14.04
11.77     29.13  22.79  24.13  20.93  18.13  1.17   7.41   7.88   13.83  11.48  9.88   0.85   3.02   6.92   11.82  12.08  14.47
3.28      8.67   17.89  14.96  13.34  20.19  1.36   3.30   6.20   8.83   13.52  13.76  1.69   2.44   5.51    4.85   7.94  11.29
0.55      3.60   6.86   11.07  15.43  20.09  0.62   2.24   5.83   8.09   9.49   13.20  0.95   1.62   2.58    5.32   7.05   9.76
1.73      4.58   8.91   8.26   9.67   9.44   1.17   4.29   8.87   6.85   11.35  11.63  0.21   3.13   6.10    5.91   9.28   9.10
1.28      2.62   2.29   2.89   4.84   6.45   1.30   3.17   3.66   2.78   3.35   4.57   1.16   2.67   2.81    4.26   4.47   4.71
0.55      2.70   1.80   1.74   4.17   4.58   0.62   1.93   2.11   2.47   3.85   3.50   0.85   2.20   0.59    2.01   2.80   3.84
0.46      3.11   6.13   8.18   8.26   10.93  0.56   2.80   4.22   7.41   7.88   10.13  0.32   2.55   5.39    6.74   7.16   9.43
0.55      1.23   1.47   2.15   3.25   3.64   0.49   1.06   1.24   1.98   3.10   3.81   0.63   0.93   0.94    1.65   2.91   3.62
];
B=[4.75
3.35
2.36
1.7
1.18
0.85
0.6
0.425
0.3
0.212
0.15
0.106
0.075
];
xvars = [];
yvars = [];
ES=[];
for i=1:18
xvars = [xvars; B];
yvars = [yvars; A(:,i)];
end
PS=[0.92*ones(13*6,1);1.84*ones(13*6,1);3.66*ones(13*6,1)];
gplotmatrix(xvars,yvars,PS,'rgb','xos')
ylabel('%');xlabel('Size (mm)');
ES= [ES; 0.1*ones(13,1); 0.5*ones(13,1); 1*ones(13,1); 1.5*ones(13,1); 2*ones(13,1); 2.5*ones(13,1)];
ES=[ES;ES;ES];
figure
gplotmatrix(xvars,yvars,ES,'rgbcmk','xos^*+')
ylabel('%');xlabel('Ecs (kWh/t)');
```

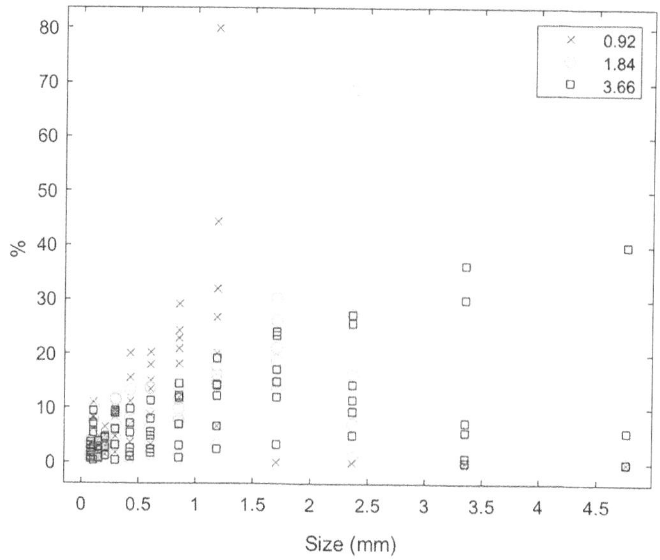

Figure 8.13 Particle size distribution (percentage retained) from Mini-DWT tests.

(2) Scatter plots for particles' cumulative percentage

```
% Based on the A and B of particle size distribution in the above code
k=length(A(:,1));
l=length(A(1,:));
A1=A;
for i=1:k-1
    for j=1:l
    A1(k-i,j)=A1(k-i+1,j)+A1(k-i,j);
    end
end
xvars = [];
yvars = [];
ES=[];
for i=1:18
xvars = [xvars; B];
yvars = [yvars; A1(:,i)];
end
PS=[0.92*ones(13*6,1);1.84*ones(13*6,1);3.66*ones(13*6,1)];
gplotmatrix(xvars,yvars,PS,'rgb','xos')
ylabel('%');xlabel('Size (mm)');
ES= [ES; 0.1*ones(13,1); 0.5*ones(13,1); 1*ones(13,1); 1.5*ones(13,1);...
 2*ones(13,1); 2.5*ones(13,1)];
ES=[ES;ES;ES];
figure
gplotmatrix(xvars,yvars,ES,'rgbcmk','xos^*+')
ylabel('%');xlabel('Ecs (kWh/t)');
```

It can be seen roughly in Figure 8.13 that the finest original particle size of 0.92 mm was broken to a higher percentage of the finest particle sizes. It is even more apparent in Figure 8.14 with a cumulative percentage that the finest product particle sizes are in the leftmost region, and the coarsest product particle sizes are in the rightmost region.

It is not easy to distinguish the trend in Figure 8.15 how the specific energy Ecs affects the percentage retained from Mini-DWT tests. However, the trend in Figure 8.16 shows that the

Table 8.4 The particle breakage data from Mini-DWT.

Sample size	0.92						1.84						3.66					
Ecs / % retained / Sieve sizes	0.10	0.50	1.00	1.50	2.00	2.50	0.10	0.50	1.00	1.50	2.00	2.50	0.10	0.50	1.00	1.50	2.00	2.50
4.75	0.00	0.00	0.00	0.00	0.00	0.00	0.00	0.00	0.00	0.00	0.00	0.00	39.96	5.68	0.00	0.00	0.00	0.00
3.35	0.00	0.00	0.00	0.00	0.00	0.00	0.00	0.00	0.00	0.00	0.00	0.00	36.26	29.93	7.27	5.56	0.89	0.00
2.36	0.00	0.00	0.00	0.00	0.00	0.00	68.60	27.38	16.19	9.26	7.01	1.00	11.42	27.15	25.67	14.30	9.28	4.93
1.7	0.00	0.00	0.00	0.00	0.00	0.00	21.08	30.24	26.12	18.77	14.76	13.45	3.28	12.06	24.03	23.29	17.00	14.80
1.18	79.84	44.35	31.86	26.61	20.10	6.54	3.03	16.18	17.68	19.75	14.21	15.07	2.43	6.61	12.19	14.30	19.13	14.04
0.85	11.77	29.13	22.79	24.13	20.93	18.13	1.17	7.41	7.88	13.83	11.48	9.88	0.85	3.02	6.92	11.82	12.08	14.47
0.6	3.28	8.67	17.89	14.96	13.34	20.19	1.36	3.30	6.20	8.83	13.52	13.76	1.69	2.44	5.51	4.85	7.94	11.29
0.425	0.55	3.60	6.86	11.07	15.43	20.09	0.62	2.24	5.83	8.09	9.49	13.20	0.95	1.62	2.58	5.32	7.05	9.76
0.3	1.73	4.58	8.91	8.26	9.67	9.44	1.17	4.29	8.87	6.85	11.35	11.63	0.24	3.13	6.10	5.91	9.28	9.10
0.212	1.28	2.62	2.29	2.89	4.84	6.45	1.30	3.17	3.66	2.78	3.35	4.57	1.16	2.67	2.81	4.26	4.47	4.71
0.15	0.55	2.70	1.80	1.74	4.17	4.58	0.62	1.93	2.11	2.47	3.85	3.50	0.85	2.20	0.59	2.01	2.80	3.84
0.106	0.46	3.11	6.13	8.18	8.26	10.93	0.56	2.80	4.22	7.41	7.88	10.13	0.32	2.55	5.39	6.74	7.16	9.43
0.075	0.55	1.23	1.47	2.15	3.25	3.64	0.49	1.06	1.24	1.98	3.10	3.81	0.63	0.93	0.94	1.65	2.91	3.62

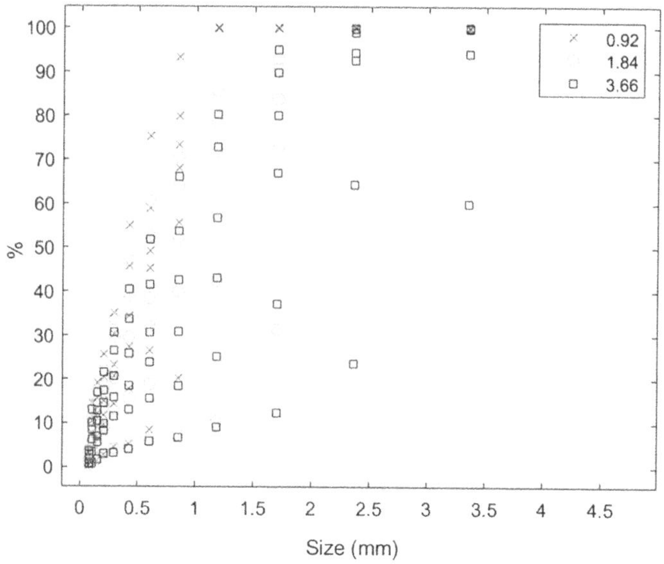

Figure 8.14 Particle size distribution (cumulative percentage) from Mini-DWT tests.

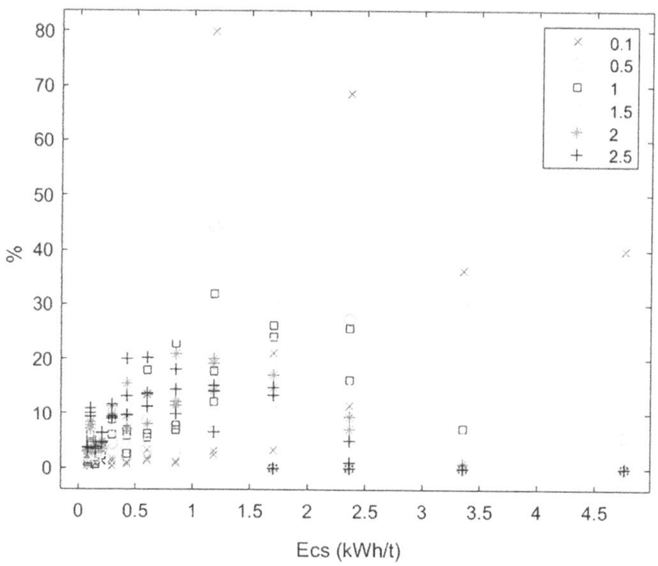

Figure 8.15 Ecs vs. percentage retained from Mini-DWT tests.

product particle sizes became finer when increasing specific energy Ecs for the breakage. The highest specific energy Ecs of 2.50 kWh/t generated the finest product particle sizes.

(3) Test whether the data of particles' percentage retained is a normal distribution.

```
% Based on the A and A1 in the above code
normplot(A)
figure
normplot(A1)
```

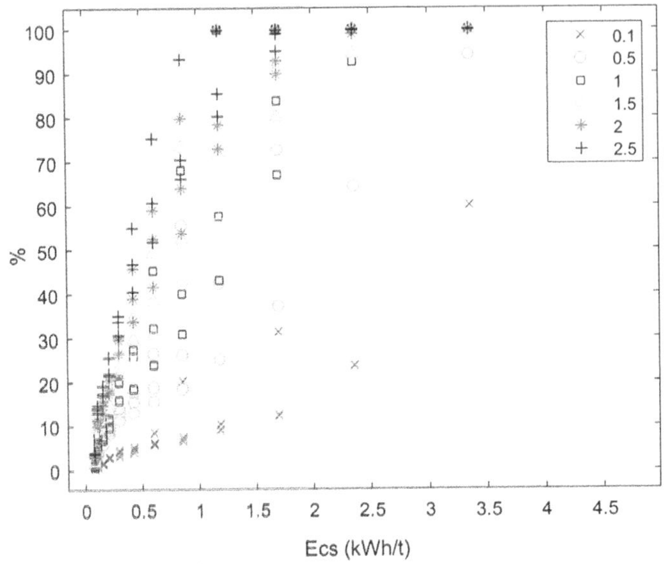

Figure 8.16 Ecs vs. cumulative percentage from Mini-DWT tests.

It can be seen in Figure 8.17 that both the particle size data (percentage retained and cumulative percentage) are not normal distributions. We need to find a better distribution model.

(4) Linear regression for the cumulative percentage of the product size distribution

```
% Based on the A, A1, B, xvars, yvars, PS, and ES in the above code
tbl = table(PS,ES,xvars,yvars,'VariableNames',{'PS','ES','xvars','yvars'});
lm  = fitlm(tbl,'yvars~PS+ES+xvars')
```

```
Results:
lm =
Linear regression model:
    yvars ~ 1 + PS + ES + xvars
Estimated Coefficients:
                    Estimate        SE          tStat        pValue
                    ________    _______       _______    __________

    (Intercept)      13.656      3.6013        3.7919    0.00019106
    PS              -4.7405      1.1438       -4.1446    4.7863e-05
    ES               11.467       1.569        7.3088    4.4122e-12
    xvars             22.95     0.93459        24.557     3.255e-66
Number of observations: 234, error degrees of freedom: 230
Root mean squared error: 19.9
R-squared: 0.745,  adjusted R-squared: 0.742
F-statistic vs. constant model: 225, p value = 4.8e-68
```

The R-squared is 0.745 for the linear regression, which did not fit well. We need to carry on finding other models.

(5) The plsregress for a cumulative percentage of the product size distribution

```
% Based on the A, A1, B, xvars, yvars, PS, and ES in the above code
X=[PS,ES,xvars];
Y=[yvars];
[XL,yl,XS,YS,beta,PCTVAR] = plsregress(X,Y,3);
```

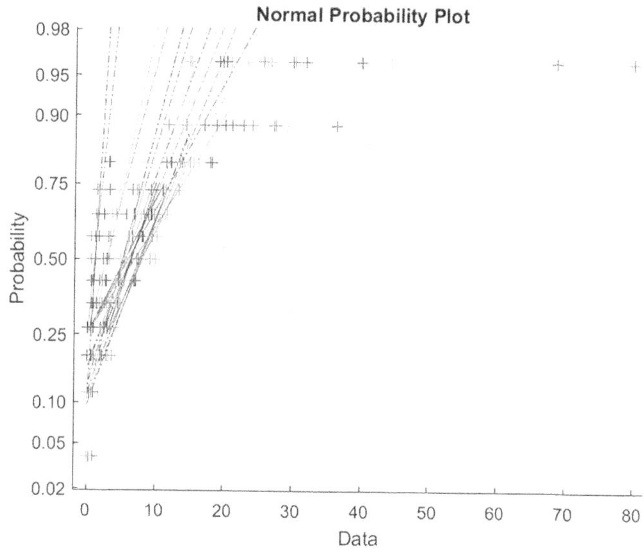

(a)

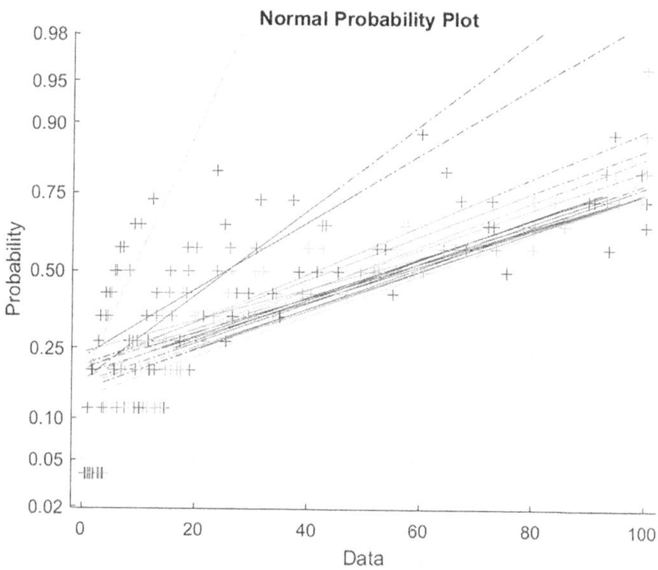

(b)

Figure 8.17 (a) The normal probability plot of the particle size data (percentage retained). (b) The normal probability plot of the particle size data (cumulative percentage) from Mini-DWT tests.

```
plot(1:3,cumsum(100*PCTVAR(2,:)),'-bo');
xlabel('Number of PLS components');
ylabel('Percent variance explained in y');
figure
yfit = [ones(size(X,1),1) X]*beta;
TSS = sum((Y-mean(Y)).^2);
RSS = sum((Y-yfit).^2);
```

```
R_squared = 1 - RSS/TSS
plot(yfit,Y,'*')
xlabel('Fitted cumulative PSD');
ylabel('Measured cumulative PSD');

Results:
R_squared =
    0.7455
```

The R-squared is 0.7455 for the plsregress (partial least-squares regression), which potentially has a linear trend but it is weak. It can also be seen in Figure 8.18a that three PLS components can only reach 74.5%, which did not show a good fit. The parity plot in Figure 8.18b shows that the linear model did not catch all features of the particle size data (cumulative percentage).

(6) Nonlinear regression for a cumulative percentage of the product size distribution

```
% Based on the A, A1, B, xvars, yvars, PS, and ES in the above code
tbl = table(PS,ES,xvars,yvars,'VariableNames',{'PS','ES','xvars','yvars'});
modelfun = @(b,x)b(1) + b(2)*x(:,1).^b(3) + b(4)*x(:,2).^b(5)+ b(6)*...
x(:,3).^b(7);
beta0 = [-50 500 -1 500 -1 500 -1];
mdl = fitnlm(tbl,modelfun,beta0)
figure
plot(mdl.Fitted, yvars,'*')
xlabel('Cumulative PSD fitted')
ylabel('Cumulative PSD')
figure
plot(yvars, yvars -mdl.Fitted,'*')
xlabel(' Cumulative PSD ')
ylabel(' Cumulative PSD - cumulative PSD fitted')

Results:
mdl =
Nonlinear regression model:
    yvars ~ b1 + b2*PS^b3 + b4*ES^b5 + b6*xvars^b7
Estimated coefficients:
            Estimate        SE           tStat        pValue
            ________    __________    _______    __________

    b1         32334       0.35871      90139             0
    b2         -1521       0.35871    -4240.1             0
    b3     0.0064913      0.001093      5.939    1.0511e-08
    b4       -5230.1       0.35871     -14581             0
    b5    -0.0016772    0.00016541     -10.14    3.2866e-20
    b6        -25515       0.35871     -71130             0
    b7    -0.0010533    2.8594e-05    -36.836    1.9472e-98
Number of observations: 234, error degrees of freedom: 230
Root mean squared error: 14.3
R-squared: 0.868,  adjusted R-squared 0.866
F-statistic vs. constant model: 505, p value = 7e-101
```

The R-squared is 0.868 for the nonlinear regression, which is better than the partial least-squares regression. It can also be seen in Figure 8.19a that data points are closer to

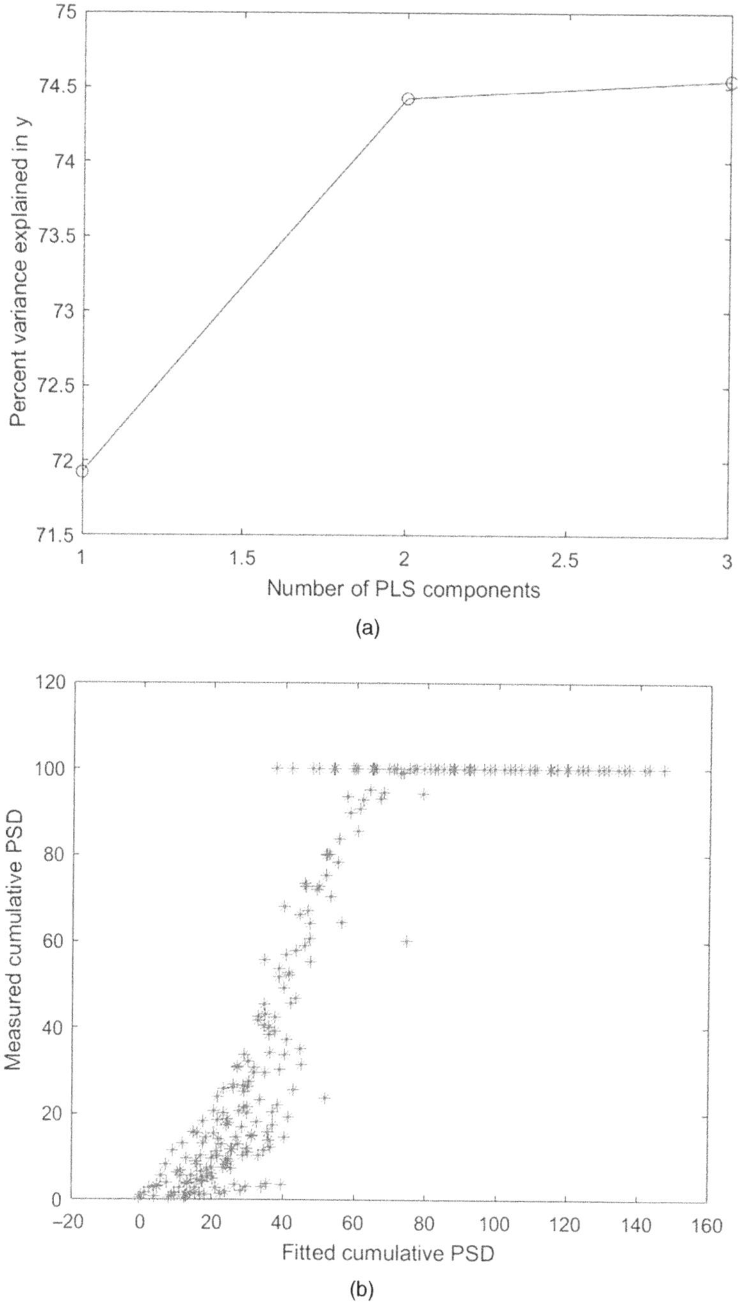

Figure 8.18 The plots for the cumulative percentage of the product size distribution: (a) the percentage variance as a function of the number of components; (b) the parity plot.

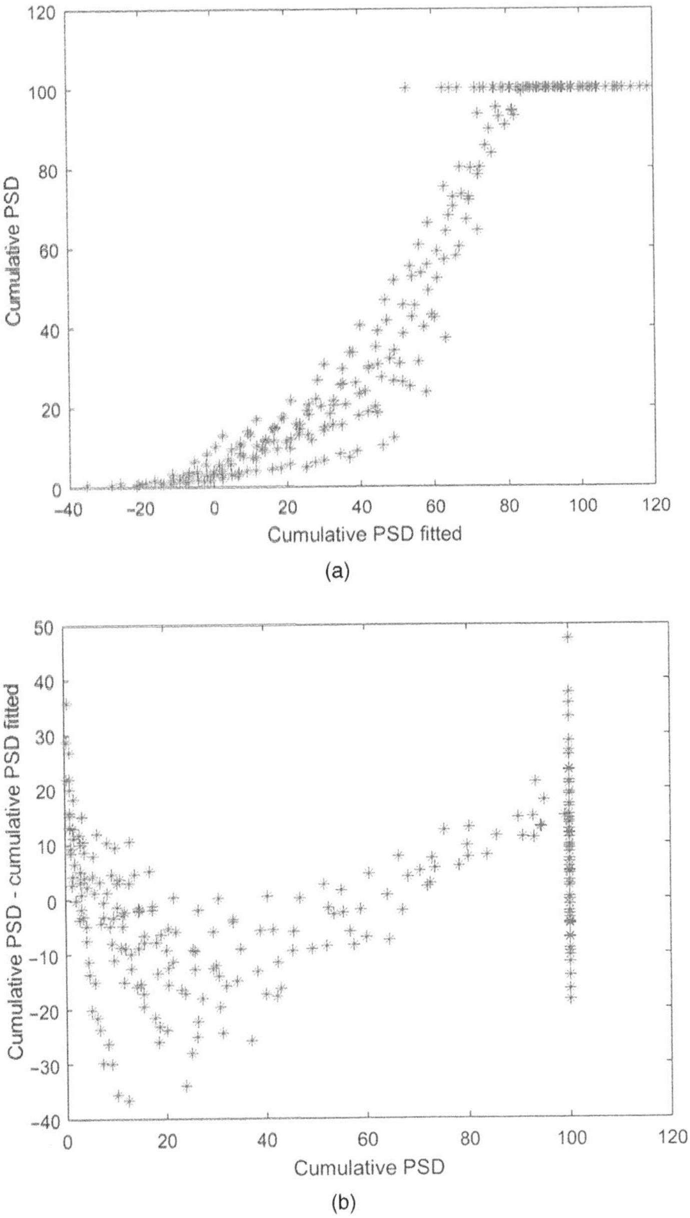

Figure 8.19 The plots for the cumulative percentage of the product size distribution: (a) the parity plot; (b) the residuals plot.

the $y = x$ line compared to Figure 8.18b. The residuals plot in Figure 8.19b shows that data points are scattered and far away from the $y = 0$ line. We need to find a better distribution model.

(7) Nonlinear regression based on the Rosin–Rammler distribution model for the cumulative percentage of the product size distribution

```
% Based on the A, A1, B, xvars, yvars, PS, and ES in the above code
tbl = table(PS,ES,xvars,yvars,'VariableNames',{'PS','ES','xvars','yvars'});
modelfun = @(b,x) 100*(1-0.2.^(((diag(x(:,1))+b(3)*eye(k*1))\...
(diag(x(:,2))+b(4)*eye(k*1)))*(x(:,3)./b(1)).^b(2)));
beta0 = [1 1 1 1];
mdl = fitnlm(tbl,modelfun,beta0)
figure
plot(mdl.Fitted, yvars,'*')
xlabel('Cumulative PSD fitted')
ylabel('Cumulative PSD')
figure
plot(yvars, yvars -mdl.Fitted,'*')
xlabel('Cumulative PSD')
ylabel('Cumulative PSD - cumulative PSD fitted')
```

```
Results:
mdl =
Nonlinear regression model:
    yvars ~ F(b,x)
Estimated coefficients:
         Estimate         SE          tStat        pValue
         ________    ________       _______    ___________

    b1     1.0446    0.053779        19.424    2.1393e-50
    b2     1.5413    0.052263         29.49    4.2057e-80
    b3    0.39325     0.14829        2.6519     0.0085616
    b4    0.37535    0.056118        6.6887    1.6928e-10
Number of observations: 234, error degrees of freedom: 230
Root mean squared error: 7.99
R-squared: 0.959,  adjusted R-squared 0.959
F-statistic vs. zero model: 3.32e+03, p value = 4.26e-202
```

The R-squared is 0.959 for the nonlinear regression based on the Rosin–Rammler distribution model, which is better than the results from the other regression method. It can also be seen in Figure 8.20a that data points are closer to the $y = x$ line compared to Figures 8.18b and 8.19a. This means the Rosin–Rammler distribution model is suitable for modeling particle size distribution. The residuals plot in Figure 8.20b shows that data points are scattered and still far away from the $y = 0$ line.

8.4 Summary

In this chapter, we have presented practical data statistics using the MATLAB built-in Statistics and Machine Learning Toolbox to perform data analysis, such as regressions, PCA, dimensional reduction, and statistical visualization. It is helpful to draw different plot types for statistical visualization, including scatter plots using grouped data, box plots, and distribution plots to see the data trends. Due to large datasets or complicated data, the data trend may not be discernible. Domain knowledge in engineering disciplines is needed to choose a better model for the data of interest.

We also introduced dimensional analysis, especially the application of Buckingham Pi, to correlate experimental data with dimensionless groupings of parameters and variables. After applying the Buckingham Pi Method, experimental data becomes dimensionless, by which mathematical models are easily built.

The operation of regression is an optimization problem. The knowledge and skills learned from previous chapters can be applied to evaluate the regression fit (the goodness of the optimization).

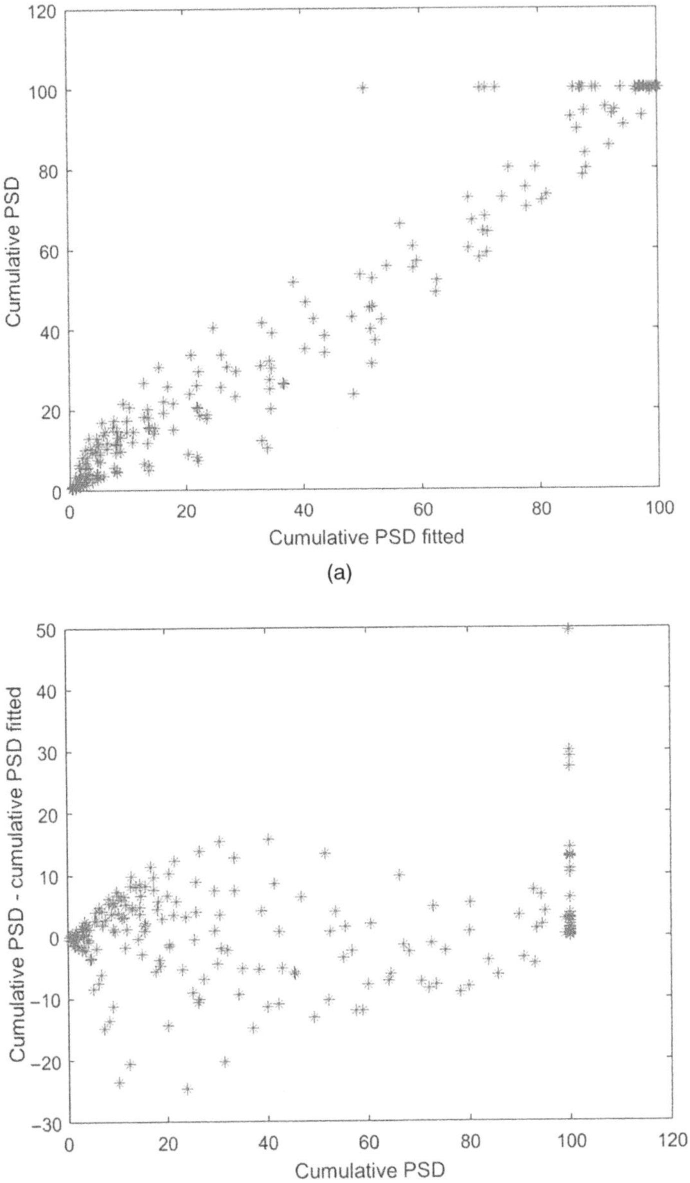

(b)

Figure 8.20 Plots based on the Rosin–Rammler distribution model for the cumulative percentage of the product size distribution: (a) parity plot and (b) residuals plot.

Problems

8.1 Decide whether the following equations are OK and why:

$$C_p = a + bT + \frac{c}{T} + dT^2 \tag{8.26}$$

$$k = k_o e^{-\frac{E_A}{RT}} \tag{8.27}$$

$$logp^{sat} = a/T \tag{8.28}$$

$$x = x_o(1 - e^{-t}) \tag{8.29}$$

8.2 Consider the standard problem of one-dimensional heat conduction in an "infinite" bar with constant thermal properties, initially heated by a point source of heat.

(1) The temperature of any point of the bar T depends on the distance from the point source x, the elapsed time after the initial heating t, the mass density of the bar ρ, the specific heat of the bar c, the thermal conductivity of the bar λ, and the strength of the heat source Q measured in energy units per (length units)^2, i.e. Joule/m^2. To give the SI unit to all those seven variables.

(2) To use x, t, ρ, λ, as basis units to apply Buckingham Pi (π) theorem to obtain the three πs (i.e. for T, c, Q).

8.3 A piezoelectric vibration sensor measured a fluid's kinetic energy fluctuation (KEF) in a flotation cell. The data is shown in Table 8.5. The distance data is the distance to the impeller. The level data shows the vertical measurement point of the fluid from the bottom of the flotation cell.

(1) Scatter plots using grouped data.
(2) Test the data for a normal distribution.
(3) Linear regression for the data.

Table 8.5 The data of fluid's kinetic energy fluctuation (KEF) was measured by a piezoelectric vibration sensor in a flotation cell.

Distance / KEF / Level (cm)	32cm	44cm	57cm	69cm
2	119.1876	45.2147	40.4176	35.9277
5	158.6847	54.6392	41.5525	32.8495
8	204.3603	61.2693	51.0602	30.8726
11	225.0062	73.6293	51.5984	30.7853
14	261.6178	83.6303	53.1313	30.8604
17	287.2098	108.4362	78.5375	27.1616
20	314.617	134.9304	74.2909	34.8954
23	399.6472	144.436	83.4605	34.6845
26	409.0237	190.5037	105.2917	45.6732
29	411.0641	194.6311	110.9736	52.4107
32	406.3152	176.592	99.2534	58.9087
42	128.6605	91.1989	60.1764	57.3923
52	22.9843	26.0638	25.5042	60.1614
72	27.9864	38.3598	33.9151	22.3508
102	12.6384	18.1693	14.794	18.6904

(4) Use plsregress to regress the data and draw the parity and residual plots.

(5) Nonlinear regression model using yvars ~ b1 + b2*PS ^ b3 + b4*xvars ^ b5 and draw the parity plot and the residuals plot.

(6) Create a nonlinear regression model based on the Gaussian model and draw the parity and the residual plots.

Hint: you need to use the matrix format of this equation: (PS ^ b(1) + b(2))\exp(−((x−b(3))/b(4)) ^ 2)*b(5).

9

Complex Equation Systems

Complex equation systems in this chapter aim to help our readers learn how to understand dynamic systems, numerical integration, and differential equations and the methods to solve those equations.

In Chapters 2–8, we have only considered systems described by sets of algebraic equations, which are relatively more straightforward than what we study in this chapter. In this chapter, we study systems described by derivatives of states. Mainly, we are interested in time evolution, i.e. derivatives with respect to time, and in some cases, we also have derivatives with respect to space. Having a simple way to solve equations with derivatives is not straightforward. There are relatively few cases that have analytical solutions. Therefore, numerical integration and numerical solutions to differential equations are practical and valuable to engineering problems.

MATLAB has many differential equation solvers, including ordinary differential equations (ODEs) posed as either initial or boundary value problems, delay differential, and partial differential equations (PDEs). In addition, there are numerical integration functions for functional expressions via quadrature or discrete data sets.

We first introduce solvers for ODEs and PDEs, and then apply them to cases in chemical engineering. Due to the complexity of some real systems, existing MATLAB solvers cannot be applied directly. We introduce a method of numerical algorithm flowchart for these difficult cases.

9.1 Introduction

9.1.1 Dynamic Systems

In dynamic systems, their physical properties tend to evolve with time. Depending on whether the physical property of interest for the system is a continuous function or not, the system can be categorized into a continuous time dynamic system (e.g. a continuously operating ball mill as shown in Figure 9.1a), or discrete dynamic system (e.g. adsorption/desorption switching operating column as shown in Figure 9.1b). If the continuous function is continuously differentiable, the system is called a differentiable dynamic system.

Chemical Engineering Analysis and Optimization Using MATLAB®, First Edition.
Weiguo Xie, Sam Toan, and Richard Davis.

Companion website: www.wiley.com/go/xie/chemicalengineering

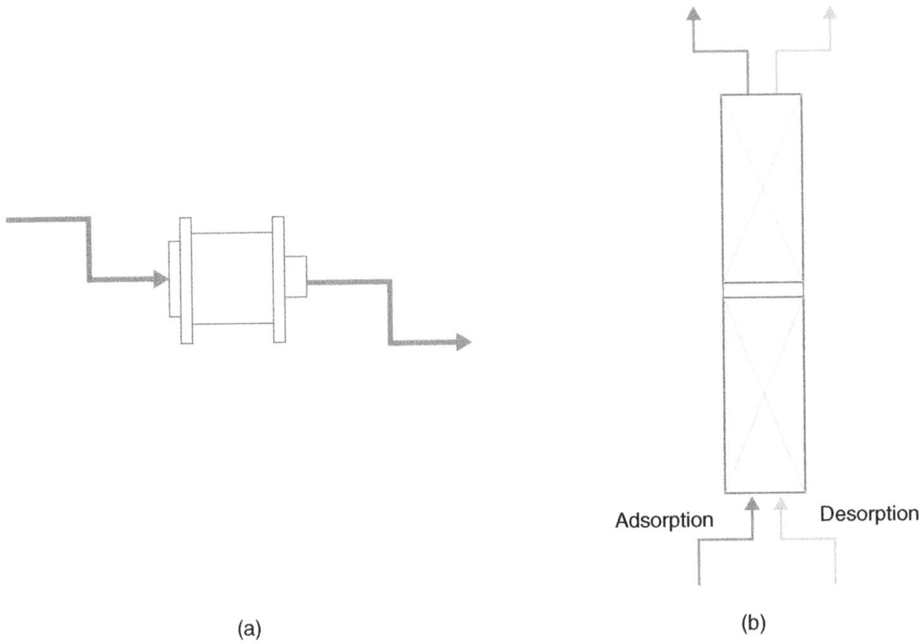

Figure 9.1 (a) A continuous operating ball mill; (b) adsorption/desorption column.

9.1.2 Lump Parameter System and Distributed Parameter System

A lump parameter system is a simplified representation of a physical system that assumes all components can be concentrated at a single point, and uniform mathematical models can describe their behavior. A perfect continuous stirred-tank reactor (CSTR), shown in Figure 9.2a, can be treated as a lump parameter system. For example, concentration C is uniform everywhere in the CSTR.

A distributed parameter system differs from a lump parameter system because its behavior is distributed spatially and cannot be considered uniform. A tubular reactor shown in Figure 9.2b is a distributed parameter system because its fluid properties have spatial distributions. For example, concentration C is different at different locations inside the tubular reactor $C(x, y, z)$, where x, y, and z show the location for the concentration.

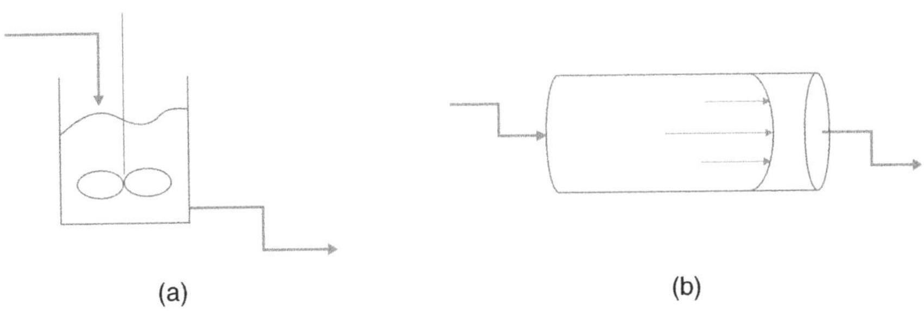

Figure 9.2 (a) A CSTR; (b) a tubular reactor.

9.2 Ordinary Differential Equations (ODEs)

9.2.1 ODEs for Dynamic Systems

ODEs, derivatives with respect to time, can be used to represent time evolution.

Example 9.1

In the following simple chemical reaction as shown in Figure 9.3, component A has a reaction rate of 3/s to produce component B, while component B has a reaction rate of 10/s to produce component A.

Solution:

We have two equations for this reaction:

$$\frac{dA}{dt} = -3A + 10B \tag{9.1}$$

$$\frac{dB}{dt} = 3A - 10B \tag{9.2}$$

Equations (9.1) and (9.2) are ODEs and can also be written in a matrix form:

$$\frac{dX}{dt} = CX \tag{9.3}$$

where $X = \begin{bmatrix} A \\ B \end{bmatrix}, C = \begin{bmatrix} -3 & 10 \\ 3 & -10 \end{bmatrix}$.

We need an initial condition for the above reaction to solve the equations. The initial condition (IC) is given as $X(0) = \begin{bmatrix} A_0 \\ B_0 \end{bmatrix}$. Then, the solution for Eq. (9.3) becomes:

$$x(t) = e^{tC}X(0) \tag{9.4}$$

There is an analytical solution for Eq. (9.3) due to its relative simplicity; however, there are no analytical solutions for most cases.

9.2.2 Higher-Order ODEs

ODEs are generally classified according to their order, linearity, and boundary conditions. The order of an ODE is the order of its highest derivative. For example, the following equation is third-order nonlinear ODE:

$$\frac{d^3y}{dx^3} + a\frac{d^2y}{dx^2} + b\left(\frac{dy}{dx}\right)^2 = kx \tag{9.5}$$

There is no analytical solution ready for Eq. (9.5). It is very challenging to obtain its numerical solution.

There is a way to solve this type of high-order solution by transforming it to canonical form, i.e. using a series of substitutions. For example, an nth-order ODE is as follows:

$$\frac{d^nz}{dx^n} = g\left(z, \frac{dz}{dx}, \frac{d^2z}{dx^2}, \ldots, \frac{d^{n-1}z}{dx^{n-1}}, x\right) \tag{9.6}$$

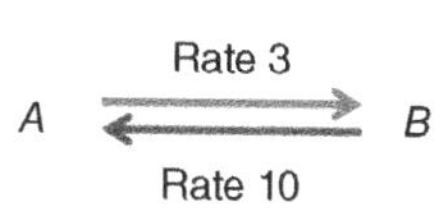

Figure 9.3 A chemical reaction between component A and component B.

Equation (9.6) can be transformed into a series of equations:

$$
\begin{aligned}
z &= y_1 \\
\frac{dz}{dx} &= \frac{dy_1}{dx} = y_2 \\
&\vdots \\
\frac{d^{n-1}z}{dx^{n-1}} &= \frac{dy_{n-1}}{dx} = y_n \\
\frac{d^n z}{dx^n} &= \frac{dy_n}{dx}
\end{aligned}
\tag{9.7}
$$

This transformation makes the nth-order ODE an equivalent set of n first-order ODEs in canonical form with n variables. Numerical methods can easily solve it.

Example 9.2

To transform the following ODE into its canonical form using three ODEs with three variables

$$\frac{d^3z}{dx^3} + 3\left(\frac{d^2z}{dx^2}\right)^2 - z^2\left(\frac{dz}{dx}\right)^3 + 5z = 0 \tag{9.8}$$

Solution:

Equation (9.8) can be transformed into a series of equations:

$$z = y_1 \tag{9.9}$$

$$\frac{dz}{dx} = \frac{dy_1}{dx} = y_2 \tag{9.10}$$

$$\frac{d^2z}{dx^2} = \frac{dy_2}{dx} = y_3 \tag{9.11}$$

$$\frac{d^3z}{dx^3} = \frac{dy_3}{dx} \tag{9.12}$$

$$\frac{dy_3}{dx} + 3y_3^2 - y_1^2y_2^3 + 5y_1 = 0 \tag{9.13}$$

Therefore, Eq. (9.8) has been transformed into three equations with three variables: y_1, y_2, and y_3.

$$\frac{dy_1}{dx} = y_2 \tag{9.14}$$

$$\frac{dy_2}{dx} = y_3 \tag{9.15}$$

$$\frac{dy_3}{dx} + 3y_3^2 - y_1^2y_2^3 + 5y_1 = 0 \tag{9.16}$$

9.2.3 Numerical Integration and Differentiation

Integration is another foundational concept in calculus to work on differential equations for analyzing and interpreting complex datasets. Finding an antiderivative of a function is not always possible, which leads to a need for numerical integration instead of analytical integration.

MATLAB has some numerical integration and differentiation functions, as shown in Table 9.1.

Example 9.3

Find the value of the gradient of a multivariate function $f(x, y) = xy^2$ at a specified point (1,1).

Solution:

```
MATLAB code:
x = -2:0.2:2;
y = x';
f = x.* y.^2;
surf(x,y,f)
xlabel('x')
ylabel('y')
zlabel('z')
[fx,fy] = gradient(f,0.2);
x0 = 1;
y0 = 1;
t = (x == x0) & (y == y0);
indt = find(t);
f_grad = [fx(indt) fy(indt)]

Results:
f_grad =
    1.0000    2.0000
```

Table 9.1 MATLAB functions for numerical integration and differentiation (https://www.mathworks.com/The MathWorks, Inc.).

MATLAB functions	Description
integral	Numerical integration
integral2	Numerically evaluate a double integral
integral3	Numerically evaluate a triple integral
quadgk	Numerically evaluate integral – Gauss–Kronrod quadrature
quad2d	Numerically evaluate double integral – tiled method
polyint	Polynomial integration
cumtrapz	Cumulative trapezoidal numerical integration for numeric data
trapz	Trapezoidal numerical integration for numeric data
del2	Finite-difference (FD) derivatives using discrete Laplacian
diff	Differences and approximate derivatives
gradient	Numerical gradient
polyder	Polynomial differentiation

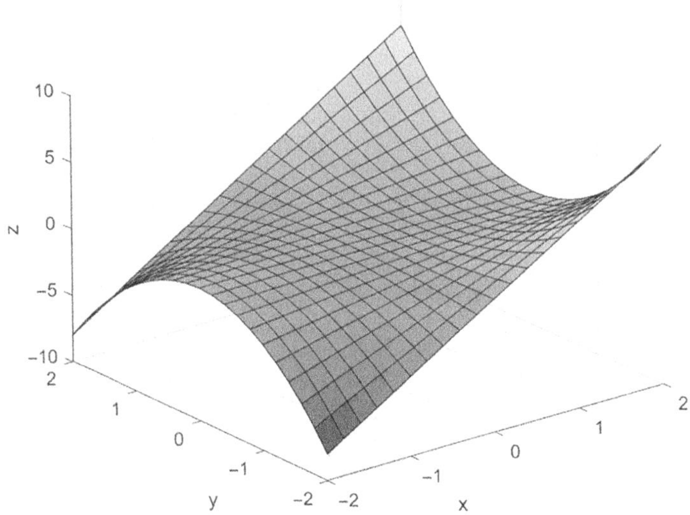

Figure 9.4 Surf plot of Example 9.5 for $f(x, y) = xy^2$.

The surf plot of Example 9.3 is shown in Figure 9.4.

The MATLAB *gradient* function has been applied in Example 9.3. MATLAB's *find* function can be used to get the gradient of a multivariate function at any specified point.

Example 9.4

(1) $f(x) = \cos(x) - 2\ln(x)$, with x range (0.01, 20). Plot the curve of $f(x)$ vs. x

Solution:

```
≫ x=linspace(0.01,20,200);
y=cos(x) - 2*log(x);plot(x,y);
```

The plot of f(x) vs. x for the Example 9.4 is shown in Figure 9.5.

(2) Use the MATLAB integral function to evaluate the integral from $x = 0.01$ to $x = 20$.

Solution:

```
≫q = integral(@(x) cos(x) - 2*log(x),0.01,20)

Result:
q =
  -79.0384
```

(3) Use the MATLAB quadgk function to evaluate the integral from $x = 0.01$ to $x = 20$.

Solution:

```
≫p = quadgk(@(x) cos(x) - 2*log(x),0.01,20)

Result:
p =
  -79.0384
```

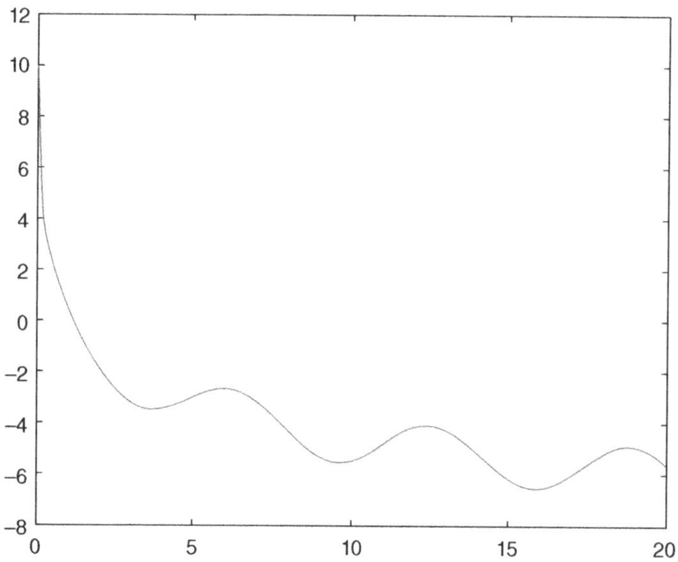

Figure 9.5 The plot of $f(x)$ vs. x for the Example 9.4.

It can be seen that both integral and quadgk functions can give the same result for the above example.

9.2.4 Numerical Method to Solve ODEs

Commonly, analytical solutions are not available for ODE problems. Then, numerical solutions with relatively reasonable accuracy are acceptable in engineering practices. This section introduces four methods: Euler's explicit method, Euler's implicit method, the Runge–Kutta method, and the FD method.

9.2.4.1 Euler's Explicit Method

Euler's explicit method is the most basic method for solving ODEs and has a long history. It is named after Leonhard Euler. It is a basic method to construct more sophisticated methods.

Consider the following differential:

$$\frac{dx}{dt} = g(t, x) \tag{9.17}$$

with $x(0) = x_0$

Euler's explicit method is to use forward FD to approximate the derivative in Eq. (9.17):

$$\left.\frac{dx}{dt}\right|_{t=t_n} = \frac{x(t_n + \Delta t) - x(t_n)}{\Delta t} = \frac{x_{n+1} - x_n}{\Delta t} \tag{9.18}$$

To substitute this FD to the above Eq. (9.17):

$$x_{n+1} = x_n + g(t_n, x_n)\Delta t \tag{9.19}$$

It approximates the derivative by neglecting the higher-order terms in the Taylor series.

Example 9.5

(1) Solve the above Example 9.1 using Euler's explicit method with time range [0, 0.5] and initial condition $X(0) = \begin{bmatrix} A_0 \\ B_0 \end{bmatrix} = \begin{bmatrix} 0 \\ 1 \end{bmatrix}$. The time step uses 0.005 seconds.

Solution:

```
MATLAB code:
%Euler explicit method
n=100;
h=0.5/n;
y=zeros(2,n+1);
y(1,1)=0;
y(2,1)=1;
for i=1:n
f = @(y) [-3*y(1)+10*y(2); 3*y(1)-10*y(2)];
y(:, i + 1) = y(:, i) + h * f(y(:, i));
end
te=linspace(0, 0.5,n);
plot(te,y(1,1:n),'g','LineWidth',2);
hold on;
plot(te,y(2,1:n),'b','LineWidth',2);
legend('Component A','Component B');
xlabel('Time (s)');
ylabel('Concentration');
title(['Chemical reactions - Euler explicit method'...
 'timestep=0.005s']);
```

Results:

The plot of concentration vs. t for Example 9.5 using Euler's explicit method with timestep = 0.005 s is shown in Figure 9.6.

(2) Use Euler's explicit method with a time step of 0.05 seconds and compare results with those of question 1.

Solution:

```
MATLAB code:
n=10;
h=0.5/n;
y=zeros(2,n+1);
y(1,1)=0;
y(2,1)=1;
for i=1:n
f = @(y) [-3*y(1)+10*y(2); 3*y(1)-10*y(2)];
y(:, i + 1) = y(:, i) + h * f(y(:, i));
end
te=linspace(0, 0.5,n);
plot(te,y(1,1:n),'k','LineWidth',2);
```

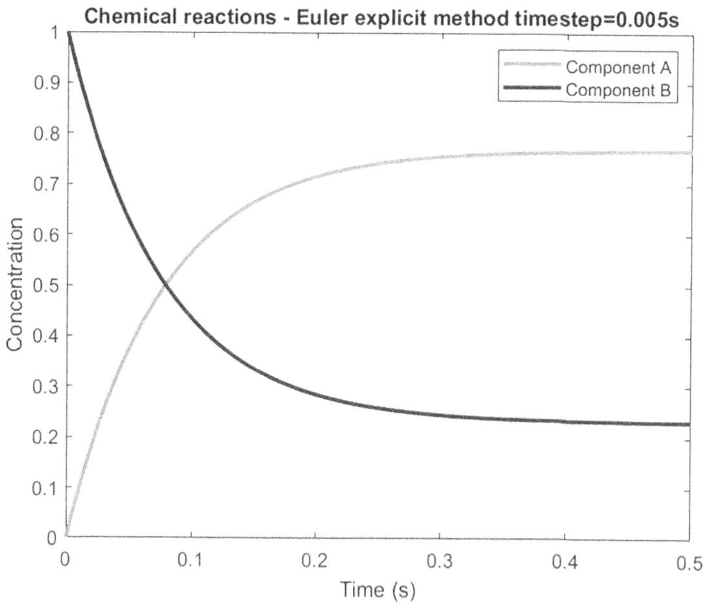

Figure 9.6 The plot of concentration vs. t for Example 9.5 using Euler's explicit method with timestep = 0.005 s.

```
hold on;
plot(te,y(2,1:n),'r','LineWidth',2);
hold on
clear y
n=100;
h=0.5/n;
y=zeros(2,n+1);
y(1,1)=0;
y(2,1)=1;
for i=1:n
f = @(y) [-3*y(1)+10*y(2); 3*y(1)-10*y(2)];
y(:, i + 1) = y(:, i) + h * f(y(:, i));
end
te2=linspace(0, 0.5,n);
plot(te2,y(1,1:n),'g','LineWidth',2);
hold on;
plot(te2,y(2,1:n),'b','LineWidth',2);
legend('Component A -timestep 0.05s',...
'Component B -timestep 0.05s',...
'Component A -timestep 0.005s',...
'Component B -timestep 0.005s');
xlabel('Time (s)');
ylabel('Concentration');
title('Chemical reactions - Euler explicit method');
```

Results:

The comparison of two results is shown in Figure 9.7.

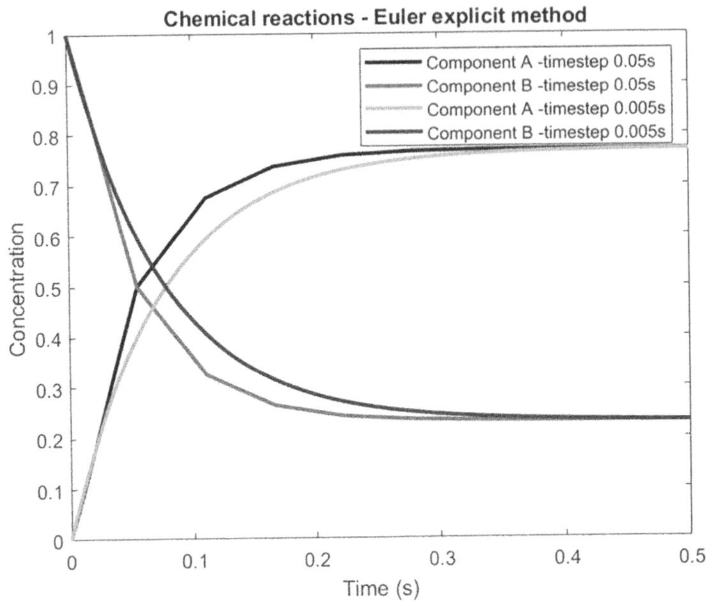

Figure 9.7 The comparison of the results between timestep = 0.05 s and timestep = 0.005 s for the Example 9.5.

It can be seen that the results depended on the selection of the timestep. Solutions with large timesteps tend to differ significantly from accurate results.

Note: Euler's explicit method is simple, but it is conditionally stable for a limited choice of small timesteps Δt, and it is slow convergence.

9.2.4.2 Euler's Implicit Method

Euler's implicit method is different from Euler's implicit method by using backward FD to approximate the derivative in the above Eq. (9.17):

$$\left.\frac{dx}{dt}\right|_{t=t_n} = \frac{x(t_n) - x(t_n - \Delta t)}{\Delta t} = \frac{x_n - x_{n-1}}{\Delta t} \tag{9.20}$$

To substitute this FD to the above Eq. (9.17):

$$x_n = x_{n-1} + g(t_n, x_n)\Delta t \tag{9.21}$$

All items in Eq. (9.21) are known. The following equation is obtained by shifting the index to the next step:

$$x_{n+1} = x_n + g(t_{n+1}, x_{n+1})\Delta t \tag{9.22}$$

The parameters of the unknown x_{n+1} are on both sides of the equation, which cannot be solved directly and need iterative methods.

Example 9.6

(1) Solve the above Example 9.1 using Euler's implicit method with time range [0, 0.5] and initial condition $X(0) = \begin{bmatrix} A_0 \\ B_0 \end{bmatrix} = \begin{bmatrix} 0 \\ 1 \end{bmatrix}$. The time step uses 0.005 seconds. The iterative method needs to be applied.

Solution:

```
MATLAB code:
%Euler implicit method
n=100;
h=0.5/n;
y=zeros(2,n+1);
y(1,1)=0;
y(2,1)=1;
M_max=1000;%max iterations
m=1;
for i=1:n
    err=1;% initial tolerance
    while (m<M_max & err>1e-8)
        f = @(y) [-3*y(1)+10*y(2); 3*y(1)-10*y(2)];
        yp = y(:, i) + h * f(y(:, i+1));
        err=max(abs(y(:, i + 1)-yp));
        y(:, i + 1) =yp;
        m=m+1;
    end
end
te=linspace(0, 0.5,n+1);
figure
plot(te,y(1,:),'k','LineWidth',2);
hold on;
plot(te,y(2,:),'r','LineWidth',2);
legend('Component A','Component B');
xlabel('Time (s)');
ylabel('Concentration');
title(['Chemical reactions - Euler implicit method'...
 'timestep=0.005s']);
```

Results:

The plot of concentration vs. t for Example 9.6 using Euler's implicit method with timestep = 0.005 s is shown in Figure 9.8.

(2) Compare results from Euler's explicit method with the time step of 0.05 seconds.

Readers, please combine the above codes, plot the results, and check how different your plot is from the following one.

Results:

The comparison of two results is shown in Figure 9.9.

It can be seen from Figure 9.9 that the results are quite different between the implicit Euler method and the explicit Euler method.

Note: Euler's implicit method is unconditionally stable and can use much larger timesteps Δt, and it can be faster than Euler's explicit method to reach convergence.

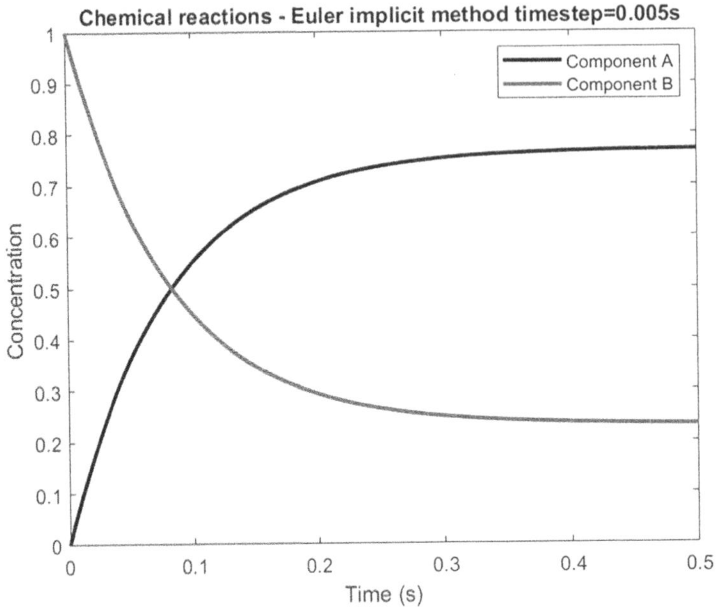

Figure 9.8 The plot of concentration vs. t for Example 9.6 using Euler's implicit method with timestep = 0.005 s.

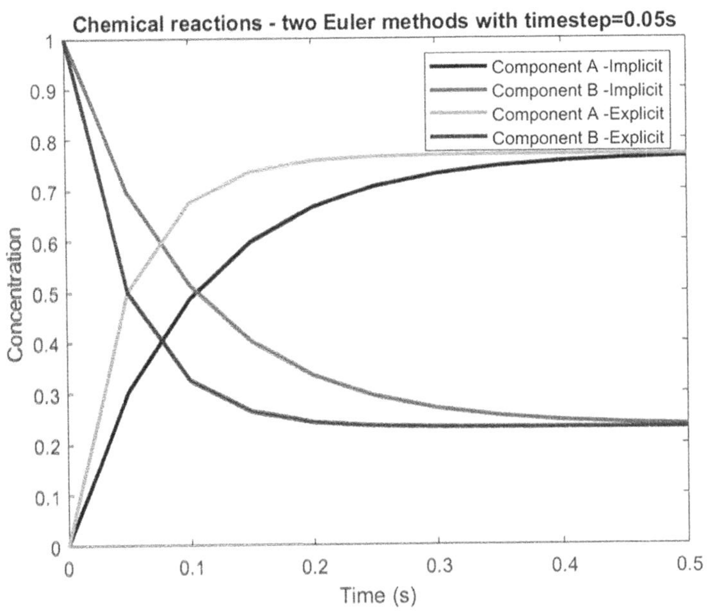

Figure 9.9 The comparison of the results between the implicit Euler method and explicit Euler method using timestep = 0.05 s for Example 9.6.

9.2.4.3 Runge–Kutta Methods

Runge–Kutta (R–K) methods are the most popular for integrating ODEs second, third, and fourth order and other variations. The fourth order R–K method with an error of $O(h^5)$ is the most widely used integration method, where h is the step size and $\frac{dx}{dt} = g(t, x)$:

$$\left.\begin{aligned} x_{n+1} &= x_n + \tfrac{1}{6}(k_1 + 2k_2 + 2k_3 + k_4) \\ k_1 &= hg(t_n, x_n) \\ k_2 &= hg\left(t_n + \tfrac{1}{2}h, x_n + \tfrac{1}{2}k_1\right) \\ k_3 &= hg\left(t_n + \tfrac{1}{2}h, x_n + \tfrac{1}{2}k_2\right) \\ k_4 &= hg(t_n + h, x_n + k_3) \end{aligned}\right\} O(h^5) \tag{9.23}$$

MATLAB has a built-in ode45 solver using an explicit Runge–Kutta–Dormand–Prince (4,5) formula, and it is easy to use. We provide examples in the MATLAB solvers section.

9.2.4.4 Finite-Difference Method

A nodal network in a Cartesian Grid is used to identify discrete points where the property of interest is determined using an (i,j) notation to designate their location. For example, the temperature of any point inside the plate in Figure 9.10 is related to its four direct neighbors along the x-direction and the y-direction. The FD method gives

$$4T_{i,j} - T_{i+1,j} - T_{i-1,j} - T_{i,j+1} - T_{i,j-1} = 0 \tag{9.24}$$

i.e.

$$\left.\begin{aligned} 4T_{1,1} - T_{2,1} - T_{1,2} &= 2T_1 \\ 4T_{2,1} - T_{1,1} - T_{2,2} &= 2T_1 \\ 4T_{1,2} - T_{1,1} - T_{2,2} &= T_1 + T_2 \\ 4T_{2,2} - T_{1,2} - T_{2,1} &= T_1 + T_2 \end{aligned}\right\} \tag{9.25}$$

Example 9.7

Simulate steady-state two-dimensional heat conduction using the FD method. The following heat equation describes the temperature profile with constant diffusivity:

$$\frac{\partial^2 T}{\partial x^2} + \frac{\partial^2 T}{\partial y^2} = 0 \tag{9.26}$$

where $T_1 = 0$ and $T_2 = 1$.

(1) Solve the problem using a grid of 4×4 nodes, as shown in Figure 9.10 with $\Delta x = \Delta y$.

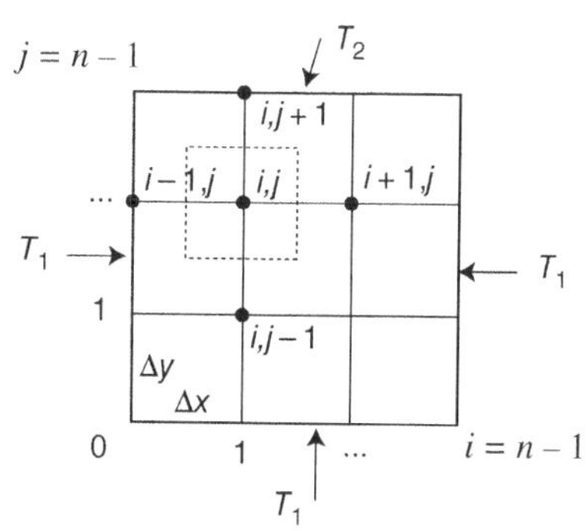

Figure 9.10 A grid of 4×4 nodes for finite-difference method.

Solution:

The problem simplifies to a system of four linear equations in Eq. (9.25) for all four internal points because the temperature values of all boundary points have already been assigned.

$$\begin{bmatrix} 4 & -1 & -1 & 0 \\ -1 & 4 & 0 & -1 \\ -1 & 0 & 4 & -1 \\ 0 & -1 & -1 & 4 \end{bmatrix} \begin{bmatrix} T_{1,1} \\ T_{2,1} \\ T_{1,2} \\ T_{2,2} \end{bmatrix} = \begin{bmatrix} 0 \\ 0 \\ 1 \\ 1 \end{bmatrix} \tag{9.27}$$

MATLAB code:

```
%Finite difference method for steady-state two-dimensional heat
%conduction
A=[4 -1 -1 0; -1 4 0 -1; -1 0 4 -1; 0 -1 -1 4];
b=[0;0;1;1];
T=A\b
```

Results:

```
T =
      0.1250
      0.1250
      0.3750
      0.3750
```

(2) Solve the problem using a grid of 5×5 nodes, as shownwith $\Delta x = \Delta y$, and draw a 2-D contour plot.

Solution:

The problem simplifies to a system of nine linear equations for all nine internal points because the temperature values of all boundary points have already been assigned.

$$\begin{bmatrix} 4 & -1 & 0 & -1 & & & & & \\ -1 & 4 & -1 & 0 & -1 & & & & \\ 0 & -1 & 4 & 0 & 0 & -1 & & & \\ -1 & 0 & 0 & 4 & -1 & 0 & -1 & & \\ & -1 & 0 & -1 & 4 & -1 & 0 & -1 & \\ & & -1 & 0 & -1 & 4 & 0 & 0 & -1 \\ & & & -1 & 0 & 0 & 4 & -1 & 0 \\ & & & & -1 & 0 & -1 & 4 & -1 \\ & & & & & -1 & 0 & -1 & 4 \end{bmatrix} \begin{bmatrix} T_{1,3} \\ T_{2,3} \\ T_{3,3} \\ T_{1,2} \\ T_{2,2} \\ T_{3,2} \\ T_{1,1} \\ T_{2,1} \\ T_{3,1} \end{bmatrix} = \begin{bmatrix} 1 \\ 1 \\ 1 \\ 0 \\ 0 \\ 0 \\ 0 \\ 0 \\ 0 \end{bmatrix} \tag{9.28}$$

MATLAB code:

```
%Finite difference method for steady-state two-dimensional heat
%conduction
A=[4 -1 0 -1 0 0 0 0 0;-1 4 -1 0 -1 0 0 0 0;...
0 -1 4 0 0 -1 0 0 0;-1 0 0 4 -1 0 -1 0 0;...
0 -1 0 -1 4 -1 0 -1 0;0 0 -1 0 -1 4 0 0 -1;0 0 0 -1 0 0 4 -1 0;...
0 0 0 0 -1 0 -1 4 -1;0 0 0 0 0 -1 0 -1 4];
```

```
b=[1;1;1;0;0;0;0;0;0];
T=A\b

Results:
T =
     0.4286
     0.5268
     0.4286
     0.1875
     0.2500
     0.1875
     0.0714
     0.0982
  0.0714

% MATLAB code for 2-D contour plot based on the above T values
T1=0;
T2=1;
Temperature(1,1:5)=T1;
Temperature(1:5,1)=T1;
Temperature(1:5,5)=T1;
Temperature(5,1:5)=T2;
Temperature(4,2:4)=T(1:3);
Temperature(3,2:4)=T(4:6);
Temperature(2,2:4)=T(7:9);
x=linspace(0,1,5);
y=linspace(0,1,5);
contourf(x,y,Temperature);

Results:
```

The 2-D contour plot of steady-state two-dimensional heat conduction is shown in Figure 9.11.

Note: It is not always to use $A\backslash b$ to obtain results for the FD method. Especially in complex geometry or shapes, the numerical algorithm flowchart with an iterative approach must be applied to solve the complex equation systems. An exercise in the question section solves a steady-state two-dimensional heat conduction problem using a numerical algorithm flowchart.

9.2.5 MATLAB ODE Solvers

MATLAB has the ODE solvers for the first-order ODEs. Table 9.2 shows the basic MATLAB ODE solvers. The ode45 solver is based on an explicit Runge–Kutta (4,5) formula, the Dormand–Prince pair. It is a single-step solver – in computing $y(t_n)$, it needs only the solution at the immediately preceding time point, $y(t_{n-1})$. The ode45 performs well with most ODE problems and should be tried first. However, ode23, ode78, ode89, and ode113 can be more efficient than ode45 for problems with looser or tighter accuracy requirements. Those

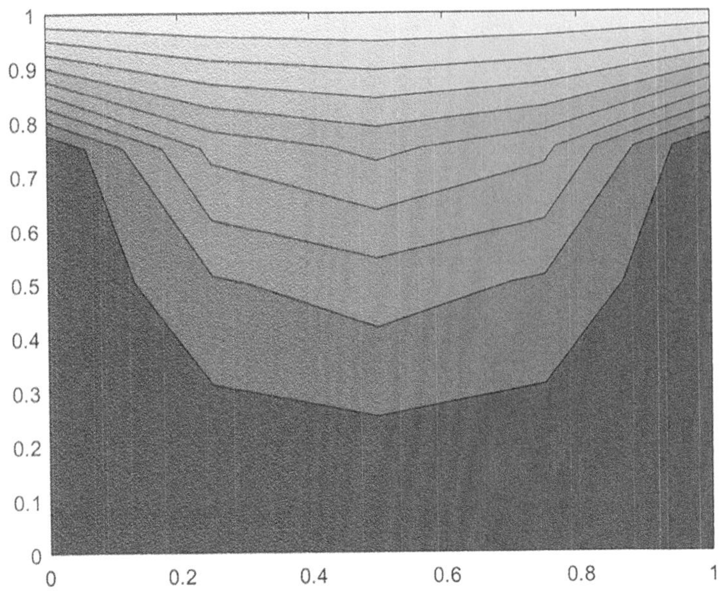

Figure 9.11 2-D contour plot of steady-state two-dimensional heat conduction.

are nonstiff solvers unable to reach a solution or extremely slow for the stiff problem. The ode15s can be used for stiff problems.

Example 9.8

Solve the above Example 9.1 with time range [0, 0.5] and initial condition $X(0) = \begin{bmatrix} A_0 \\ B_0 \end{bmatrix} = \begin{bmatrix} 0 \\ 1 \end{bmatrix}$.

Solution:

```
MATLAB code:
[t,y]=ode45(@chemReaction, [0 0.5], [0 1]);
plot(t,y(:,1),'k','LineWidth',2);
hold on;
plot(t,y(:,2),'r','LineWidth',2);
legend('Component A','Component B');
xlabel('Time (s)');
ylabel('Concentration');
title('Chemical reactions');
function dydt=chemReaction (t,y)
dydt=zeros(2,1);
dydt(1)=-3*y(1)+10*y(2);
dydt(2)=3*y(1)-10*y(2);
end
```

Results:

Figure 9.12 shows the concentrations of components A and B varying with time.

Table 9.2 The basic MATLAB ODE solvers (https://www.mathworks.com/The MathWorks, Inc.).

MATLAB ODE solvers	Description
ode45	It is used for nonstiff problems with medium accuracy.
ode23	It is used for nonstiff problems with low accuracy. It can be more efficient than ode45 at problems with crude tolerances or in the presence of moderate stiffness.
ode113	It is used for nonstiff problems with low to high accuracy. It can be more efficient than ode45 at problems with stringent error tolerances or when the ODE function is expensive to evaluate.
ode78	It is used for nonstiff problems with high accuracy. It can be more efficient than ode45 at solving problems with smooth solutions and high accuracy requirements.
ode89	It is used for nonstiff problems with high accuracy. It can be more efficient than ode78 on very smooth problems when integrating over long intervals or with tight tolerances.
ode15s	It is used for stiff problems with low to medium accuracy. It can be used when ode45 fails or inefficient, and the problem may be stiff. It can be used when solving differential-algebraic equations (DAEs).
ode23s	It is used for stiff problems with low accuracy. It can be more efficient than ode15s at problems with crude error tolerances. It can solve some stiff problems where ode15s is not adequate.
ode23t	It is used for stiff problems with low accuracy. It can be used if the problem is only moderately stiff and to find a solution without numerical damping. It can also solve DAEs.
ode23tb	It is used for stiff problems with low accuracy. It might be more efficient than ode15s at problems with crude error tolerances.
ode15i	It is a fully implicit method with low accuracy. It can be used for fully implicit problems $f(t,y,y') = 0$ and DAEs of index 1.
dsolve	Solves a differential equation symbolically. Use diff and == to represent differential equations. For example, diff(y,x)==y represents the equation $dy/dx = y$. Solve a system of differential equations by specifying them as a vector of those equations.
bvp4c	Solve boundary value problem – fourth-order method.
bvp5c	Solve boundary value problem – fifth-order method.
bvpinit	Form initial guess for boundary value problem solver.

The ode45 has been applied in Example 9.8. Readers interested in other ODE solvers should try to modify the above MATLAB code using different ODE solvers and compare the results.

Example 9.9
Solve the above Example 9.2 using Eqs.(9.14)–(9.16) with x range [0, 0.5] and initial condition $Y(0) = \begin{bmatrix} y_1(0) \\ y_2(0) \\ y_3(0) \end{bmatrix} = \begin{bmatrix} 0 \\ 1 \\ 0 \end{bmatrix}$.

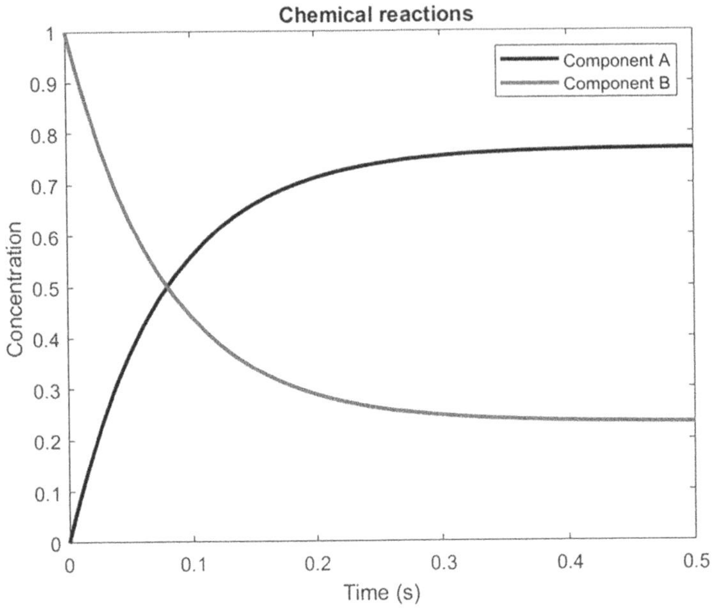

Figure 9.12 A chemical reaction between components *A* and *B* varies with time.

Solution:

```
MATLAB code:
[x,y]=ode45(@ODE_Code, [0 0.5], [0 1 0]);
plot(x,y(:,1),'k','LineWidth',2);
hold on;
plot(x,y(:,2),'r','LineWidth',2);
hold on;
plot(x,y(:,3),'g','LineWidth',2);
legend('y1','y2','y3');
xlabel('x');
ylabel('y');
title('Higher order ODE');
function dydx=ODE_Code (x,y)
dydx=zeros(3,1);
dydx(1)=y(2);
dydx(2)=y(3);
dydx(3)=-3*y(3)^2+y(1)^2*y(2)^3-5*y(1);
end
```

```
Results:
```

Figure 9.13 shows the plot of Example 9.9 for a higher-order ODE case.

The ode45 has been applied in Example 9.9. Readers interested in other ODE solvers should try to modify the above MATLAB code using different ODE solvers and compare the results.

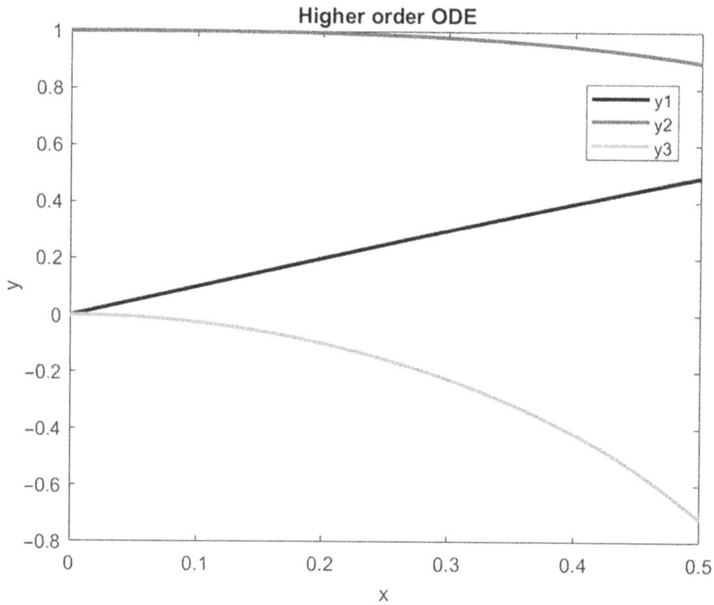

Figure 9.13 A plot of Example 9.9 for a higher-order ODE case.

Learning Tips

- The ode45 can be first choice to solve ODEs problems.
- The ode15s can be tried if ode45 fails or is inefficient.
- Due to their simplicity, both Euler's explicit method and Euler's implicit method can be tried for ODEs problems.

9.3 Partial Differential Equation Toolbox in MATLAB (PDE)

PDEs have more than one independent variable. The independent variables can be different spatial coordinates or spatial and time coordinates. The highest order of the derivative is the order of the PDE. The MATLAB PDE toolbox provides functions for solving structural mechanics, heat transfer, and general PDEs using the finite-element method.

9.3.1 Forms of Equations Can Be Solved

The MATLAB PDE toolbox can solve some forms of equations:

Scalar equations:

$$m\frac{\partial^2 u}{\partial t^2} + d\frac{\partial u}{\partial t} - \nabla \cdot (c\nabla u) + au = f \tag{9.29}$$

Eigenvalue equations:

$$-\nabla \cdot (c\nabla u) + au = \lambda du \tag{9.30}$$

or

$$-\nabla \cdot (c\nabla u) + au = \lambda^2 mu \tag{9.31}$$

Systems of equations:

$$\mathbf{m}\frac{\partial^2 \mathbf{u}}{\partial t^2} + \mathbf{d}\frac{\partial \mathbf{u}}{\partial t} - \nabla \cdot (\mathbf{c} \otimes \nabla \mathbf{u}) + \mathbf{au} = \mathbf{f} \tag{9.32}$$

Eigenvalue systems of equations:

$$-\nabla \cdot (\mathbf{c} \otimes \nabla \mathbf{u}) + \mathbf{au} = \lambda \mathbf{du} \tag{9.33}$$

or

$$-\nabla \cdot (\mathbf{c} \otimes \nabla \mathbf{u}) + \mathbf{au} = \lambda^2 \mathbf{mu} \tag{9.34}$$

There are two kinds of boundary conditions:
Dirichlet boundary conditions on the edge or face for scalar equations:

$$hu = r \tag{9.35}$$

Generalized Neumann boundary conditions – on the edge or face for scalar equations:

$$\vec{n} \cdot (c\nabla u) + qu = g \tag{9.36}$$

For systems of PDEs, Dirichlet boundary conditions on the edge or face for systems of equations:

$$\mathbf{hu} = \mathbf{r} \tag{9.37}$$

Generalized Neumann boundary conditions – on the edge or face for systems of equations:

$$\mathbf{n} \cdot (\mathbf{c} \otimes \nabla \mathbf{u}) + \mathbf{qu} = \mathbf{g} \tag{9.38}$$

The bold letters in the Eqs. (9.29)–(9.38) are matrices or vectors. MATLAB has a special function to specify the above coefficients, **m**, **d**, **c**, **a**, and **f**:

```
specifyCoefficients(model,m=m,d=d,c=c,a=a,f=f)
```

To define the PDE coefficients and include them in the model.

```
specifyCoefficients(model,m=m,d=d,c=c,a=a,f=f,...
RegionType=RegionID)
```

To assign coefficients for a specified geometry region.

9.3.2 MATLAB PDE Solver *pdepe*

The MATLAB PDE solver *pdepe* can solve parabolic and elliptic PDEs and systems of PDEs in one spatial variable and time.

Its syntax in MATLAB is as follows:

```
sol = pdepe(m,pdefun,icfun,bcfun,xmesh,tspan)
sol = pdepe(m,pdefun,icfun,bcfun,xmesh,tspan,options)
[sol,tsol,sole,te,ie] = pdepe(m,pdefun,icfun,bcfun,xmesh,...
tspan,options)
```

where

- m represents symmetry constant with $m = 0$ for 1-D Cartesian coordinates with no symmetry; $m = 1$ for 2-D cylindrical coordinates with azimuthal angular symmetry; $m = 2$ for 3-D spherical coordinates with azimuthal and zenith angular symmetries.
- *pdefun* stores equations to be solved.
- *icfun* stores the initial condition functions.
- *bcfun* stores the initial condition functions.
- *xmesh* stores spatial mesh as a vector [x0 x1 ... xn]. Specifying the points at which a numerical solution is requested for every value in *tspan*. Second-order approximations to the solution are made on the mesh specified in *xmesh*.
- *Tspan* is the time span of integration, specified as a vector [t0 t1 ... tf]. Specify the points at which a solution is requested for every value in *xmesh*. The *pdepe* function performs the time integration with an ODE solver that dynamically selects both the time step and the formula.

More information about *pdepe* can be seen at http://www.mathworks.com.

Note: The standard form that *pdepe* expects is as follows:

$$c\left(x, t, u, \frac{\partial u}{\partial x}\right)\frac{\partial u}{\partial t} = x^{-m}\frac{\partial}{\partial x}\left(x^m f\left(x, t, u, \frac{\partial u}{\partial x}\right)\right) + s\left(x, t, u, \frac{\partial u}{\partial x}\right) \tag{9.39}$$

with an initial condition:

$$u(x, t_0) = u_0(x) \tag{9.40}$$

with boundary conditions at $x = a$ and $x = b$:

$$p(x, t, u) + q(x, t)f\left(x, t, u, \frac{\partial u}{\partial x}\right) = 0 \tag{9.41}$$

Example 9.10

(1) Solve the following heat transfer PDE using the *pdepe* function:

$$4\frac{\partial u}{\partial t} = \frac{\partial^2 u}{\partial x^2} \tag{9.42}$$

with the following initial and boundary conditions:

$$u(x, 0) = \sin(2\pi x) \tag{9.43}$$

$$u(0, t) = 0 \tag{9.44}$$

$$u(1, t) = 0 \tag{9.45}$$

Solution:

To compare, Eq. (9.42) with Eqs.(9.39) and (9.30) can be re-written as:

$$4\frac{\partial u}{\partial t} = x^{-0}\frac{\partial}{\partial x}\left(x^0\frac{\partial u}{\partial x}\right) + 0 \tag{9.46}$$

We can get $m = 0, c = 4, f = DuDx$ and $s = 0$.

The MATLAB code is as follows:

```
%m=0 specifies no symmetry in the problem.
m = 0;
%Define the solution mesh
x = linspace(0,1,100);
t = linspace(0,0.2,5);
%Solve the PDE
u = pdepe(m,@pdefun,@pdeic,@pdebc,x,t);
% A surface plot
surf(x,t,u);
title('Surface plot of solution.');
xlabel('Distance x');
ylabel('Time t');
zlabel('u');
function [c,b,s] = pdefun(x,t,u,DuDx)
c = 4;
b = DuDx;
s = 0;
end
function u0 = pdeic(x)
u0 = sin(2*pi*x);
end
function [pl,ql,pr,qr] = pdebc(xl,ul,xr,ur,t)
pl = ul;
ql = 0;
pr = ur;
qr = 0;
end
```

```
Results:
```

A surface plot of solution for Example 9.10 is shown in Figure 9.14.

(2) Plot u vs. x at different values of $t = 0, 0.05, 0.1, 0.15$, and 0.2.

Solution:

```
% A plot of u vs. x at different values of t
figure;
plot(x,u);
axis([0,1,-1,1]);
legend('t = 0','t = 0.05','t = 0.1','t = 0.15','t = 0.2');
xlabel('Distance x');
ylabel('u');
```

```
Results:
```

A plot of u vs. x at different values of t for Example 9.10 is shown in Figure 9.15.

(3) Calculate u and $DuDx$ at $x = 0.5$ and $t = 0.2$ by using the *pdeval* function:

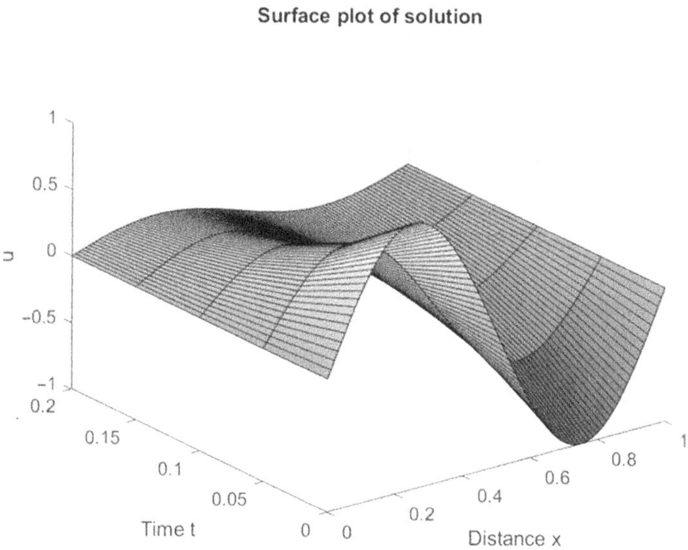

Figure 9.14 A surface plot of solution for Example 9.10.

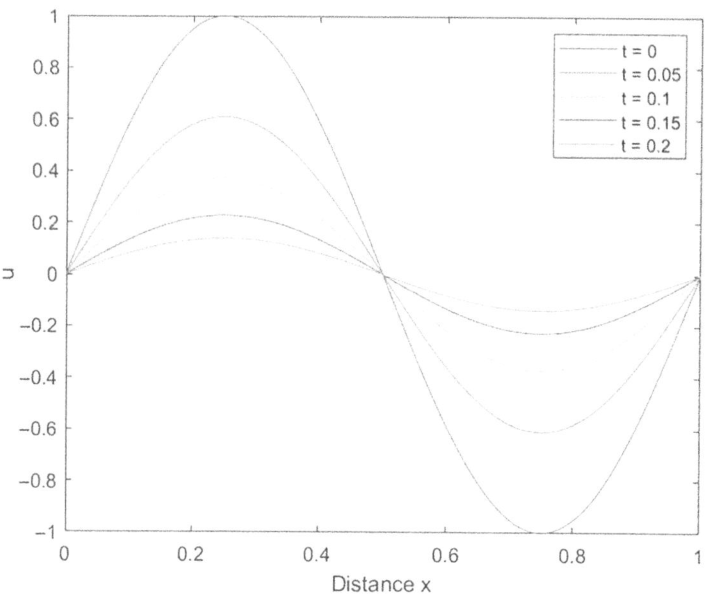

Figure 9.15 A plot of u vs. x at different values of t for Example 9.10.

Solution:

```
% Values of u and DuDx at x = 0.5 & t = 0.2
[uint,DuDxint] = pdeval(m,x,u(5,:),0.5)
```

```
Results:
uint =
  -1.5542e-12
```

```
DuDxint =
   -0.8705
```

The analytical result of u at $x = 0.5$ and $t = 0.2$ should be zero, which can also be seen from the almost zero simulation result uint = −1.5542e−12.

9.4 Application in Chemical Engineering

9.4.1 Complex Chemical Reactions

Example 9.11

In the following complex chemical reactions as shown in Figure 9.16, components A, B, and C have the following reactions: component A has a reaction rate of 5/s to produce component B, while component B has a reaction rate of 20/s to produce component A; component A has a reaction rate 2/s to produce component C, while component C has a reaction rate $3 \times C$ to produce component A; component B has a reaction rate 10/s to produce component C, while component C has a reaction rate 3/s to produce component B.

We have three equations for these complex chemical reactions:

$$\frac{dA}{dt} = -5A + 20B - 2A + 3C^2 \tag{9.47}$$

$$\frac{dB}{dt} = 5A - 20B - 10B + 3C \tag{9.48}$$

$$\frac{dC}{dt} = 2A + 10B - 3C - 3C^2 \tag{9.49}$$

Equations (9.47)–(9.49) are ODEs. If initial values are provided (i.e. initial value problem, the initial concentrations $C_A = 0$, $C_B = 1$, and $C_C = 0$), they can also be solved by MATLAB ODE solvers.

Solution:

```
MATLAB code:
[t,y]=ode45(@chem3, [0 0.5], [0 1 0]);
figure
p=plot(t,y(:,1),'k','LineWidth',2);
hold on;
```

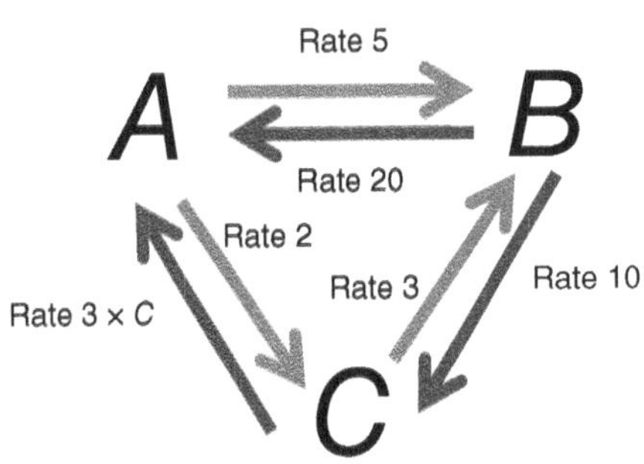

Figure 9.16 Complex chemical reactions among components A, B, and C.

```
plot(t,y(:,2),'r','LineWidth',2);
hold on;
plot(t,y(:,3),'g','LineWidth',2);
legend('Component A','Component B','Component C');
xlabel('Time (s)');
ylabel('Concentration');
title('Chemical reactions');
function dydt=chem3 (t,y)
dydt=zeros(3,1);
dydt(1)=-5*y(1)+20*y(2)-2*y(1)+3*y(3)^2;
dydt(2)=5*y(1)-20*y(2)-10*y(2)+3*y(3);
dydt(3)=2*y(1)+10*y(2)-3*y(3)-3*y(3)^2;
end
```

```
Results:
```

Figure 9.17 shows the concentrations of components A, B, and C varying with time.

The ode45 has been applied in Example 9.11. Readers interested in other ODE solvers should try to modify the above MATLAB code using different ODE solvers and compare the results.

9.4.2 Liquid Level Response

Example 9.12

Figure 9.18 shows two interacting square tanks: the first tank is 4 m tall with a cross-sectional area (A_1) of $2 \times 2 = 4$ m^2, and the second tank is 3 m tall with a cross-sectional area (A_2) of

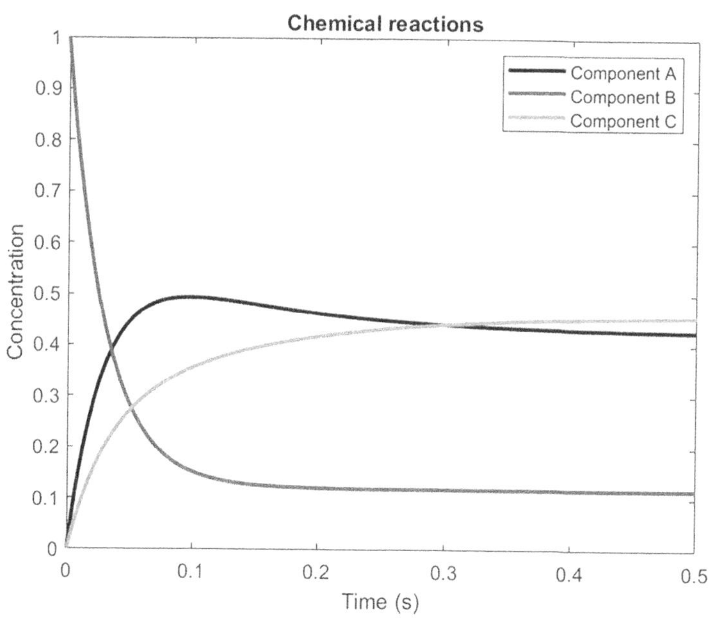

Figure 9.17 A chemical reaction between components *A*, *B*, and *C* varies with time.

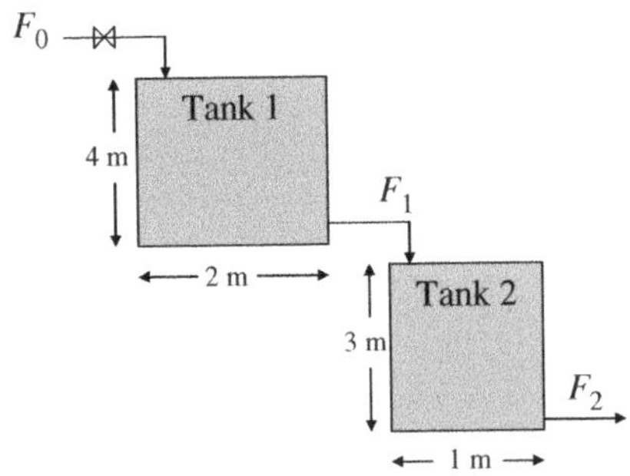

Figure 9.18 A schematic of two interacting tanks.

$1 \times 1 = 1$ m^2. The feed to the first tank is $F_0 = 16\text{m}^3/\text{hour}$. F_1 is the discharge from the first tank directly fed to the second tank. F_2 is the discharge from the second tank. At $t = 0$, both tanks are empty.

Mass balance of tank 1:

$$A_1 \frac{dh_1}{dt} = F_0 - F_1 \tag{9.50}$$

Mass balance of tank 2:

$$A_2 \frac{dh_2}{dt} = F_1 - F_2 \tag{9.51}$$

Linear model:

$$F_1 = c_1 h_1 \tag{9.52}$$

$$F_2 = c_2 h_2 \tag{9.53}$$

The value of $c_1 = 16$ and $c_2 = 16/3$.

(1) Solve the dynamic equations using ode45 for a time span [0 10] and plot the liquid level response.

Solution:

```
%MATLAB code
% Liquid level using ode45
global A1 A2 C1 C2 F0
A1=4;
A2=1;
C1=4;
C2=16/3;
F0=16;
[t,y]=ode45(@liquidlevel, [0 10], [0 0]);
plot(t,y(:,1),'k','LineWidth',2);
hold on;
plot(t,y(:,2),'r','LineWidth',2);
legend('Tank 1','Tank 2');
xlabel('Time (hour)');
ylabel('Liquid level (m)');
title('Liquid level response');
```

```
function dydt=liquidlevel (t,y)
global A1 A2 C1 C2 F0
dydt=zeros(2,1);
dydt(1)=(F0-C1*y(1))/A1;
dydt(2)=(C1*y(1)-C2*y(2))/A2;
end
```

```
Results:
```

Figure 9.19 shows the liquid level of two interacting square tanks that was solved using ode45.

(2) Solve the dynamic equations using Euler's explicit method for a time span [0 10] and plot the liquid level response.

Solution:

```
%MATLAB code
%Euler's explicit method for liquid level case
A1=4;
A2=1;
C1=4;
C2=16/3;
F0=16;
n=100;
h=10/n;
y=zeros(2,n+1);
y(1,1)=0;
```

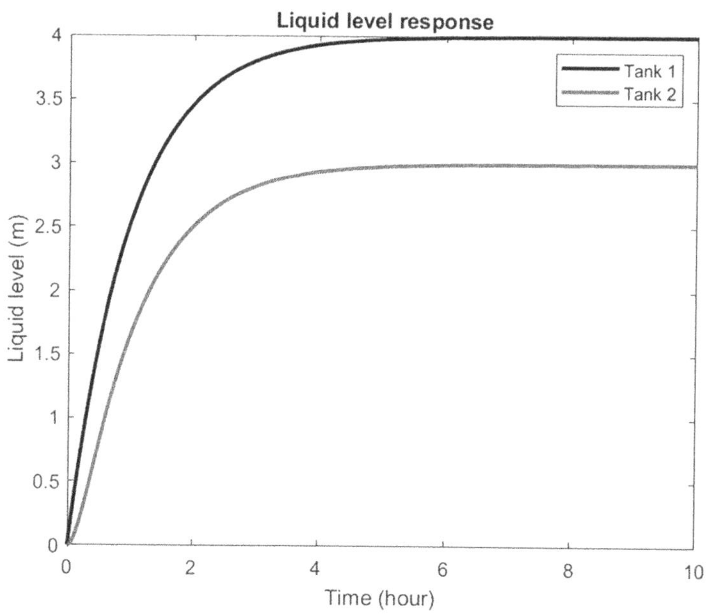

Figure 9.19 The liquid level of two interacting square tanks was solved using ode45.

```
y(2,1)=0;
for i=1:n
f = @(y) [(F0-C1*y(1))/A1; (C1*y(1)-C2*y(2))/A2];
y(:, i + 1) = y(:, i) + h * f(y(:, i));
end
te=linspace(0, 10,n);
plot(te,y(1,1:n),'k','LineWidth',2);
hold on;
plot(te,y(2,1:n),'r','LineWidth',2);
legend('Tank 1','Tank 2');
xlabel('Time (hour)');
ylabel('Liquid level (m)');
title('Liquid level response -Euler explicit method');
```

```
Results:
```

Figure 9.20 shows the liquid level of two interacting square tanks that was solved using Euler's explicit method.

It can be seen that both ode45 and Euler's explicit method give similar results for the liquid level case in Example 9.12. For this case, readers may try to use other ode solvers in MATLAB or Euler's implicit method.

9.4.3 Plane Poiseuille Flow

Example 9.13

Plane Poiseuille flow is the flow between two infinitely long parallel plates (with a distance

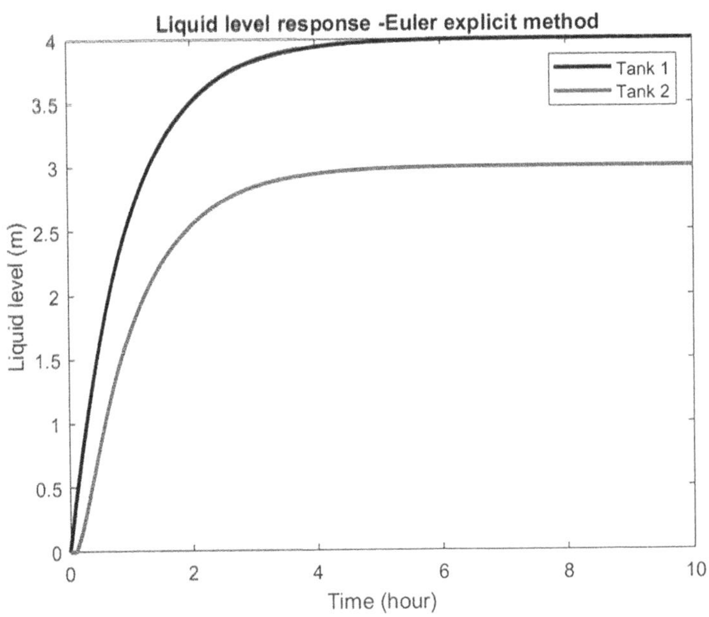

Figure 9.20 The liquid level of two interacting square tanks was solved using Euler's explicit method.

of h) driven by a constant pressure gradient $G = -dp/dx$ applied in the direction of flow. The Navier–Stokes equations are reduced to

$$\frac{d^2u}{dy^2} = -\frac{G}{\mu} \tag{9.54}$$

with the no-slip condition on both walls

$$u(0) = 0 \text{ and } u(h) = 0 \tag{9.55}$$

(1) Solve the equation using bvp4c and plot fluid velocity (u) vs. distance between plates (h) with conditions: $h = 0.1$, $G = -100$, and $\mu = 1$.

Solution:

```
%MATLAB code
%Plane Poiseuille flow
global mu G
h = 0.1;
mu = 1;
G = -100;
x = linspace(0,h);
solinit = bvpinit(x,@guess);
sol = bvp4c(@odefcn,@bcfcn,solinit);
y = sol.x;
u1 = sol.y(1,:);
u2 = sol.y(2,:);
plot(y,u1);
xlabel('Distance between plates')
ylabel('Fluid velocity')
title('Plane Poiseuille flow using bvp4c');
function Uinit = guess(x)
Uinit = [0; 0];
end
function dUdy = odefcn(y,U)
global mu G
u1 = U(1);
u2 = U(2);
du1dy = u2;
du2dy = 1/mu*G;
dUdy = [du1dy; du2dy];
end
function res = bcfcn(U_0,U_d)
res = [0 - U_0(1);0 - U_d(1)];
end
```

Results:

Figure 9.21 shows the plane Poiseuille flow that was solved using the bvp4c method.

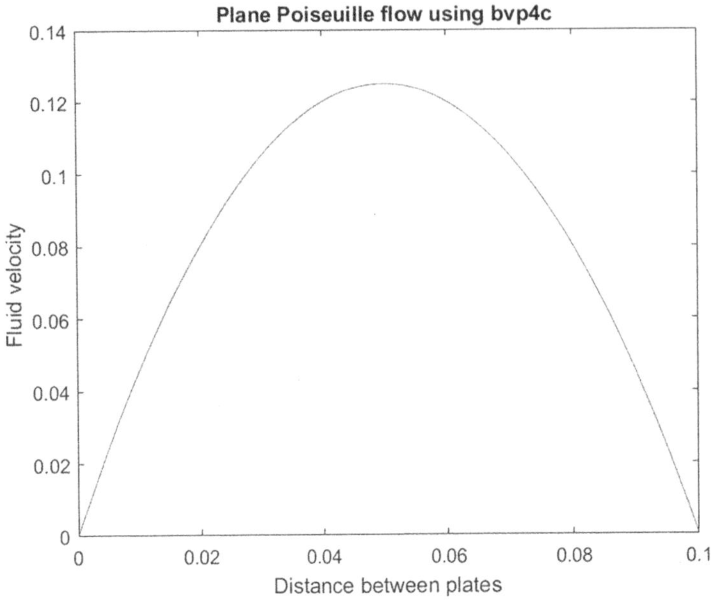

Figure 9.21 Plane Poiseuille flow solved using the bvp4c method.

(2) Solve the equation using dsolve and plot fluid velocity (u) vs. distance between plates (h) with conditions: $h = 0.1$, $G = -100$, and $\mu = 1$.

Solution:

```
%MATLAB code
%Plane Poiseuille flow using dsolve
syms y(x)
h = 0.1;
mu = 1;
G = -100;
eqn = diff(y,x,2) == G/mu;
cond1 = y(0)==0;
cond2 = y(h)==0;
y(x) = dsolve(eqn,[cond1,cond2]);
vx = linspace(0,h,100);
vy = double(y(vx));
plot(vx,vy);
xlabel('Distance between plates');
ylabel('Fluid velocity');
title('Plane Poiseuille flow using dsolve function');
```

```
Results:
```

Figure 9.22 shows the plane Poiseuille flow that was solved using the dsolve method.

It can be seen from Figures 9.21 and 9.22 that both bvp4c and dsolve can solve the plane Poiseuille flow boundary value problem and obtain similar results.

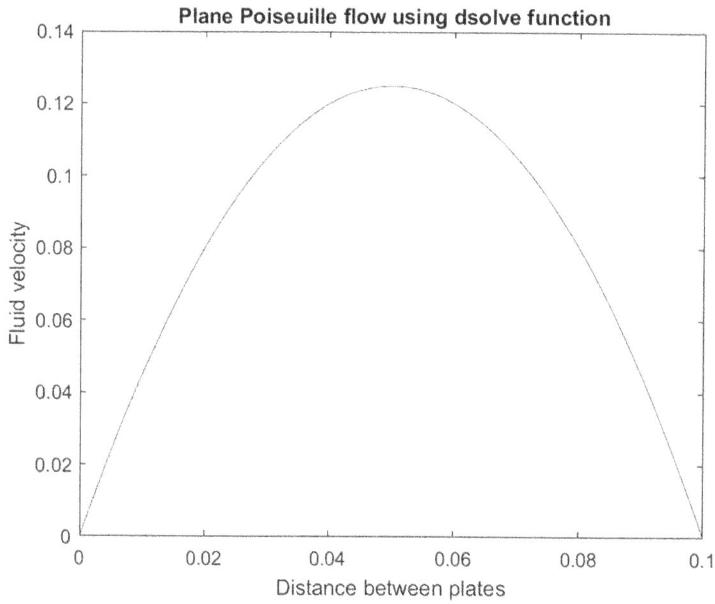

Figure 9.22 Plane Poiseuille flow solved using the dsolve method.

> **Learning Tips**
>
> - The *dsolve* can be useful for initial value problems and boundary value problems.
> - The *bvp4c* or *bvp5c* can be used for boundary value problems, but *ode45* and its variations cannot be used for boundary value problems directly.

9.5 Summary

This chapter presents complex equation systems and their potential solutions. Due to difficulty arising from complexity, it can be very challenging to solve some problems, even using numerical methods.

Some MATLAB functions exist to solve ODEs and particular PDEs. The ode45 and its variations can be used to solve initial value problems but not for boundary value problems. Where ode45 cannot work for stiff problems, ode15s or Euler's explicit and Euler's implicit methods should be tried.

The *pdepe* can be used for 1-D parabolic and elliptic PDEs or a system of those kinds of PDEs. However, these are quite limited. The FD method can be chosen for challenging ODEs or PDEs.

Problems

9.1 In the following complex chemical reactions as shown in Figure 9.23, components A, B, C, and D have the following reactions: component A has a reaction rate 5/s to produce component B, while component B has a reaction rate 2/s to produce component

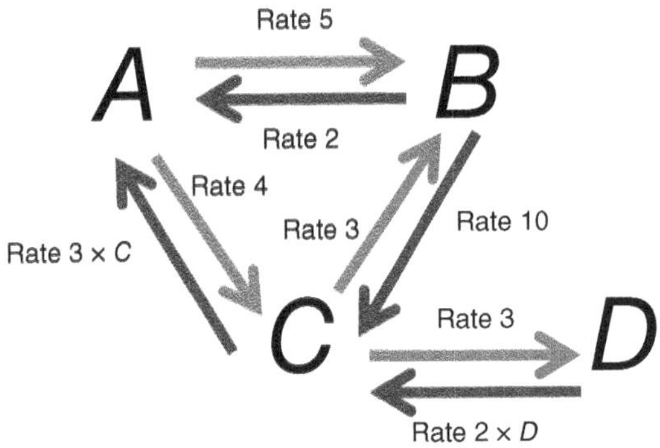

Figure 9.23 Complex chemical reactions among components A, B, C, and D.

A; component A has a reaction rate 4/s to produce component C, while component C has a reaction rate $3 \times C$ to produce component A; component B has a reaction rate 10/s to produce component C, while component C has a reaction rate 3/s to produce component B; component C has a reaction rate 3/s to produce component D, while component D has a reaction rate $2 \times D$ to produce component C. The initial concentrations are $C_A = 1$, $C_B = 0$, $C_C = 0$, and $C_D = 0$.

(1) List all ODEs for the chemical reactions.
(2) Use ode45 to solve the equations with a time range [0, 1].
(3) Use ode15s to solve the equations with a time range [0, 1].
(4) Use Euler's explicit method to solve the equations with time range [0, 1] and timestep = 0.01 s. Compare these results with the above results.
(5) Use Euler's implicit method to solve the equations with time range [0, 1] and timestep = 0.05 s. Compare these results with the above results.

9.2 There is a second-order ODE:

$$\frac{d^2y}{dt^2} + 5\frac{dy}{dt} - 2y = \sin(t) \tag{9.56}$$

with boundary condition:

$$\frac{dy}{dt} = 0 \text{ when } t = 0 \tag{9.57}$$

$$y = 1 \text{ when } t = 5 \tag{9.58}$$

(1) Solve the equation using dsolve and calculate y at $t = 2$.
(2) Solve the equation using bvp4c and calculate y at $t = 2$.

9.3 There is a system of PDEs:

$$\frac{\partial u_1}{\partial t} = \frac{\partial^2 y}{\partial x^2} - u_1 e^{-\frac{1}{u_2}} + u_2 e^{-\frac{1}{u_1}} \tag{9.59}$$

$$\frac{\partial u_1}{\partial t} = \frac{\partial^2 y}{\partial x^2} + u_1 e^{-\frac{1}{u_2}} - u_2 e^{-\frac{1}{u_1}} \tag{9.60}$$

The equation holds in the interval $0 \leq x \leq 1$ for times $t \geq 0$. The initial conditions are

$$u_1(x, 0) = 1 \tag{9.61}$$

$$u_2(x, 0) = 1 \tag{9.62}$$

with boundary conditions:

$$\frac{\partial u_1(0,t)}{\partial t} = 0 \tag{9.63}$$

$$u_2(0,t) = 0 \tag{9.64}$$

$$\frac{\partial u_2(1,t)}{\partial t} = 0 \tag{9.65}$$

$$u_1(1,t) = 0 \tag{9.66}$$

(1) Solve the equations using *pdepe* and create surface plots of the solutions for u_1 and u_2 plotted at the selected mesh points for x and t.
(2) Calculate u_1, $\frac{\partial u_1}{\partial x}$, u_2, and $\frac{\partial u_1}{\partial x}$ at $x = 0.5$ and $t = 0.8$ by using the *pdeval* function.

9.4 Simulate steady-state two-dimensional heat conduction using the FD method. The following heat equation describes the temperature profile with constant diffusivity:

$$\frac{\partial^2 T}{\partial x^2} + \frac{\partial^2 T}{\partial y^2} = 0 \tag{9.67}$$

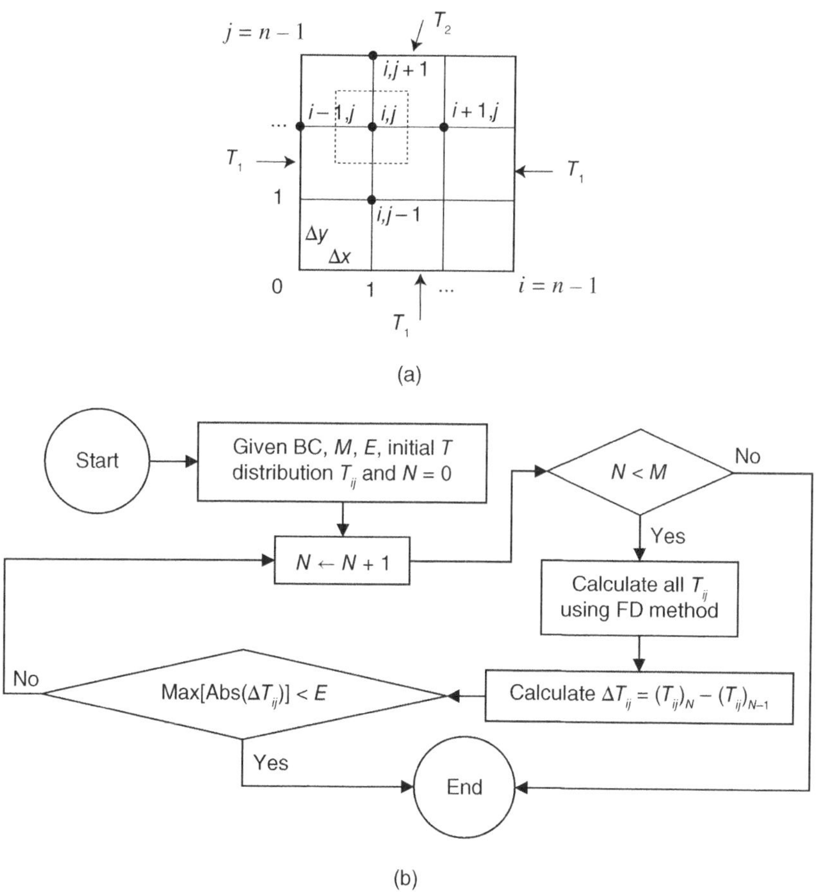

Figure 9.24 (a) The nodal network in a Cartesian grid; (b) the numerical algorithm flowchart using finite-difference method.

where $T_1 = 0$ and $T_2 = 1$. Figure 9.24a shows the nodal network in a Cartesian grid. Figure 9.24b shows the numerical algorithm flowchart using the FD method, where N is the iteration counter, M is the maximum iterations, and E is the error tolerance for convergence.

(1) Solve the problem using the numerical FD method for a grid of 5×5 nodes, and draw a 2-D contour plot.

(2) Increase the number of nodes to $n \times n$ to solve the problem using the FD method, compare to the following analytical solution to reach an error tolerance of less than 0.01, and draw a 2-D contour plot.

$$\frac{T_{i,j} - T_1}{T_2 - T_1} = \frac{2}{\pi} \sum_{k=1}^{\infty} \frac{1 + (-1)^{k+1}}{k \sinh(k\pi)} \sin\left(\frac{k\pi i}{n-1}\right) \sinh\left(\frac{k\pi j}{n-1}\right) \tag{9.68}$$

10

Process Integration and Optimization

Process integration and optimization presented in this chapter aim to expand our readers' capability to view process design in a bigger picture with a holistic approach. With the surging availability of massive data in chemical engineering and emerging advanced artificial intelligence, it is time to explore further the potential benefit of process integration and optimization. Achieving a whole process optimization is more important than just single-unit optimization.

MATLAB is a general programming language that provides a numeric computing environment not explicitly tailored for chemical process integration and optimization. However, it can be a helpful tool to assist us in performing optimization with other software packages. There are many existing chemical process simulators, such as Aspen Plus, Aspen HYSYS, CHEMCAD, GAMS, gPROMS, and HSC Sim, which can readily be applied in chemical process simulations. Experts with domain knowledge in chemical engineering can play a critical role in this area.

Previous Chapters 2–9 focused on optimizing a single process or single-unit operation. In contrast, this chapter considers the interactions between different unit operations and how they contribute to the whole process circuit. This chapter uses two types of solve extraction circuits (cross-current and counter-current) as illustrative examples. We apply skills and knowledge from previous Chapters 2–9, such as orthogonal experimental design (OED), model fitting, and optimization. It is a new experience for optimizing the whole process.

10.1 Introduction

10.1.1 Process Simulators

With the development of computers, more and more software packages emerge to perform simulations of the material and energy balances of chemical process plants. They can be used for process simulation, data reconciliation, engineering design, control system design, equipment scheduling, pipeline management, production management, and digital twins.

Table 10.1 lists a selection of popular process simulators. Some of them are general process simulators (e.g. Aspen Plus, Aspen HYSYS, CHEMCAD, CHEMPRO, and PRO/II), which can be used to perform mass and energy balances, calculation of thermal, physical and chemical properties, equipment sizing and costing. Some of them are specialized simulators (e.g. BatchColumn, BatchReactor, BioSTEAM, ProPhyPlus, and ProSim HEX) for

Chemical Engineering Analysis and Optimization Using MATLAB®, First Edition.
Weiguo Xie, Sam Toan, and Richard Davis.

Companion website: www.wiley.com/go/xie/chemicalengineering

batch process modeling and simulation, heat exchanger network design and optimization, and conceptual design of distillation processes. Some can be used for multiphase flows (e.g. IDEAS, JKSimMet, METSIM, OLGA, UniSim Design Suite, USIM PAC, and VMG Symmetry). Few open-source simulators exist (e.g. Advanced Simulation Library, ASCEND, BioSTEAM, COCO simulator, OpenModelica, and Visual MINTEQ).

10.1.2 Flowsheet Simulation and Design

There are many simulators shown in Table 10.1. Meanwhile, there is no one simulator fit for all-purpose chemical engineering design. Choosing a suitable simulator or combining some simulators depends on the actual processes, materials, equipment, and systems. Flowsheet simulation and design can analyze existing processes and retrofit design options. There are a few things that need to be considered for flowsheet simulation and design:

(1) Start with a simplified version of the flowsheet, such as initially omitting pressure and temperature changing devices (valves, pumps, and heat exchangers), and use simplified reaction models first.
(2) Use steady state operation first and consider process dynamics and control at a later stage.
(3) Apply process simulators for calculating heat duties, power requirements, phase and chemical equilibrium, performance of distillation columns, etc.
(4) Perform process synthesis using process simulators for potential reasonable results.
(5) Compare with the existing processes on analyzing the effect of changing operating conditions on the performance of the process.
(6) Retrofit design options by examining different ways of using existing equipment.

10.2 Process Integration and Optimization

10.2.1 Process Models

Using good models for every equipment is crucial, as it lays a solid foundation for further process integration and optimization. Many built-in models in different software packages and mathematical models can be referenced from existing literature. It is also possible to build your model by experiments or the methods shown in Chapters 7–9.

There are a few realities:

(1) Most models, historically, were built based on petrochemicals.
(2) Available models are for standard chemical processing unit operations.
(3) New models, such as polymerization and fermentation reactions, emerge.
(4) Some areas still lack models, such as solids crushing, grinding, and sorting equipment.

10.2.2 Simulating Novel Processes and Equipment

Under particular circumstances, novel processes that cannot be modeled using standard unit operations are sometimes needed (e.g. thermally coupled distillation columns, counter-current multiphase reactors).

Novel equipment is often encountered due to the faster innovation process nowadays. In this case, most of the time, standard chemical processing unit models are unsuitable. Therefore, it is necessary to build or combine new models.

Table 10.1 List of a selection of current process simulators.

Software	Developers	Applications
Advanced Simulation Library (ASL)	Avtech Scientific	ASL is an open source and can be used to model various coupled physical and chemical phenomena, especially in computational fluid dynamics.
ASCEND	ASCEND team	It is an open source for mathematical modeling chemical process modeling system.
Aspen HYSYS	Hyprotech, Aspen Technology	It is used for steady-state and dynamic simulation, process design, performance modeling, and optimization.
Aspen Plus	Aspen Technology	It is used for process simulation and optimization for bulk chemicals, specialty chemicals, and pharmaceutical industries.
BatchColumn	Fives ProSim	It is used for simulation and optimization of batch distillation columns.
BatchReactor	Fives ProSim	It is used to simulate and optimize chemical reactors running in batch mode.
BioSTEAM	BioSTEAM Development Group	It is an open-source platform that streamlines the design, simulation, techno-economic analysis (TEA), and life-cycle assessment (LCA) of biorefineries.
CADSIM Plus	Aurel Systems Inc.	It is used to perform mass and energy balances and simulate dynamic conditions.
CHEMCAD	Chemstations	It is used for process simulation in chemical, petrochemical, pharmaceutical, and thermodynamic models and contains a database of thousands of components for gases, liquids, solids, and electrolytes.
CHEMPRO	EPCON	It is used for process flow simulation, fluid flow simulation, & process equipment sizing.
COCO simulator	AmsterCHEM	It is a CAPE-OPEN-based steady-state flowsheet simulation environment.
DYNSIM	AVEVA	It is a dynamic simulator that predicts the time-dependent behavior of industrial processes.
GAMS	GAMS	It is used for modeling and solving linear, nonlinear, and mixed-integer optimization problems.
gPROMS	PSE Ltd	It is used for process modeling and optimization using static or dynamic models.
HSC Sim	Outotec Oyj	It primarily simulates static processes in the mining and mineral industries as well as biochemical and organic chemistry.
IDEAS	Andritz Automation	It is used for steady-state or dynamic modeling for pulp and paper, mineral processing, oil sands operations, potash operations, and power industry.
iiSE Simulator	iiSE company	It is used to simulate chemical, petrochemical, and energy conversion processes.

Table 10.1 (Continued)

Software	Developers	Applications
INOSIM	INOSIM Software GmbH	It is a discrete-event dynamic simulation software for mixed batch-continuous processes in the chemical and pharmaceutical industry.
JADE	GSE Systems	It is used for dynamic process simulation for nuclear, oil, and gas electricity generators and the chemical process industries.
JKSimMet	JK Tech	It analyzes and simulates comminution and classification circuits in mineral processing operations.
JModelica.org	Modelon AB	It is used for process simulation in robotics, vehicle systems, energy systems, CO_2 separation, and polyethylene production.
METSIM	Metsim International	It is used for dynamic and steady-state simulation for metallurgical and chemical engineering processes.
OLGA	Schlumberger	It models multiphase oil, natural gas, and water transportation in the same pipeline.
OpenModelica	Open-Source Modelica Consortium	It is an open-source simulator for modeling, simulating, optimizing, and analyzing complex dynamic systems.
Petro-SIM	KBC Advanced Technologies	It is used for dynamic process simulation for oil, gas, refining, and petrochemicals.
ProMax	Bryan Research & Engineering	It is used for process troubleshooting and design and is considered the industry standard for designing amine gas treating and glycol dehydration units.
ProPhyPlus	Fives ProSim	It is used for thermodynamic calculations.
ProSim HEX	Fives ProSim	It is used to analyze, simulate, and optimize heat exchangers.
ProSimPlus	Fives ProSim	It is used for steady-state simulation and optimization of processes.
PRO/II	AVEVA	It is used for steady-state process simulation.
AVEVA Process Simulation (former SimCentral)	AVEVA	It is used for steady state, fluid flow, and dynamic process simulation.
UniSim Design Suite	Honeywell	It is used for process simulation and optimization for oil and gas, LNG, refining, petrochemicals, metals, mining, and minerals.
USIM PAC	CASPEO	It is used for steady-state simulation and optimization of processes.
Visual MINTEQ	Jon Petter Gustafsson at KTH	It is open-source software and can simulate the chemical composition of solutions in contact with gases, solid compounds, and particle surfaces.
VMG Symmetry	Schlumberger	It is used for steady-state simulation, dynamic process simulation, and transient multiphase flowline simulation.

Example 10.1 Simulating a dividing wall column as shown in Figure 10.1:

The dividing wall column can be simulated using three conventional columns, as shown in Figure 10.2. The first column, C1, has three feeds and two outputs. The second and third columns have three feeds and three side outputs. After this conversion, many software packages can simulate the three conventional columns.

10.2.3 Physical and Chemical Property Estimation

Process simulators generally have built-in models and parameters for an extensive range of individual components and binary mixtures, which can be used to perform the material and energy balances. The built-in models come from the development cases' scope and

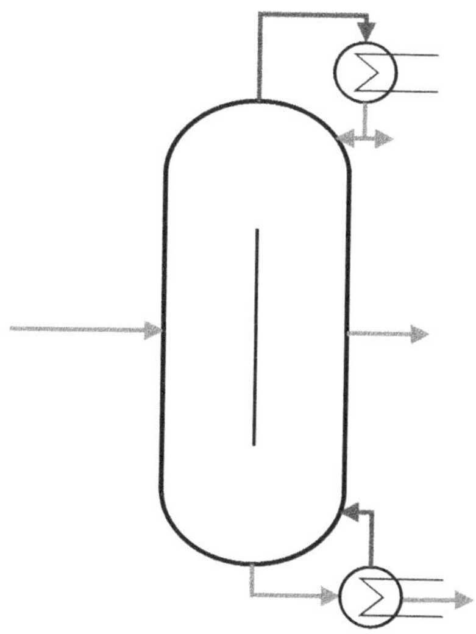

Figure 10.1 A dividing wall column with blue represents liquid streams, and red represents vapor streams.

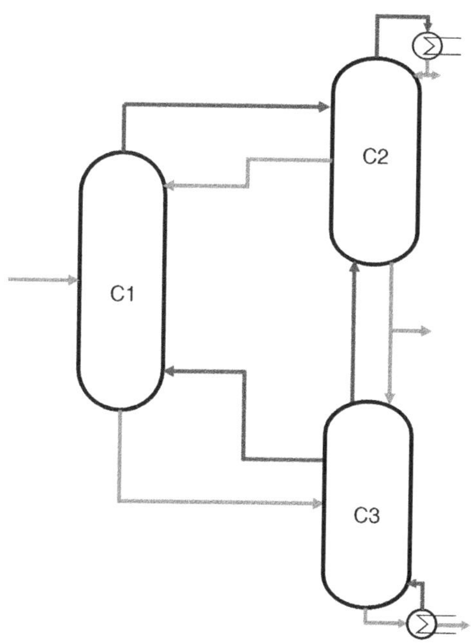

Figure 10.2 A dividing wall column can be simulated using three conventional columns with blue represents liquid streams, and red represents vapor streams.

parameters regressed from their experimental data. However, sometimes, the built-in parameters are unsuitable for a given application. Therefore, double-checking with the lab experimental pilot plant or published data is worthwhile. It can also be helpful to check by using different simulators.

Most process simulators can accommodate user-specified values, which is where your data are used to replace built-in parameters. It is often the case that we are not just limited by the existing model parameters from the process simulators.

Ideal cases following Raoult's law (Dahm and Visco 2014) are often assumed in process simulators, e.g. an ideal vapor–liquid equilibrium diagram as shown in Figure 10.3. However, nonideal cases deviating from Raoult's law are often encountered when estimating physical and chemical properties, such as those with a positive azeotrope as shown in Figure 10.4.

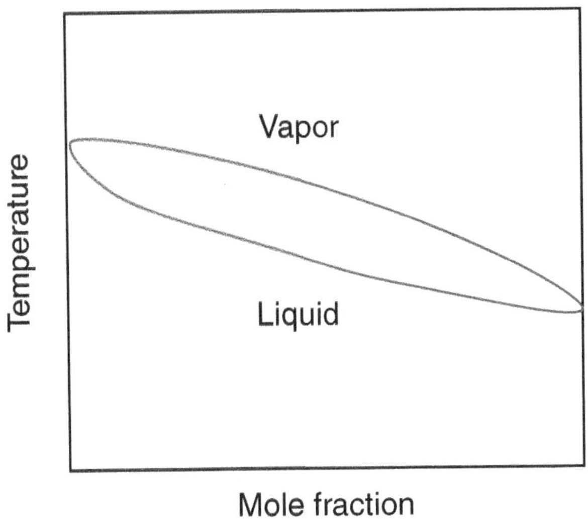

Figure 10.3 An ideal vapor–liquid equilibrium diagram.

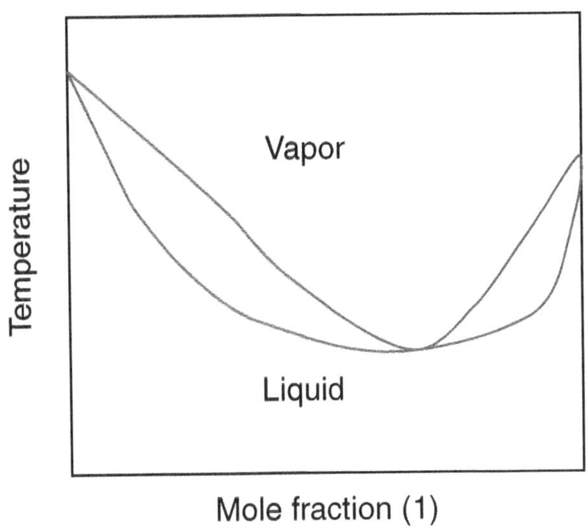

Figure 10.4 A nonideal vapor–liquid equilibrium diagram shows the mixture with a positive azeotrope.

The azeotrope is a constant heating point mixture in fluidic states whose proportions cannot be changed by simple distillation. This happens because when an azeotrope is boiled, the vapor has the exact proportions of constituents as the unboiled mixture. Each azeotrope has a characteristic boiling point. The boiling point of an azeotrope is either less than the boiling point temperatures of any of its constituents (a positive azeotrope) or greater than any of its constituents (a negative azeotrope). Separating the components by fractional distillation is impossible for both positive and negative azeotropes, and azeotropic distillation or adsorption–distillation can be used instead (Hu and Xie, 2001).

10.2.4 Process Optimization

Process optimization is a big topic and relates to specific processes. In general, we need to have a good optimization mindset and apply the optimization techniques of previous chapters for individual operational units, as well as the OED for whole flowsheet optimization. In the following section, we use extraction circuits (cross-current and counter-current) as illustrative examples of process optimization.

Learning Tips

- Use good conceptual design methods to initialize process integration and not just rely on simulators.
- Divide a large flowsheet into sub-sections and study them first for a baseline.
- Work on gradual changes to parameters and conditions, and perform sensitivity analysis.
- Apply OED (Orthogonal Experimental Design) for process optimization.

10.3 Application in Chemical Engineering

10.3.1 Introduction to Solvent Extraction

Many readers may have learned solvent extraction (SX) previously and get some review in this section. SX, also known as liquid–liquid extraction, is a method to separate compounds based on their solution preferences for two different immiscible liquids, usually water and an organic solvent. An actual application uses a nonaqueous phase as a medium to transfer solute from one aqueous solution to another.

SX is exclusively used to separate and purify copper, zinc, nickel, cobalt, PGMs (platinum-group metals), uranium, other nuclear materials, and rare earth elements. It is also widely used to produce fine organic compounds, process perfumes, produce vegetable oils and biodiesel, and other industries.

Figure 10.5a shows a separatory funnel with two different immiscible liquids to perform SX of solute in aqueous solution. Figure 10.5b shows a brief SX process in a lab batch operation to add immiscible solvent to an aqueous solution, shake the funnel, and settle it into two layers. The equilibrium is reached, and the distribution of solute between the two layers may be described by the distribution coefficients (the extract and raffinate phases)

$$\alpha_{BA,j} = \frac{c_{j,A}}{c_{j,B}} \tag{10.1}$$

where component j is distributed between two nonmiscible liquid phases, A and B, at the equilibrium state.

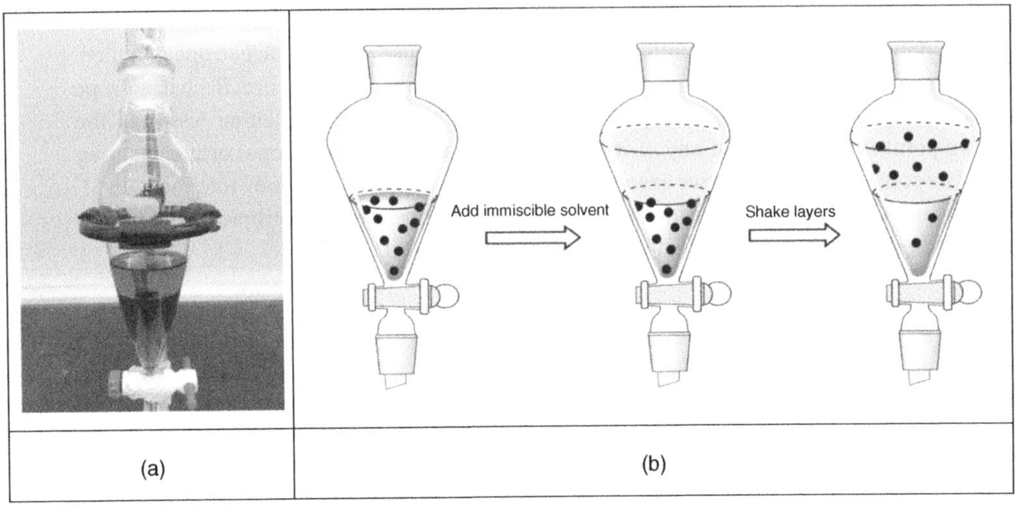

Figure 10.5 (a) Separatory funnel with two immiscible liquids; (b) a brief solvent extraction process.

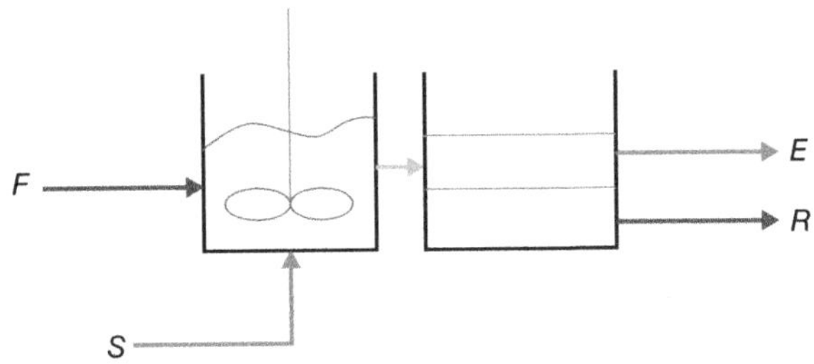

Figure 10.6 A schematic of a typical industrial equipment mixer-settler: *F* represents feed, *S* represents solvent, *E* represents extract, and *R* represents raffinate.

Figure 10.5 shows a batch operation using a separatory funnel, which is unsuitable for industrial applications. A typical industrial equipment, a mixer-settler, is shown in Figure 10.6, which can perform continuous operation.

10.3.2 Ternary Diagram (Partially Soluble Ternary Systems)

Figure 10.7 shows a typical ternary diagram for partially soluble ternary systems (Seader et al. 2011). It is an equilateral triangle with three vertices representing *A* – solute, *S* – solvent, and *C* – carrier. The three edges represent *SA* – a mass fraction of solute, *CS* – a mass fraction of solvent, and *AC* – a mass fraction of carrier.

In Figure 10.7, the binodal curve separates the two-phase and single-phase regions. Any points in the two-phase region have two-phase layers (extract phase layer and raffinate phase layer). A tie line links an extract with its corresponding raffinate. When extract and raffinate collapse into one point, it is called a plait point. There are six existing tie lines in Figure 10.7, which generally come from experimental tests. There are many ternary diagrams from historical databases.

10.3.3 Ternary Diagram – One Equilibrium Stage

Figure 10.8 shows one equilibrium stage: a batch operation in a separatory vessel or a continuous operation in a mixer-settler. The feed data and solvent data are given. The feed point is

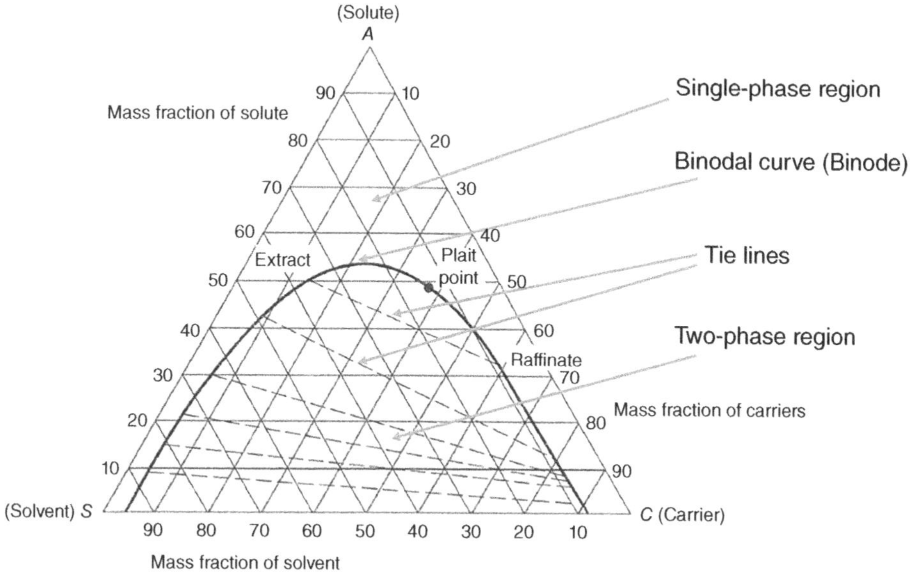

Figure 10.7 A ternary diagram showing single-phase region, binodal curve (binode), tie lines, and two-phase region.

shown on the *AC* edge with $x_{F,A} = 0.26$ and $x_{F,C} = 0.74$. The solvent point is shown at point *S* with $x_{S,S} = 1.00$.

The overall stream mass balance is:

$$M = F + S = E_1 + R_1 \tag{10.2}$$

The mixture of feed and solvent is marked at the *M* point. The *M* point is lying on the *SF* line. The component mass balances are:

$$x_{M,A}M = x_{F,A}F + x_{S,A}S \tag{10.3}$$

$$x_{M,C}M = x_{F,C}F + x_{S,C}S \tag{10.4}$$

$$x_{M,S}M = x_{F,S}F + x_{S,S}S \tag{10.5}$$

M point can also be located by distance relationships, which are lever rules:

$$S/F = \overline{MF}/\overline{MS} = 100/250 = 0.4 \tag{10.6}$$

or

$$S/M = \overline{MF}/\overline{SF} = 100/(100 + 250) = 0.286 \tag{10.7}$$

Because the *M* point is not lying on an existing tie line, we need to find a tie line to pass the *M* point. The interpolation between *M* point's two nearby tie lines can be performed to find E_1 and R_1 points. For example, Figure 10.9 shows that the tie line E_1R_1 passes the *M* point. Both E_1 and R_1 points are on the binodal curve.

The interpolation between two tie lines is quite difficult to perform, and it is hard to code using computational software. There are two ways to find the E_1R_1 line using a method of the conjugate line, as shown in Figure 10.10 (Seader et al. 2011). The points of the conjugate line in Figure 10.10a are generated by the extract and raffinate points of existing tie lines using

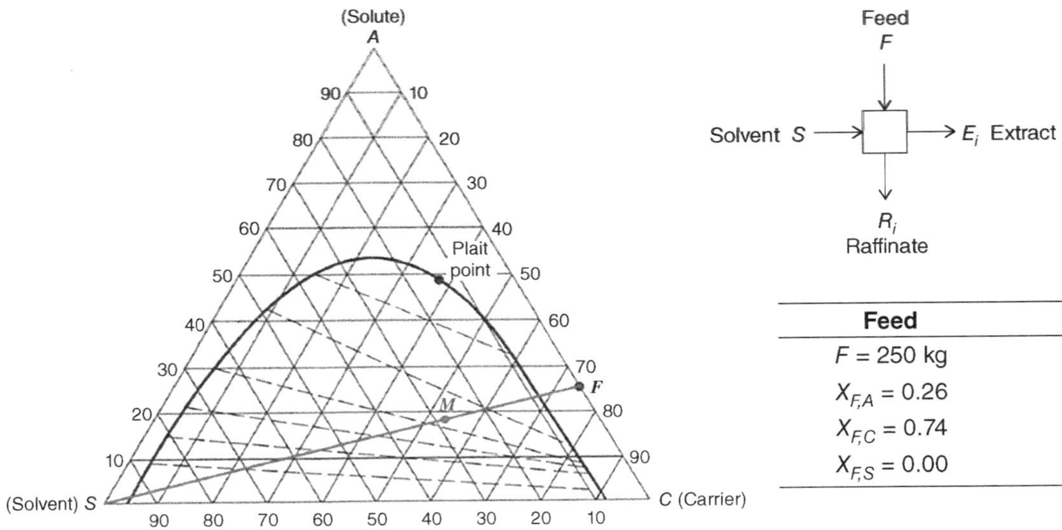

Feed	Solvent
$F = 250$ kg	$S = 100$ kg
$X_{F,A} = 0.26$	$X_{S,A} = 0.00$
$X_{F,C} = 0.74$	$X_{S,C} = 0.00$
$X_{F,S} = 0.00$	$X_{S,S} = 1.00$

Figure 10.8 One equilibrium stage in a ternary diagram with feed data and solvent data.

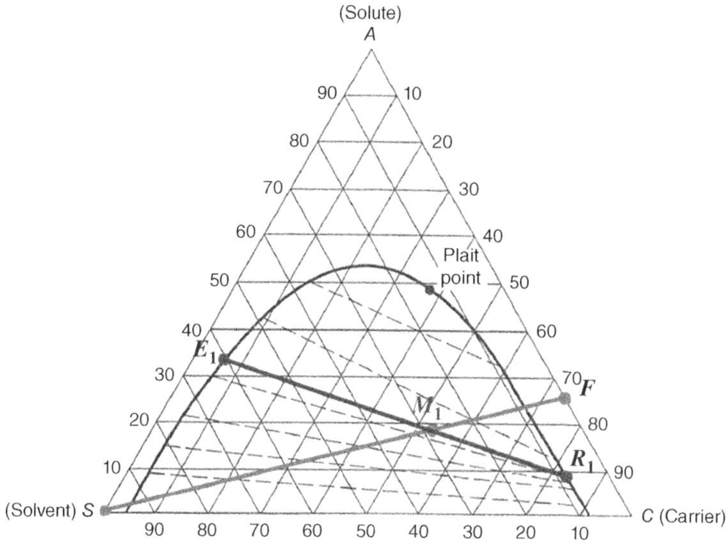

Figure 10.9 One equilibrium stage in a ternary diagram with feed data and solvent data.

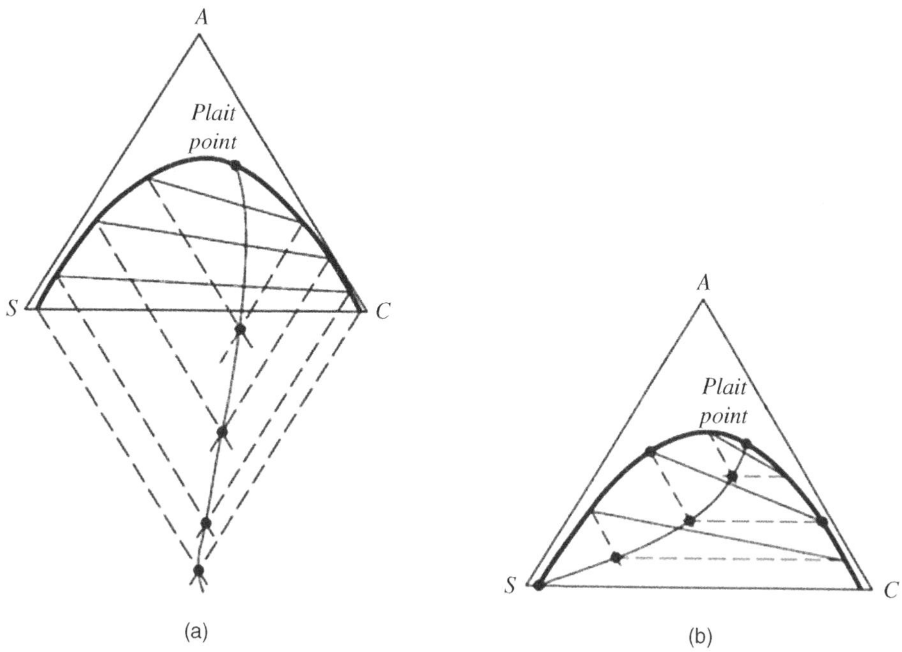

Figure 10.10 Using a conjugate line method, there are two ways to find the E_1R_1 tie line (Seader et al. 2011).

their parallel line to the two side edges of the equilateral triangle. One end of the conjugate line is the plait point, while another end from this method is extended to the outside of the equilateral triangle. The points of the conjugate line in Figure 10.10b are generated by the extract and raffinate points of existing tie lines using their parallel line to the one side edge and the bottom edge of the equilateral triangle. One end of the conjugate line is the plait point, while another end from this method is an intersection point of the binodal curve and the bottom edge of the equilateral triangle.

After generating the conjugate line, new tie lines can be generated by any point on the conjugate line to draw parallel lines (two different ways in Figure 10.10a,b) and obtain their intersection points of the binodal curve. Figure 10.10a shows the lines paralleled to AS edge and AC edge; while Figure 10.10b shows the lines paralleled to SC edge and AC edge The trend of the new tie lines can be seen by moving points on the conjugate line. When the new tie line passes the M point, two intersection points of the binodal curve are E_1 and R_1 points. We can optimize the minimum distance between the M point and a tie line to find the E_1R_1 tie line.

10.3.4 Ternary Diagram – Cross-Current Two Stages Solvent Extraction

Figure 10.11 shows two cross-current SX stages: batch operation in two separatory vessels or continuous operation in two mixer-settler (separator) sets. The feed and solvent data are the same as shown in Figure 10.8. Fresh solvent has been added to two mixers. The two outputs from the first mixer-settler are Extract 1 and Raffinate 1. The Raffinate 1 from the first settler (separator) is fed to mixer 2 and mixed with a fresh solvent to produce Extract 2 and Raffinate 2. The solute concentrations have the following order: Feed > Raffinate 1 > Raffinate 2.

10.3.5 Ternary Diagram – Counter-Current Two Stages Solvent Extraction

There is another configuration different from cross-current SX for the equipment arrangement in SX. Figure 10.12 shows counter-current two SX stages: batch operation in two separatory vessels or continuous operation in two mixer-settlers (separator) sets. The feed and solvent data are the same as shown in Figure 10.8. Fresh solvent has only been added to the second mixer-settlers set. The feed is mixed with Extract 2 from the second set of mixer-settler to produce Extract 1 and Raffinate 1. No fresh solvent is fed into the first set of mixer-settlers. The Raffinate 1 is fed to mixer-settler 2 and mixed with fresh solvent to produce Extract 2 and Raffinate 2. The solute concentrations also have the following order: Feed > Raffinate 1 > Raffinate 2.

The stream mass balance for the first set of mixer-settler is:

$$F + E_2 = E_1 + R_1 \tag{10.8}$$

The stream mass balance for the second set of mixer-settler is:

$$S + R_1 = E_2 + R_2 \tag{10.9}$$

Rearrange the above two equations:

$$F - E_1 = R_1 - E_2 \tag{10.10}$$

The stream mass balance for the second set of mixer-settler is:

$$R_2 - S = R_1 - E_2 \tag{10.11}$$

We assign $P = (F - E_1)$, so

$$P = (F - E_1) = (R_1 - E_2) = (R_2 - S) \tag{10.12}$$

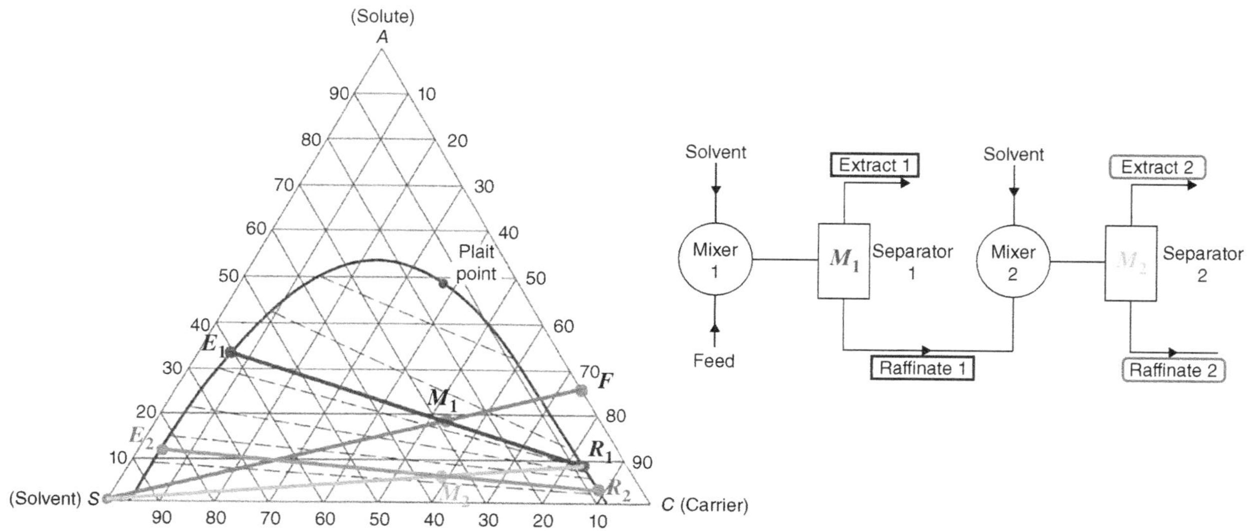

Figure 10.11 Cross-current two stages solvent extraction.

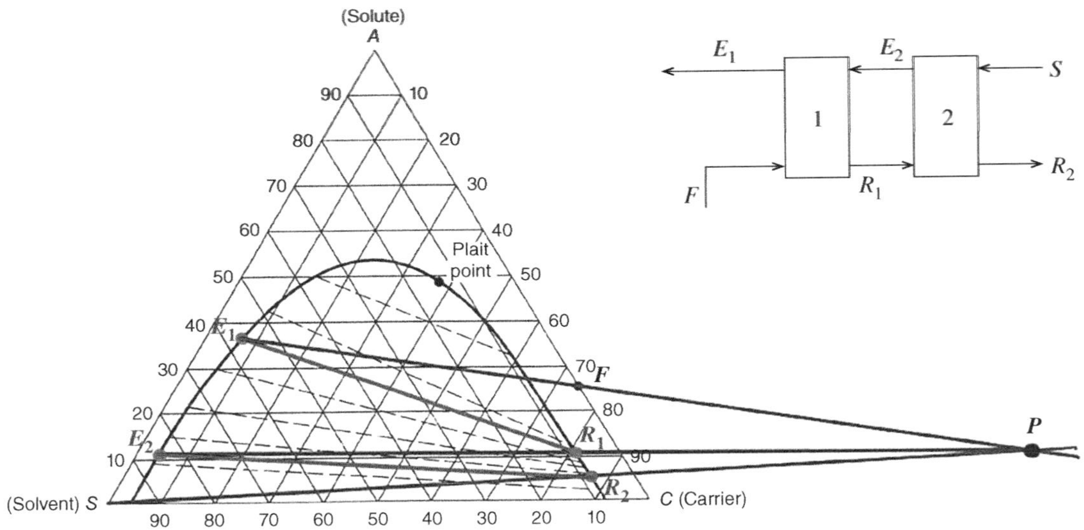

Figure 10.12 Counter-current two stages solvent extraction.

Therefore, there is a P point, as shown in Figure 10.12, which is the intersection point of all three lines: E_1F, E_2R_1, and SR_2. Meanwhile, E_1R_1 and E_2R_2 are tie lines. This P point can be used for mathematical modeling (or computational modeling) of the counter-current two SX stages.

10.3.6 Ternary Diagram – Counter-Current *N* Stages Solvent Extraction

The above counter-current two stages SX can be extended to counter-current N stages SX as shown in Figure 10.13. Fresh solvent has only been added to the mixer-settler's last set (Nth). The feed is mixed with Extract 2 from the second set of mixer-settlers to produce Extract 1 and Raffinate 1. No fresh solvent is fed into the first set of mixer settlers. The Raffinate 1 is fed to mixer-settler 2 and mixed with Extract 3 from the second set of mixer-settlers to produce Extract 2 and Raffinate 2. The inputs of any ith mixer-settler are Extract $(i + 1)$th and Raffinate $(i - 1)$th, while their outputs are Extract ith and Raffinate ith. The inputs of the last number Nth mixer-settler are fresh solvent S and Raffinate $(N-1)$th, while their outputs are Extract Nth and Raffinate Nth. The concentrations of the solute also have the following order: Feed > Raffinate 1 > Raffinate 2 > · · ·> Raffinate i > · · ·> Raffinate $N-1$ > · · ·> Raffinate N.

We can use a similar idea as the above counter-current two stages SX, and the stream mass balance for the first set of mixer-settler is:

$$F + E_2 = E_1 + R_1 \tag{10.13}$$

The stream mass balance for the second set of mixer-settler is:

$$E_3 + R_1 = E_2 + R_2 \tag{10.14}$$

The stream mass balance for the ith set of mixer-settler is:

$$E_{i+1} + R_{i-1} = E_i + R_i \tag{10.15}$$

The stream mass balance for the last Nth set of mixer-settler is:

$$S + R_{N-1} = E_N + R_N \tag{10.16}$$

Rearrange the above Eqs. (10.13)–(10.16) and assign $P = (F - E_1)$, so:

$$P = (F - E_1) = (R_1 - E_2) = \cdots = (R_{i-1} - E_i) = (R_i - E_{i+1}) = \cdots = R_N - S \tag{10.17}$$

Therefore, there is a P point, the intersection point of all lines: E_1F, $E_2R_1, \cdots E_iR_{i-1}$, $E_{i+1}R_i \cdots$, and SR_N. Meanwhile, E_1R_1, E_2R_2, $\cdots E_iR_i \cdots E_NR_N$ are tie lines. This P point can be used for mathematical modeling (or computational modeling) of the counter-current N stages SX.

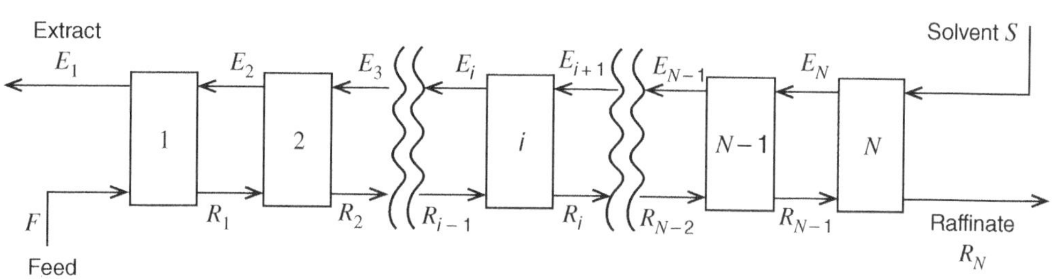

Figure 10.13 Counter-current N stages SX configuration.

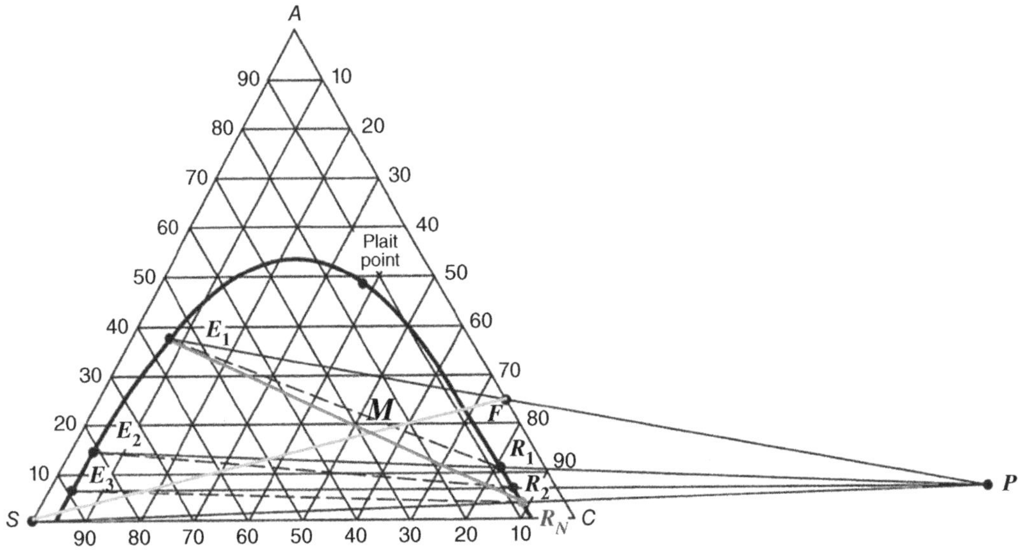

Figure 10.14 Counter-current N stages solvent extraction.

Hunter–Nash graphical equilibrium-stage method can simulate counter-current N stages SX. Its procedure is as follows:

(1) We know F and S; connect with a line and locate "mixture" M
(2) Either specify E_1 or R_N (we always know one of them)
(3) Connect a straight line through M, passing through the one specified, and its intersection point of the Binodal curve is the unspecified one.
(4) Connect S through R_N and extrapolate.
(5) Connect E_1 through F and extrapolate; cross lines at P
(6) Locate P by the intersection of two lines
(7) In general, connect E_n and R_n via equilibrium tie lines

Figure 10.14 shows the Hunter–Nash graphical equilibrium-stage method results with counter-current N stages. There are points shown in the figure: F, M, E_1, R_1, E_2, R_2, E_3, and R_N. Meanwhile, E_1R_1 and E_2R_2 are tie lines.

10.3.7 Case Study – Cross-Current Multistages Solvent Extraction

Example 10.2

Figure 10.15 shows three cross-current SX stages: a batch operation in two separatory vessels or a continuous operation in three mixer-settler (separator) sets. The feed data and solvent data are shown in Table 10.2. Fresh solvent has been added to three mixers. The two outputs from the first mixer-settler are Extract 1 and Raffinate 1. The Raffinate 1 from the first settler (separator) is fed to the second mixer. The two outputs from the second mixer-settler are Extract 2 and Raffinate 2. The Raffinate 2 from the second settler (separator) is fed to the third mixer. The two outputs from the third mixer-settler are Extract 3 and Raffinate 3.

(1) To code Hunter–Nash graphical equilibrium-stage method in MATLAB to get all data points on the ternary diagram: E_1, E_2, E_3, R_1, R_2, R_3, M_1, M_2, and M_3.

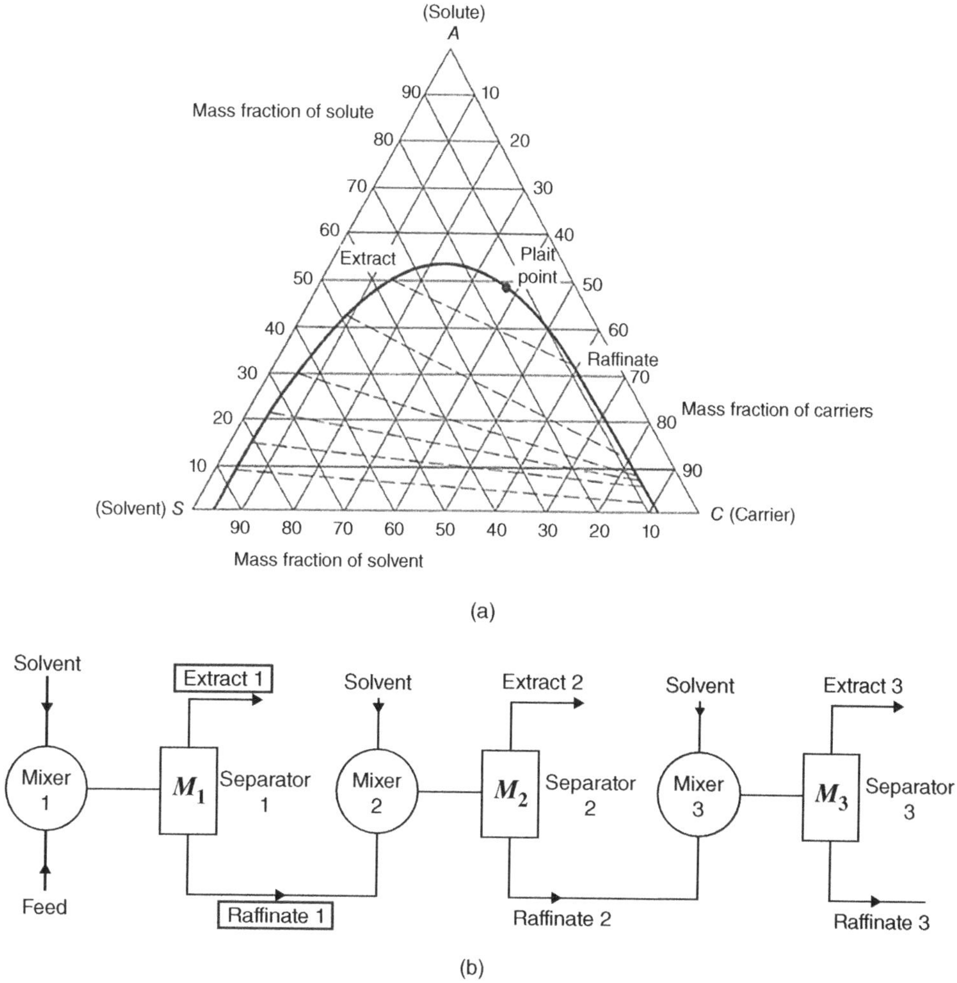

Figure 10.15 (a) Ternary diagram of the system; (b) cross-current three-stages solvent extraction.

Table 10.2 The feed and solvent data.

Feed	Solvent
F = 300kg	S = 100kg
$X_{F,A} = 0.40$	$X_{S,A} = 0.00$
$X_{F,C} = 0.60$	$X_{S,C} = 0.00$
$X_{F,S} = 0.00$	$X_{S,S} = 1.00$

Solution with all Steps

(a) Use grabit.m to obtain the endpoints' data of the existing tie-lines as shown in Table 10.3 and three vertex points: $S(0,0)$, $C(1,0)$, and $A\left(\frac{1}{2}, \frac{\sqrt{3}}{2}\right)$.

```
MATLAB code for triangle plot and binodal curve with the Plait point. *Note: the global
variables defined here are used in the main and sub-function codes.
```

Table 10.3 The endpoints' data of the existing tie-lines saved to Grabit_Data001.mat.

0.0466	0
0.0879	0.0772
0.1157	0.1272
0.1538	0.1837
0.2078	0.2584
0.3046	0.3673
0.3935	0.4332
0.4990	0.4642
0.6188	0.4229
0.7520	0.2790
0.8578	0.1044
0.8792	0.0742
0.8847	0.0646
0.8919	0.0519
0.9109	0.0217
0.9228	0

```
global p mx my xe xr ye yr
%Triangle plot
x=(100:100:100)';
h=TrianglePlot(abs([x x(end:-1:1).*sin(x) cos(x).*x]),'');
h=TrianglePlot(h,'xlabel','S');
h=TrianglePlot(h,'ylabel','C');
h=TrianglePlot(h,'zlabel','A');
h=TrianglePlot(h,'grid',0.1);
x=linspace(0,1);
y=linspace(0,1);
hold on
load('Grabit_Data001.mat')
pt=Grabit_Data001;
plot([pt(7,1) pt(10,1)],[pt(7,2) pt(10,2)],'black') %tieline 1
plot([pt(6,1) pt(11,1)],[pt(6,2) pt(11,2)],'black') %tieline 2
plot([pt(5,1) pt(12,1)],[pt(5,2) pt(12,2)],'black') %tieline 3
plot([pt(4,1) pt(13,1)],[pt(4,2) pt(13,2)],'black') %tieline 4
plot([pt(3,1) pt(14,1)],[pt(3,2) pt(14,2)],'black') %tieline 5
plot([pt(2,1) pt(15,1)],[pt(2,2) pt(15,2)],'black') %tieline 6
plot([pt(1,1) pt(16,1)],[pt(1,2) pt(16,2)],'black') %tieline 7 (y=0)
xlim([0 1]) %Border
ylim([0 1]) %Border
% polyfit for binodal curve n=5
a=pt(:,1); b=pt(:,2);
p = polyfit(a,b,5);
fplot(@(x) p(1)*x^5 + p(2)*x^4 + p(3)*x^3 + p(4)*x^2 + p(5)*x + p(6),...
[pt(1,1) pt(16,1)],'black')
ppx=pt(9,1);%Plait point
ppy=pt(9,2);%Plait point
plot(ppx,ppy,'*');
text(ppx,ppy,'Plait point')
```

```
Results:
Figure 10.16 shows the ternary diagram of the system plotted by our MATLAB code.
```

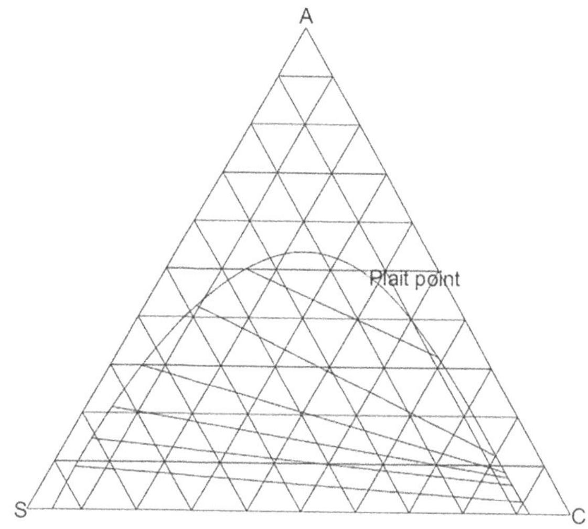

Figure 10.16 Ternary diagram of the system plotted by MATLAB code with digitalized data from Figure 10.15a using grabit.m.

(b) Use the existing tie-lines and plait point to generate the conjugate line and plot the curve.

The method shown in Figure 10.10a is used to generate the conjugate line. The conjugate line points are produced by extracting and raffinate points of existing tie lines using their parallel line to the two side edges of the equilateral triangle. One end of the conjugate line is the plait point. The other end of the conjugate line is produced by the two intersect points between the binodal curve and the *SC* line. This method's end of the conjugate line is extended outside the equilateral triangle.

The left parallel line to the *AC* line has the linear function:

$$y = -\sqrt{3}x + b_1 \tag{10.18}$$

We use the left extract points represented by *Ex* and *Ey*, which are:

$$Ey = -\sqrt{3}Ex + b_1 \tag{10.19}$$

The b_1 can be solved:

$$b_1 = Ey + \sqrt{3}Ex \tag{10.20}$$

Then, the left parallel line function becomes:

$$y = -\sqrt{3}x + Ey + \sqrt{3}Ex \tag{10.21}$$

The right parallel line to the *AS* line has the linear function:

$$y = \sqrt{3}x + b_2 \tag{10.22}$$

We use the right raffinate points represented by *Rx* and *Ry*, which are:

$$Ry = \sqrt{3}Rx + b_2 \tag{10.23}$$

The b_2 can be solved:

$$b_2 = Ry - \sqrt{3}Rx \tag{10.24}$$

Then, the right parallel line function becomes:

$$y = \sqrt{3}x + Ry - \sqrt{3}Rx \tag{10.25}$$

Therefore, any points on the conjugate line can be solved by the above two Eqs. (10.21) and (10.25) using existing tie-lines data:

$$x = \frac{1}{2\sqrt{3}}[(Ey + \sqrt{3}Ex) - (Ry - \sqrt{3}Rx)] \tag{10.26}$$

$$y = \frac{1}{2}[(Ey + \sqrt{3}Ex) + (Ry - \sqrt{3}Rx)] \tag{10.27}$$

MATLAB code for generating the conjugate line using the above Grabit_Data001 data of extract and raffinate points and plot the curve.

```
for i=1:7
    cx(i)=1/2*((pt(i,2)+sqrt(3)*pt(i,1))-(pt(16-i+1,2)-sqrt(3)*pt(16-i+1,1)))/sqrt(3);
    cy(i)=1/2*((pt(i,2)+sqrt(3)*pt(i,1))+(pt(16-i+1,2)-sqrt(3)*pt(16-i+1,1)));
end
cx(8)=ppx;% Plait point
cy(8)=ppy;% Plait point
hold on
plot(cx,cy,'Marker','o')
xlim([0 1]) %Border
ylim([-1 1]) %Border
Results:
Figure 10.17 shows the plot of the conjugate line.
```

MATLAB code for model-fitting of the conjugate line: (*Note: it is challenging to obtain a relationship of $y = f(x)$, instead it is relatively easy to obtain a relationship of $x = g(y)$.) Therefore, we use cx as the y data and cy as the x data and use the Rational Curve Fit (rat32) as shown in Figure 10.18. The Goodness of Fit, R-square, is 0.9967, which is a good result.

```
Results of the fitted model:
val =
     General model Rat32:
     val(x) = (p1*x^3 + p2*x^2 + p3*x + p4) /
                  (x^2 + q1*x + q2)
     Coefficients (with 95% confidence bounds):
       p1 =        0.1087  (-0.03166, 0.249)
       p2 =        0.6098  (0.5516, 0.6679)
       p3 =         0.316  (-0.129, 0.7609)
       p4 =       0.08389  (-0.009326, 0.1771)
       q1 =        0.5375  (-0.1543, 1.229)
       q2 =        0.1348  (-0.01477, 0.2843)
```

We can plot the fitted conjugate curve and compare it with the original data (*Note: we use cx as the y data and cy as the x data). It can be seen from Figure 10.19 that the model-fitted conjugate curve matches very well with the smooth curve original data points.

```
% General model Rat32:
% val(x) = (p1*x^3 + p2*x^2 + p3*x + p4) / (x^2 + q1*x + q2)
% Coefficients (with 95% confidence bounds):
% *Note that we use cp and cq here because the other place uses p.
       cp1 =       0.1087;
       cp2 =       0.6098;
       cp3 =        0.316;
       cp4 =      0.08389;
       cq1 =       0.5375;
       cq2 =       0.1348;
%the conjugate line, note cx->y and cy->x
yz=linspace(min(cy),max(cy),100);
xz=(cp1*yz.^3 + cp2*yz.^2 + cp3*yz + cp4)./(yz.^2 + cq1*yz + cq2);
plot(xz,yz)
```

Results:

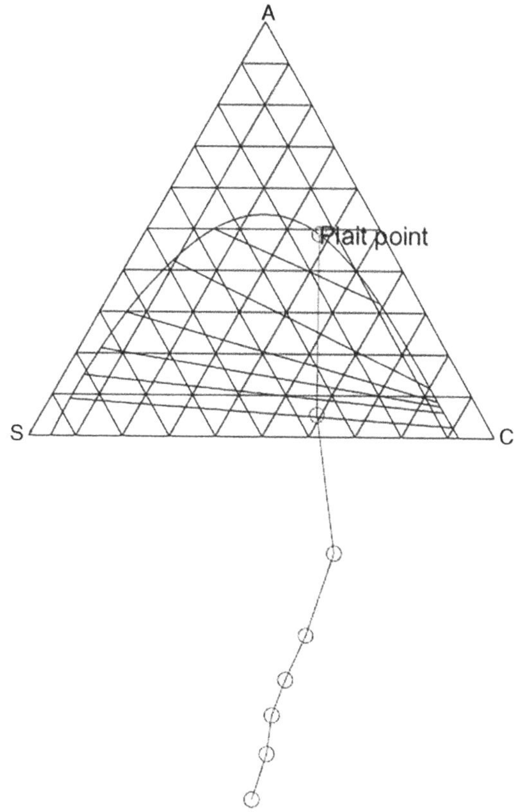

Figure 10.17 The plot of the conjugate line was generated using the above Grabit_Data001 data of extract and raffinate points.

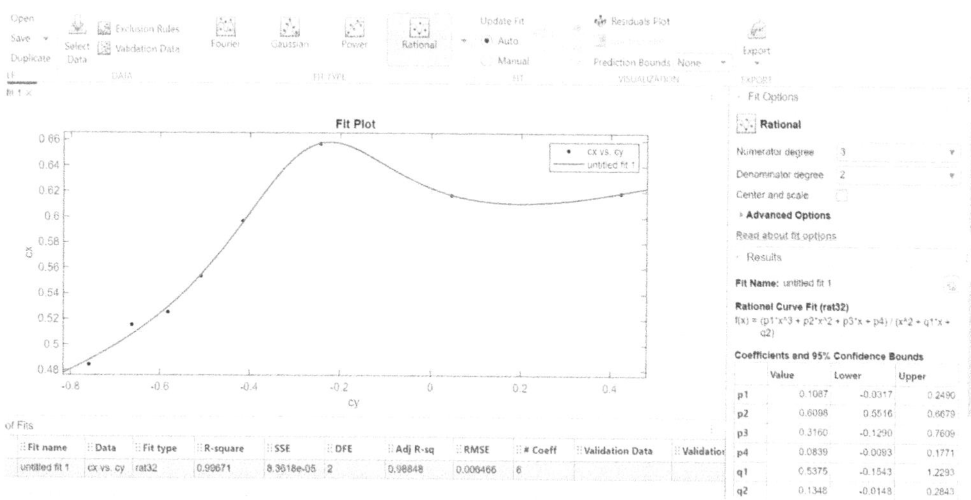

Figure 10.18 The conjugate line's model-fitting was performed using the above Grabit_Data001 data of extract and raffinate points.

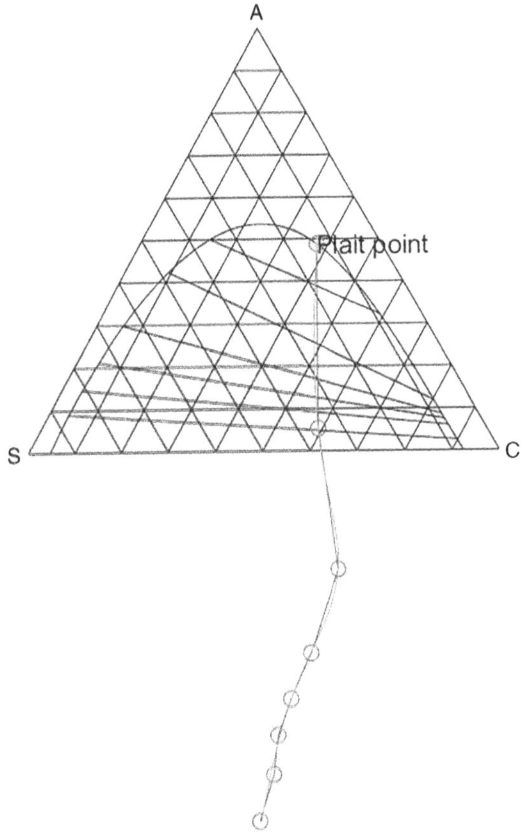

Figure 10.19 The model-fitted conjugate curve compared to the smooth curve of the original data points.

(c) Input the feed and solvent data and plot them on the diagram.

```
MATLAB code:
y1=input('enter ternary percentages as decimals [Af,Cf,Sf]');
S=input('enter Solvent value [S]');
F=input('enter Feed value ');
```

(d) Locate the feed and solvent data points on the diagram.

```
MATLAB code:
    Af=y1(1)/(sum(y1));
    Cf=y1(2)/(sum(y1));
    Sf=y1(3)/(sum(y1));
xf=(1/2)*(Af+(2*Cf))/(Sf+Af+Cf);
yf=(sqrt(3)/2)*Af/(Sf+Af+Cf);
%Solvent point
y2=[0,0,1];
    As=y2(1)/(sum(y2));
    Cs=y2(2)/(sum(y2));
    Ss=y2(3)/(sum(y2));
xs=(1/2)*(As+(2*Cs))/(Ss+As+Cs);
ys=(sqrt(3)/2)*As/(Ss+As+Cs);
```

```
plot(xf,yf,'*') %Feed point
text(xf,yf,'F')
plot(xs,ys,'*') %Solvent point
```

```
Results:
```

The feed data point and solvent data point are shown in Figure 10.20.

(e) Locate the mixture data point on the diagram.

Equations (10.2)–(10.7) can be used with the feed and solvent data to calculate the mixture components and locate the mixture point on the diagram.

```
MATLAB code:
n=1; % It is stage 1, the first mixer-settler.
%calculating Distance Ratio
DR=S/(S+F);
%calculating the distance from F to M1
M1F=DR*sqrt((xs-xf)^2+(ys-yf)^2);
%plotting line from S to F
plot([xs xf],[ys yf],'r')
%calculating coordinates of M1
k=(ys-yf)/(xs-xf);%slope of the line
xm(n)=xf-(DR*(xf-xs));
ym(n)=k*xm(n);
plot(xm(n),ym(n),'Marker','o')
text(xm(n),ym(n),"M"+n)
```

The mixture data point is shown in Figure 10.21.

(f) Locate the extract data point E1 and the raffinate data point R1 on the diagram.

An optimization method is needed to find a tie-line passing the mixture point M1, and the two endpoints of the tie-line are E1 and R1. The object function is to minimize the absolute difference of $|y^* - ym|$, where ym is the y-axis of the mixture data point, and y^* is the y-axis of a tie-line with the corresponding x-axis being xm, the x-axis of the mixture data point.

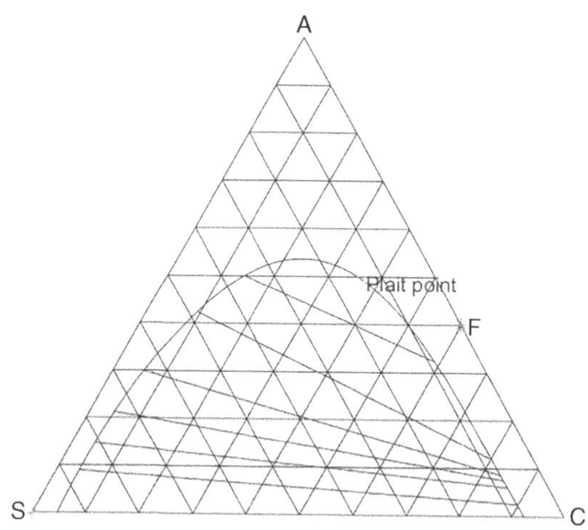

Figure 10.20 The feed data point and solvent data point are shown in the diagram.

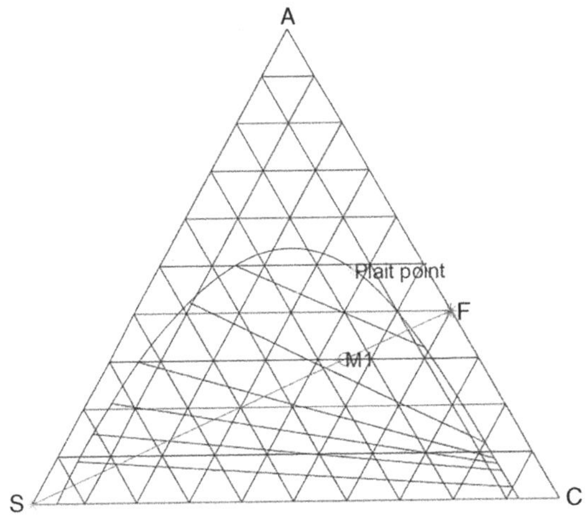

Figure 10.21 The mixture data point is shown in the diagram.

In this section, we use any point on the conjugate line to generate a new tie-line. Any point on the conjugate line, i.e. (xc, yc), which satisfies the fitted model in the above section b:

$$xc = (cp1 * yc^3 + cp2 * yc^2 + cp3 * yc + cp4)/(yc^2 + cq1 * yc + cq2) \quad (10.28)$$

The left parallel line function becomes:

$$yc = -\sqrt{3}xc + Ey + \sqrt{3}Ex \quad (10.29)$$

The right parallel line function becomes:

$$yc = \sqrt{3}xc + Ry - \sqrt{3}Rx \quad (10.30)$$

The binodal curve function from the above section a is

$$Ey = p(1) * Ex^5 + p(2) * Ex^4 + p(3) * Ex^3 + p(4) * Ex^2 + p(5) * Ex + p(6) \quad (10.31)$$

$$Ry = p(1) * Rx^5 + p(2) * Rx^4 + p(3) * Rx^3 + p(4) * Rx^2 + p(5) * Rx + p(6) \quad (10.32)$$

Due to the difficulty of solving the above Eqs. (10.28)–(10.32), the optimization method fmincon is used twice to solve those equations, as shown in the function file opt_y.m in the following MATLAB code.

```
MATLAB code:
mx=xm(n);
my=ym(n);
[yc,fval]=fmincon(@opt_y,0,[],[],[],[],min(cy),max(cy))
nxe(n)=xe;
nxr(n)=xr;
nye(n)=ye;if nye(n)<0; nye(n)=0;end
nyr(n)=yr;if nyr(n)<0; nyr(n)=0;end
plot([nxe(n) nxr(n)],[nye(n) nyr(n)],'b');
str = {"E"+n,"R"+n};
text([nxe(n) nxr(n)],[nye(n) nyr(n)],str)

function sm=opt_y(yc)
global p mx my xe xr ye yr
% General model Rat32 for the conjugate line: cx->y and cy->x
%     val(x) = (p1*x^3 + p2*x^2 + p3*x + p4) / (x^2 + q1*x + q2)
%     Coefficients (with 95% confidence bounds):
       cp1 =       0.1087;
       cp2 =       0.6098;
       cp3 =         0.316;
       cp4 =      0.08389;
```

```
        cq1 =       0.5375;
        cq2 =       0.1348;
%the conjugate line, note cx->y and cy->x
xc=(cp1*yc^3 + cp2*yc^2 + cp3*yc + cp4)/(yc^2 + cq1*yc + cq2);
%Raffinate lower bound is the Plait point, and the upper bound is the right side
intersect
%point of the binodal curve to the SC line.
[xr]=fmincon(@(xr1)(p(1)*xr1^5+p(2)*xr1^4+p(3)*xr1^3+p(4)*xr1^2+p(5)*xr1+p(6)-sqrt(3)*
(xr1-xc)-yc)^2,0.7,[],[],[],[],0.6188,0.9228);
yr=sqrt(3)*(xr-xc)+yc;
% extract lower bound is the right side intersect point of the binodal curve to the SC
line, and the upper bound is the Plait point.
[xe]=fmincon(@(xe1)(p(1)*xe1^5+p(2)*xe1^4+p(3)*xe1^3+p(4)*xe1^2+p(5)*xe1+p(6)+sqrt(3)*
(xe1-xc)-yc)^2,0.1,[],[],[],[],0.0466,0.6188);
ye=-sqrt(3)*(xe-xc)+yc;
p1=polyfit([xe xr],[ye yr],1);
f1=polyval(p1,mx);
sm=abs(f1-my);
end

Results:
The E1 data point and R1 data point are shown in Figure 10.22.
```

(g) Compute the composites of M1, E1, and R1.

```
MATLAB code:
%percentages for M1
AM(n)=(Af*F)/(S+F);
SM(n)=(S)/(S+F);
CM(n)=1-AM(n)-SM(n);
%Finding E1 and R1 Percentages
AE(n)=nye(n)*2/sqrt(3);
SE(n)=1-(nye(n)+sqrt(3)*nxe(n))/sqrt(3);
CE(n)=1-AE(n)-SE(n);
AR(n)=nyr(n)*2/sqrt(3);
SR(n)=1-(nyr(n)+sqrt(3)*nxr(n))/sqrt(3);
CR(n)=1-AR(n)-SR(n);
R(n)=(S+F)*(AM(n)-AE(n))/(AR(n)-AE(n));
```

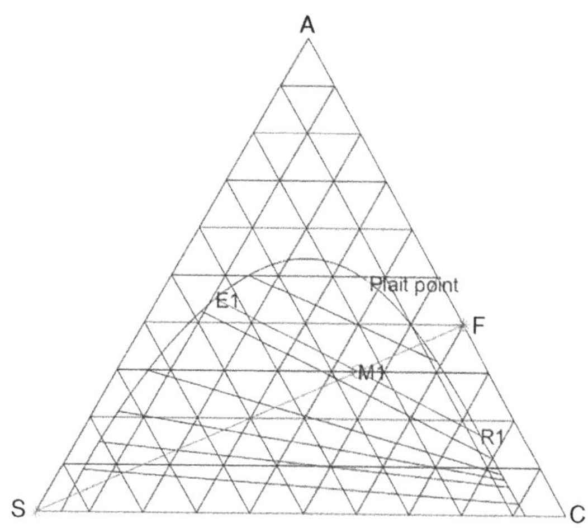

Figure 10.22 The E1 data point and R1 data point are shown in the diagram.

(h) Prepare for the next SX stage, i.e. R1 feeds the second mixer-settler and adds a new solvent.

```
MATLAB code:
%The new feed is the last raffinate
    F=R(n);
    xf=nxr(n);
    yf=nyr(n);
    Af=AR(n);
    Sf=SR(n);
    Cf=CR(n);
```

(i) Extend to cross-current N stages SX. The code from the above sections e-h can be combined to suit multistage SX. We use three stages as an example. The results are shown in Figure 10.23 for the $E2$ data point and $R2$ data point on the diagram, and Figure 10.24 for the $E3$ data point and $R3$ data point on the diagram.

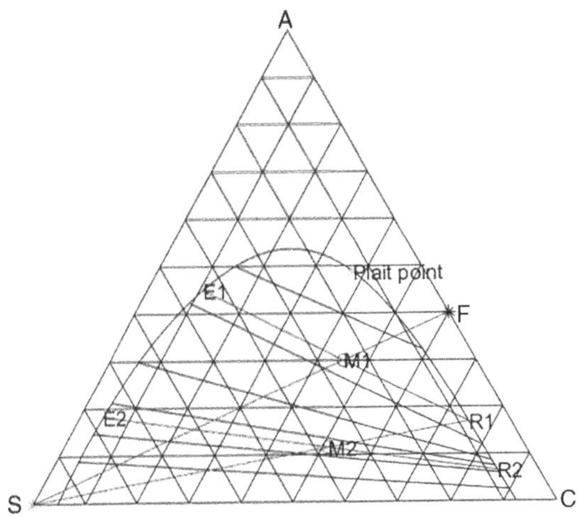

Figure 10.23 The E2 and R2 data points are shown in the diagram.

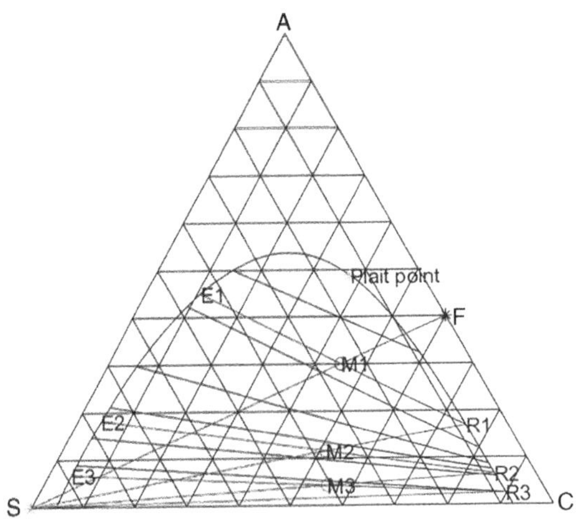

Figure 10.24 The E3 and R3 data points are shown in the diagram.

```
MATLAB code:
% n-the number of stages
for n=1:3
%calculating Distance Ratio
DR=S/(S+F);
%calculating the distance from F to M1
M1F=DR*sqrt((xs-xf)^2+(ys-yf)^2);
%plotting line from S to F
plot([xs xf],[ys yf],'r')
%calculating coordinates of M1
k=(ys-yf)/(xs-xf);%slope of the line
% xm1=xf-((M1F*(xf-xs)/sqrt((xs-xf)^2+(ys-yf)^2)));%old one
xm(n)=xf-(DR*(xf-xs));
ym(n)=k*xm(n);
plot(xm(n),ym(n),'Marker','o')
text(xm(n),ym(n),"M"+n)
%Finding E1 and R1 using optimization
mx=xm(n);
my=ym(n);
[yc,fval]=fmincon(@opt_y,0,[],[],[],[],min(cy),max(cy))
nxe(n)=xe;
nxr(n)=xr;
nye(n)=ye;if nye(n)<0; nye(n)=0;end
nyr(n)=yr;if nyr(n)<0; nyr(n)=0;end
plot([nxe(n) nxr(n)],[nye(n) nyr(n)],'b');
str = {"E"+n,"R"+n};
text([nxe(n) nxr(n)],[nye(n) nyr(n)],str)
%plotting line from S to R1
plot([nxr(n) xs],[nyr(n) ys],'r')
%percentages for M1
AM(n)=(Af*F)/(S+F);
SM(n)=(S)/(S+F);
CM(n)=1-AM(n)-SM(n);
%Finding E1 and R1 Percentages
AE(n)=nye(n)*2/sqrt(3);
SE(n)=1-(nye(n)+sqrt(3)*nxe(n))/sqrt(3);
CE(n)=1-AE(n)-SE(n);
AR(n)=nyr(n)*2/sqrt(3);
SR(n)=1-(nyr(n)+sqrt(3)*nxr(n))/sqrt(3);
CR(n)=1-AR(n)-SR(n);
R(n)=(S+F)*(AM(n)-AE(n))/(AR(n)-AE(n));
%The new feed is the last raffinate
    F=R(n);
    xf=nxr(n);
    yf=nyr(n);
    Af=AR(n);
    Sf=SR(n);
    Cf=CR(n);
end
for n=1:3
formatSpec = 'M(%d) location: XM(%d)=%5.4f, and YM(%d)=%5.4f\n';
fprintf(formatSpec,n,n,xm(n),n,ym(n))
formatSpec = 'M(%d) concentrations: AM(%d)=%5.4f, CM(%d)=%5.4f, and SM(%d)=%5.4f\n';
fprintf(formatSpec,n,n,AM(n),n,CM(n),n,SM(n))
formatSpec = 'E(%d) location: XE(%d)=%5.4f, and YE(%d)=%5.4f\n';
fprintf(formatSpec,n,n,nxe(n),n,nye(n))
formatSpec = 'E(%d) concentrations: AE(%d)=%5.4f, CE(%d)=%5.4f, and SE(%d)=%5.4f\n';
fprintf(formatSpec,n,n,AE(n),n,CE(n),n,SE(n))
formatSpec = 'R(%d) location: XR(%d)=%5.4f, and YR(%d)=%5.4f\n';
fprintf(formatSpec,n,n,nxr(n),n,nyr(n))
formatSpec = 'R(%d) concentrations: AR(%d)=%5.4f, CR(%d)=%5.4f, and SR(%d)=%5.4f\n';
fprintf(formatSpec,n,n,AR(n),n,CR(n),n,SR(n))
end
```

Results:

```
M(1) location: XM(1)=0.6000, and YM(1)=0.2598
M(1) concentrations: AM(1)=0.3000, CM(1)=0.4500, and SM(1)=0.2500
E(1) location: XE(1)=0.3309, and YE(1)=0.3895
E(1) concentrations: AE(1)=0.4498, CE(1)=0.1060, and SE(1)=0.4442
R(1) location: XR(1)=0.8353, and YR(1)=0.1464
R(1) concentrations: AR(1)=0.1691, CR(1)=0.7507, and SR(1)=0.0802
M(2) location: XM(2)=0.5688, and YM(2)=0.0997
M(2) concentrations: AM(2)=0.1151, CM(2)=0.5658, and SM(2)=0.3191
E(2) location: XE(2)=0.1362, and YE(2)=0.1579
E(2) concentrations: AE(2)=0.1823, CE(2)=0.0450, and SE(2)=0.7727
R(2) location: XR(2)=0.8887, and YR(2)=0.0567
R(2) concentrations: AR(2)=0.0654, CR(2)=0.8560, and SR(2)=0.0786
M(3) location: XM(3)=0.5715, and YM(3)=0.0364
M(3) concentrations: AM(3)=0.0421, CM(3)=0.6010, and SM(3)=0.3569
E(3) location: XE(3)=0.0768, and YE(3)=0.0578
E(3) concentrations: AE(3)=0.0667, CE(3)=0.0434, and SE(3)=0.8899
R(3) location: XR(3)=0.9100, and YR(3)=0.0219
R(3) concentrations: AR(3)=0.0252, CR(3)=0.8973, and SR(3)=0.0774
```

(2) To apply OED to find optimal parameters to minimize the operation costs and reach a final raffinate of no greater than 2.5%.

- Mass of solvent (US$15/kg)
- Addition of carrier (US$5/kg) for dilution (carrier usage is limited to max 200 kg)
- Number of stages (US$1000/stage)

Three factors, as shown in Table 10.4, are considered: the mass of solvent (Factor A), the addition of carrier (Factor B), and the number of stages (Factor C).

If we use a full factorial design for three factors with three levels, we will need to perform $3^3 = 27$ runs. However, if we use an OED method such as in Table 10.5, we only need to run nine specifically designed experiments instead.

The data analysis using the OED table for the SX case is shown in Table 10.6. The maximum *DE* implies the most critical factor is factor C, the number of stages, and the second largest *DE* implies the second important factor is factor A, the mass of solvent. It can be seen from Figure 10.25 that the number of stages is the most critical factor affecting the total cost; the mass of solvent is the second important factor affecting the total cost, while the addition of carrier has the least impact on the total cost. The optimal scheme A1–B2–C1 that was obtained from OED method is outside the 9 tests and it has also been tested, and its results are also shown in the table. Compared to Test No. 1, A1–B1–C1, the optimal scheme is worse than that of Test No. 1. It can happen because the OED method is based on discretized points, and the optimal point may be near an existing test point instead of the optimal scheme.

To reduce the costs, we need to use fewer stages and less mass of solvent. We also need to achieve a final raffinate not greater than 2.5%. The number of stages can be fixed to 2,

Table 10.4 Design of experiments for three factors with three levels on solvent extraction case.

Level no.	Factor A/kg	Factor B/kg	Factor C
1	100	0	2
2	150	100	3
3	200	200	4

Table 10.5 The OED (Orthogonal Experimental Design) table for three factors with three levels.

Test no.	Factor A	Factor B	Factor C
1	A1	B1	C1
2	A1	B2	C2
3	A1	B3	C3
4	A2	B1	C2
5	A2	B2	C3
6	A2	B3	C1
7	A3	B1	C3
8	A3	B2	C1
9	A3	B3	C2

a) Note: The factor B (the addition of carrier for dilution) affects the feed concentration: when factor B is 0 kg, the solute concentration is 0.4; when factor B is 100 kg, the solute concentration is $0.4 \times 300/(300+100) = 0.3$; when factor B is 200 kg, the solute concentration is $0.4 \times 300/(300+200) = 0.24$.

Table 10.6 The data analysis using the OED table for this case.

Test no.	Factor A	Factor B	Factor C	Cost ($)	Raffinate concentration
1	A1	B1	C1	5000	0.0654
2	A1	B2	C2	8000	0.0366
3	A1	B3	C3	11,000	0.0256
4	A2	B1	C2	9750	0.0077
5	A2	B2	C3	13,500	0.0049
6	A2	B3	C1	7500	0.0476
7	A3	B1	C3	16,000	0
8	A3	B2	C1	8500	0.0314
9	A3	B3	C2	13,000	0.0149
K1	24,000	30,750	21,000		
K2	30,750	30,000	30,750		
K3	37,500	31,500	40,500		
M1 = (K1/3)	8000	10,250	7000		
M2 = (K2/3)	10,250	10,000	10,250		
M3 = (K3/3)	12,500	10,500	13,500		
DE	4500	500	6500		
Optimal scheme	A1	B2	C1	5500	0.0714

a) Note: The total mass of solvent for a test is the mass of solvent multiplied by the number of stages.

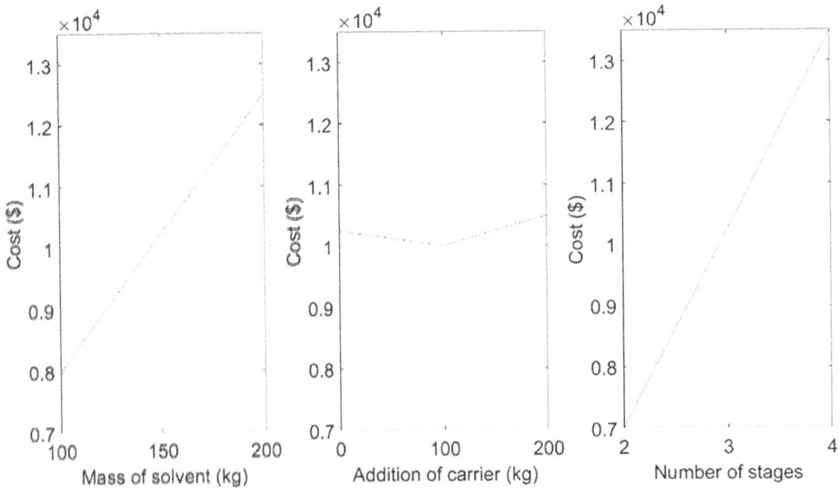

Figure 10.25 The plots of M values vs. different factors.

and no carrier can be added. Then, the only variable to optimize is the mass of the solvent. Because it is a complicated computation case, any local optimization solver may fall into local optimal; we use a global optimization method. The MATLAB code with the global optimization method, MultiStart, is as follows: (*Note: the opt_y.m and data preparation in the previous section are also used in this optimization. The only difference between opt_y_Q2.m and opt_y.m is first line of code: global p mx my xe xr ye yr cx cy xm ym nxe nxr nye nyr xs ys)

```
MATLAB code:
%Cost optimization
x0=[200];
rng default % For reproducibility
func=@(x)x(1)*2*15+0*5+2*1000;
problem = createOptimProblem('fmincon','objective',func,'x0',x0, 'lb',100,'ub',200,...
'nonlcon',@non_liquid);
ms = MultiStart;
[xw,fval] = run(ms,problem,20)
F=300;
Af=300*0.4/F;
Cf=1-Af;
Sf=0;
xf=(1/2)*(Af+(2*Cf))/(Sf+Af+Cf);
yf=(sqrt(3)/2)*Af/(Sf+Af+Cf);
plot(xf,yf,'*') %Feed point
text(xf,yf,'F')
%Solvent point
y2=[0,0,1];
    As=y2(1)/(sum(y2));
    Cs=y2(2)/(sum(y2));
    Ss=y2(3)/(sum(y2));
xs1=(1/2)*(As+(2*Cs))/(Ss+As+Cs);
ys1=(sqrt(3)/2)*As/(Ss+As+Cs);
plot([xs1 xf],[ys1 yf],'r')
% n-the number of stages
for n=1:2
plot([nxe(n) nxr(n)],[nye(n) nyr(n)],'b');
str = {"E"+n,"R"+n};
text([nxe(n) nxr(n)],[nye(n) nyr(n)],str)
    xf=nxr(n);
```

```
    yf=nyr(n);
plot(xf,yf,'*') %Feed point
text(xf,yf,'F')
plot(xs,ys,'*') %Solvent point
plot([xs xf],[ys yf],'r')
plot(xm(n),ym(n),'Marker','o')
text(xm(n),ym(n),"M"+n)
end

function [c,ceq]=non_liquid(vy)
global p mx my xe xr ye yr cx cy xm ym nxe nxr nye nyr xs ys
S=vy(1);
F=300; % F=300+vy(2);
Af=300*0.4/F;
Cf=1-Af;
Sf=0;
xf=(1/2)*(Af+(2*Cf))/(Sf+Af+Cf);
yf=(sqrt(3)/2)*Af/(Sf+Af+Cf);
y2=[0,0,1]; %Solvent point
    As=y2(1)/(sum(y2));
    Cs=y2(2)/(sum(y2));
    Ss=y2(3)/(sum(y2));
xs=(1/2)*(As+(2*Cs))/(Ss+As+Cs);
ys=(sqrt(3)/2)*As/(Ss+As+Cs);
%plot([xs1 xf],[ys1 yf],'r')
for n=1:2 % n-the number of stages
%calculating Distance Ratio
DR=S/(S+F);
%calculating the distance from F to M1
M1F=DR*sqrt((xs-xf)^2+(ys-yf)^2);
%calculating coordinates of M1
k=(ys-yf)/(xs-xf);%slope of the line
xm(n)=xf-(DR*(xf-xs));
ym(n)=k*xm(n);
%Finding E1 and R1 using optimization
mx=xm(n);
my=ym(n);
[yc,fval]=fmincon(@opt_y_Q2,0,[],[],[],[],min(cy),max(cy));
nxe(n)=xe;
nxr(n)=xr;
nye(n)=ye;if nye(n)<0; nye(n)=0;end
nyr(n)=yr;if nyr(n)<0; nyr(n)=0;end
%percentages for M1
AM(n)=(Af*F)/(S+F);
SM(n)=(S)/(S+F);
CM(n)=1-AM(n)-SM(n);
%Finding E1 and R1 Percentages
AE(n)=nye(n)*2/sqrt(3);
SE(n)=1-(nye(n)+sqrt(3)*nxe(n))/sqrt(3);
CE(n)=1-AE(n)-SE(n);
AR(n)=nyr(n)*2/sqrt(3);
SR(n)=1-(nyr(n)+sqrt(3)*nxr(n))/sqrt(3);
CR(n)=1-AR(n)-SR(n);
R(n)=(S+F)*(AM(n)-AE(n))/(AR(n)-AE(n));
%The new feed is the last raffinate
    F=R(n);
    xf=nxr(n);
    yf=nyr(n);
    Af=AR(n);
    Sf=SR(n);
    Cf=CR(n);
end
for n=1:2
formatSpec = 'M(%d) location: XM(%d)=%5.4f, and YM(%d)=%5.4f\n';
fprintf(formatSpec,n,n,xm(n),n,ym(n))
```

```
formatSpec = 'M(%d) concentrations: AM(%d)=%5.4f, CM(%d)=%5.4f, and SM(%d)=%5.4f\n';
fprintf(formatSpec,n,n,AM(n),n,CM(n),n,SM(n))
formatSpec = 'E(%d) location: XE(%d)=%5.4f, and YE(%d)=%5.4f\n';
fprintf(formatSpec,n,n,nxe(n),n,nye(n))
formatSpec = 'E(%d) concentrations: AE(%d)=%5.4f, CE(%d)=%5.4f, and SE(%d)=%5.4f\n';
fprintf(formatSpec,n,n,AE(n),n,CE(n),n,SE(n))
formatSpec = 'R(%d) location: XR(%d)=%5.4f, and YR(%d)=%5.4f\n';
fprintf(formatSpec,n,n,nxr(n),n,nyr(n))
formatSpec = 'R(%d) concentrations: AR(%d)=%5.4f, CR(%d)=%5.4f, and SR(%d)=%5.4f\n';
fprintf(formatSpec,n,n,AR(n),n,CR(n),n,SR(n))
end
ceq=AR(2)-0.025
c=[];
end

The results are as follows:
ceq =
   8.9063e-07
MultiStart completed some of the runs from the start points.
7 out of 20 local solver runs converged with a positive local solver exitflag.
xw =
  181.3065
fval =
   7.4392e+03

M(1) location: XM(1)=0.4986, and YM(1)=0.2159
M(1) concentrations: AM(1)=0.2493, CM(1)=0.3740, and SM(1)=0.3767
E(1) location: XE(1)=0.2472, and YE(1)=0.3068
E(1) concentrations: AE(1)=0.3542, CE(1)=0.0701, and SE(1)=0.5757
R(1) location: XR(1)=0.8747, and YR(1)=0.0801
R(1) concentrations: AR(1)=0.0924, CR(1)=0.8285, and SR(1)=0.0791
M(2) location: XM(2)=0.4509, and YM(2)=0.0413
M(2) concentrations: AM(2)=0.0477, CM(2)=0.4678, and SM(2)=0.4845
E(2) location: XE(2)=0.0765, and YE(2)=0.0573
E(2) concentrations: AE(2)=0.0661, CE(2)=0.0434, and SE(2)=0.8905
R(2) location: XR(2)=0.9101, and YR(2)=0.0217
R(2) concentrations: AR(2)=0.0250, CR(2)=0.8976, and SR(2)=0.0774
```

Figure 10.26 shows the results of the two stages for the optimal case.

After the OED method and performance of global optimization using MultiStart, the optimal parameters to minimize the operation costs to reach final raffinate not greater than 2.5%: mass of solvent = 181.3065 kg, addition of carrier = 0 kg, number of stages = 2, and the minimum operation cost is $7,439.

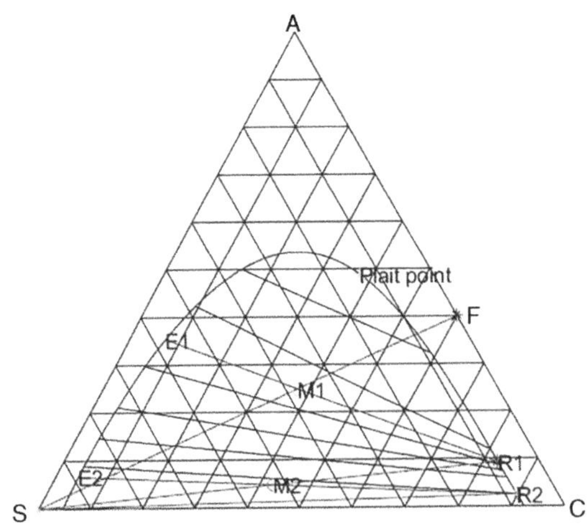

Figure 10.26 The first stage, E1, M1, and R1 data points and the second stage, E2, M2, and R2 data points for the optimal case are shown in the diagram.

10.4 Summary

This chapter presents practical process integration and optimization to expand our readers' knowledge and skills learned from Chapters 1–9. Achieving a whole process optimization is more important than just single-unit optimization. A whole process optimization can help us explore the whole process's potential benefit.

MATLAB is not specific for chemical process integration and optimization because it is a general programming language and provides a numeric computing environment. The domain knowledge in chemical engineering is essential to perform process integration and optimization with the assistance of MATLAB computational power.

The case study on the cross-current multistages solvent extraction has been used to demonstrate process integration and optimization using MATLAB. The knowledge and skills from previous chapters have been applied to this case study. It is a comprehensive case study with many aspects covered, including digitalization of ternary diagram, model-fitting for binodal curve and conjugate curve, equation solving to determine intersect points between tie-lines and binodal curve, searching the tie-line passing mixture point using optimization method, process optimization using OED method, and coding to implement all the above operations.

With the fast development of computers, more software packages emerge to perform simulations of the material and energy balances of chemical process plants. It can be envisaged that the knowledge and skills of process integration and optimization play a vital role in chemical process plants.

Problems

10.1 Use the above Example 10.2, the cross-current three stages of SX shown in Figure 10.15, with the second construction method for the conjugate line in Figure 10.10b. Then, code Hunter–Nash graphical equilibrium-stage method in MATLAB to get all data points on the ternary diagram: E1, E2, E3, R1, R2, R3, M1, M2, and M3.

(1) Compare the conjugate line with the above Example 10.2 using the first construction method for the conjugate line in Figure 10.10a.

(2) Code Hunter–Nash graphical equilibrium-stage method in MATLAB to get all data points on the ternary diagram: E1, E2, E3, R1, R2, R3, M1, M2, and M3.

(3) If solvent S1 = 50 kg for the first stage, S2 = 100 kg for the second stage, and S3 = 150 kg for the third stage, to get all new data points on the ternary diagram: E1, E2, E3, R1, R2, R3, M1, M2, and M3. Compare the results to the question (2) and explain.

10.2 Figure 10.27 shows the three counter-current stages of SX. The feed data and solvent data are shown in Table 10.7. The feed is added to the first mixer, and the fresh solvent is added to the third mixer.

(1) To code Hunter–Nash graphical equilibrium-stage method in MATLAB to get all data points on the ternary diagram: M, P, E_1, E_2, E_3, R_1, R_2, and R_3. (*Hint: It differs from cross-current searching for a new tie-line by optimizing and passing

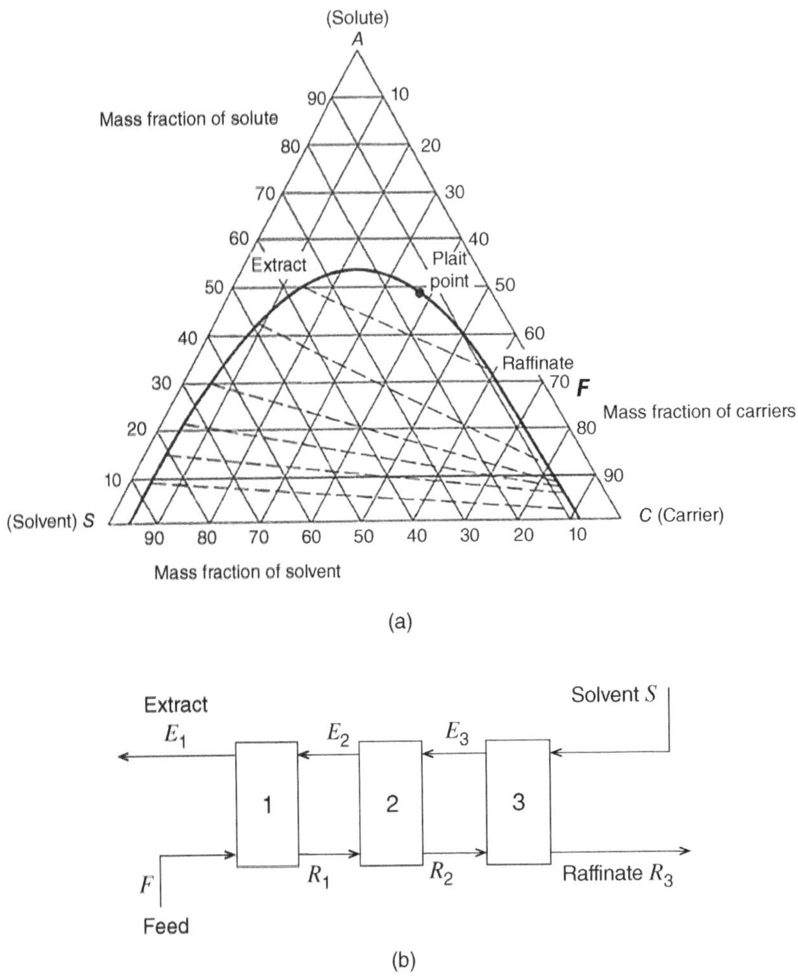

Figure 10.27 (a) Ternary diagram of the system; (b) counter-current three-stages solvent extraction.

Table 10.7 The feed and solvent data.

Feed	**Solvent**
F = 300kg	S = 100kg
$X_{F,A} = 0.40$	$X_{S,A} = 0.00$
$X_{F,C} = 0.60$	$X_{S,C} = 0.00$
$X_{F,S} = 0.00$	$X_{S,S} = 1.00$

the mixture point. Instead, you need to search the R_3, the Raffinate 3, by minimizing the difference between it and the newly calculated R_3 after following the procedure in the Section 10.3.6.)

(2) Optimize the mass of solvent ($15/kg) to minimize the operation costs to reach the final raffinate, not greater than 2.5%.

References

Aris, R. (1993). Ends and beginnings in mathematical modeling of chemical engineering systems. *Chemical Engineering Science* 48 (14): 2507–2517.

Bolboacă, S.D. and Jäntschi, L. (2007). Design of experiments: useful orthogonal arrays for number of experiments from 4 to 16. *Entropy* 9 (4): 198–232.

Buckingham, E. (1914). On physically similar systems; illustrations of the use of dimensional equations. *Physical Review* 4 (4): 345–376.

Dahm, K.D. and Visco, D.P. (2014). Fundamentals of chemical engineering thermodynamics, SI Edition. ISBN: 9781305178168.

Davis, R.A. (2021). Practical numerical methods for chemical engineers: using excel with VBA, 5th Edition. ISBN: 978-1694584687.

Farhadi, M., Azadi, P., and Zarinpanjeh, N. (2009). First principles modeling of the performance of a hydrogen-peroxide-driven chem-e-car. *Chemical Engineering Education* 43 (1): 65–69.

Gehman, H.W. (2003). Columbia accident investigation board report, vol. 1. http://www.nasa.gov/columbia/home/CAIB_Vol1.html (accessed 25 October, 2013).

Holdich, R.G. (2002). *Fundamentals of Particle Technology*. Shepshed, Leicestershire: Midland Information Technology and Publishing.

Hu, X. and Xie, W. (2001). Fixed-bed adsorption and fluidized-bed regeneration for breaking the azeotrope of ethanol and water. *Separation Science and Technology* 36 (1): 125–136.

Kim, J.J., Stafford, G.R., Beauchamp, C., and Kim, S.A. (2020). Development of a dental implantable temperature sensor for real-time diagnosis of infectious disease. *Sensors* 20 (14): 3953.

Kowalczyk, L.S. (1963). There are no “small” mathematical errors in engineering work. *Chemical Engineering Education* 2 (2): 28–32.

MathWorks Inc (2022). *Optimization toolbox version: 9.4 (R2022b)*. Natick, MA: The MathWorks Inc https://www.mathworks.com.

Lagrange, J.L. (1811). *Mécanique Analytique*, 2e, vol. 1. Paris: Courcier Complete Edition, joining the notes of the 3rd Edition revised, corrected, and annotated by Joseph Bertrand, and those of the 4th Edition published under the direction of Gaston Darboux; Albert Blanchard, Paris, France, 1965.

Levenspiel, O. (2002). Modeling in chemical engineering. *Chemical Engineering Science* 57: 4691–4696.

Meng, J., Xie, W., Tabosa, E. et al. (2016). Turbulence model development for flotation cells based on piezoelectric sensor measurements. *International Journal of Mineral Processing* 156: 116–126.

Chemical Engineering Analysis and Optimization Using MATLAB®, First Edition.
Weiguo Xie, Sam Toan, and Richard Davis.

Companion website: www.wiley.com/go/xie/chemicalengineering

Menten, L. and Michaelis, M.I. (1913). Die Kinetik der Invertinwirkung. *Biochem Z* 49: 333–369.

Napier-Munn, T.J., Morrel, S., Morrison, R.D., and Kojovic, T. (1996). Mineral comminution circuits: their operation and optimization, JKMRC Monograph Series, Queensland.

Press, W.H., Teukolsky, S.A., Vetterling, W.T., and Flannery, B.P. (1992). *Numerical Recipes in Fortran 77: The Art of Scientific Computing*, 2nde, 178. New York: Cambridge University Press.

Rasmuson, A., Adersson, B., Olsson, L., and Andersson, R. (2014). *Mathematical Modeling in Chemical Engineering*. Cambridge: Cambridge University Press.

Seader, J.D., Henley, E.J., Roper, D., and D.K. (2011). *Separation Process Principles*, 3e. Hoboken, NJ: Wiley.

Sherwood, T.K. (1963). *A Course in Process Design*. Cambridge: MIT Press.

Thompson, H.B. (1987). Good numerical techniques in chemistry: the quadratic equation. *Journal of Chemical Education* 64: 1009–1010.

Thygeson, J.R. Jr. and Grossmann, E.D. (1970). Optimization of a continuous through-circulation dryer. *AIChE Journal* 16 (5): 749–754.

Toan, S., Adidharma, H., and Nojabaei, B. (2017). Introduction to MATLAB for chemical & petroleum engineering, 2nd Edition. ISBN: 978-1548004873.

Walas, S.M. (1990). *Chemical Process Equipment: Selection and Design*. Elsevier.

Xie, W. and Davis, R.A. (2022). Chemical engineering analysis through systematic optimization. *Chemical Engineering Education* 56 (4): 1–7.

Xie, X., Deng, Y., Peng, J. et al. (2020). Nanoparticle-reinforced silica gels with enhanced mechanical properties and excellent pH-sensing performance. *Particle and Particle Systems Characterization* 37: 1900404.

Yu, P., Xie, W., Liu, L.X. et al. (2018). Evolution of a generic, dynamic and multicomponent tumbling mill model structure incorporating a wide-range 4D appearance function. *Powder Technology* 339: 396–407.

Index

Chemical Engineering Analysis and Optimization Using MATLAB®, First Edition.
Weiguo Xie, Sam Toan, and Richard Davis.

Companion website: www.wiley.com/go/xie/chemicalengineering

Printed and bound by CPI Group (UK) Ltd, Croydon, CR0 4YY
17/09/2026
14964313-0003